PINN with Python

An Introduction

G.R. Liu
University of Cincinnati, USA

2025

Scientech Publisher, Mason, USA

Published by:

ScienTech Publisher, Mason, USA

Website: https://www.sci-en-tech.com/

ISBN: 978-1-946018-04-5 (hardback)

ISBN: 978-1-946018-05-2 (paperback)

ISBN: 978-1-946018-03-8 (eBook)

First Edition

Printed in the United States of America

About the Author

Dr. Gui-Rong Liu received his Ph.D. from Tohoku University, Japan, in 1991. He was a Postdoctoral Fellow at Northwestern University, USA, from 1991 to 1993. He was a Professor at the National University of Singapore until 2010. He is currently a Professor at the Department of Aerospace Engineering and Engineering Mechanics, University of Cincinnati, USA. He was the Founder of the Association for Computational Mechanics (Singapore) (SACM) and served as the President of SACM until 2010. He served as the President of the Asia-Pacific Association for Computational Mechanics (APACM) (2010–2013) and an Executive Council Member of the International Association for Computational Mechanics (IACM) (2005–2010; 2020–2026). He authored a large number of journal papers and books, including two best-sellers, Mesh Free Method: Moving Beyond the Finite Element Method and Smoothed Particle Hydrodynamics: A Meshfree Particle Methods. He is the Editor-in-Chief of the International Journal of Computational Methods and served as an Associate Editor for IPSE and MANO. He is the recipient of numerous awards, including the Singapore Defence Technology Prize, NUS Outstanding University Researcher Award, NUS Best Teacher Award, APACM Computational Mechanics Award, JSME Computational Mechanics Award, ASME Ted Belytschko Applied Mechanics Award, Zienkiewicz Medal from APACM, AJCM Computational Mechanics Award, Humboldt Research Award, SACM Medal from the Association of Computational Mechanics (Singapore), and the Master Engineering Educator Award (UC, CEAS). He has been listed as one among the world's top 1% most influential scientists (Highly Cited Researchers) by Thomson Reuters for a number of years.

Contents

Chapter 1

Introduction

Computational methods are widely used in science, engineering, and technology across various disciplines. There are mainly two classes of computational methods. One is the standard or traditional method that solves the physical laws governing problems in science and engineering through various numerical approaches. The other builds neural networks (NNs), aiming to mimic the human brain, and uses data to model the problem. These two methods can be combined, leading to physics-informed neural networks , or PINNs, which is the topic of this book.

1.1 On Standard Law-based Numerical Methods

There are two major types of standard methods for solving partial differential equations (PDEs) along with the boundary and initial conditions: analytical methods and numerical methods. The former can only be applied to relatively simple cases of two-dimensional (2D) and three-dimensional (3D) problems, often producing exact or analytical solutions [14, 15].

Currently, the most widely used methods for practical problems in science and engineering are numerical methods. Typical examples include the finite element method (FEM) [21], the smoothed finite element method (S-FEM) [20], and, more generally, meshfree methods [11].

Most of these methods employ domain discretization techniques to approximate the solution functions in **piecewise** forms. This results in discretized algebraic system equations that can be solved using standard numerical routines. Various techniques are available for formulating these discretized systems, using either the strong form (collocation-based) or the weak form (energy-based) of the governing PDEs.

1.2 On Physics-Informed Neural Networks

A PINN is a neural network (NN) framework that incorporates physical laws into the loss function [12], typically expressed as partial differential equations (PDEs) along with boundary and/or initial conditions. The basic idea was formalized by Raissi, Perdikaris, and Karniadakis (2019) [26]. In their seminal work, they introduced the use of neural networks as universal function approximators constrained by the residuals of PDEs and the associated boundary and initial conditions.

This methodology is particularly useful for solving forward and inverse problems governed by differential equations, without relying heavily on large labeled datasets. It offers a compelling alternative to conventional PDE solvers. A PINN uses a neural network model composed of affine transformations and elementary nonlinear activation functions, arranged in a chained structure [12], as the basis for function approximation. The affine transformations are equipped with learnable parameters, which are optimized by minimizing the residuals of the PDEs.

The resulting approximated function is a **whole-piece** and **smooth** function, which allows for various orders of differentiation required in strong-form PDEs. Furthermore, differentiation can be performed automatically within the neural network using the powerful technique known as `autograd` in machine learning frameworks [12]. The derivatives computed using `autograd` are numerical, but exact to machine precision, leveraging the differentiability of the chained elementary functions in the NN. Additionally, these computations of the derivatives are highly efficient.

This book details how PINNs work, presenting fundamental formulations, basic techniques, and practical Python code for PINNs in various configurations. Both collocation-based approaches and energy methods will be discussed. Our applications will focus on three major types of PDEs commonly encountered in science and engineering:

- Poisson's equations (static),
- Heat transfer equations (time-dependent parabolic diffusion-type problems),
- Wave equations (time-dependent hyperbolic-type problems),
- Helmholtz equations (eigenvalue problems).

Intensive studies will explore how and why different types of PINN models should be constructed for specific problem types and configurations. The critical differences between using collocation and energy approaches in PINN modeling are analyzed both theoretically and through examples and case studies. Issues related to the difficulties and effective training of PINN models are also examined.

Both NumPy and SymPy are employed in developing the demonstration codes. Due to its efficiency and ease of use, PyTorch is used throughout this book for building PINN models. Discussions with ChatGPT, Gemini, and DeepSeek have also contributed significantly to the development of code and the preparation of this book.

1.2.1 Who May Read This Book

This book is intended for beginners who are interested in learning how to use PINNs to solve PDEs arising in science, engineering, and natural phenomena. It may also serve researchers who would like to further develop new techniques using PINNs. The target audience includes university students, graduate students, researchers, and professionals in any STEM discipline. Engineers and practitioners may also find this book valuable for building a solid foundation in machine learning methods, particularly in the development and application of PINN models.

Readers are encouraged to first read the book entitled *Machine Learning with Python* [12], or any other introductory book on machine learning, to gain a basic understanding of how a multilayer perceptron (MLP) works.

1.2.2 Use of the Book and Codes

Readers who purchase the book may contact the author at liugr100@gmail.com to request a free soft copy in Jupyter Notebook format, including codes (which may be updated), for academic use after registration.

Conditions of Use

The following conditions apply to the use of the book and the codes developed by the author in hard copy, soft copy, or any other medium or format:

- **User Responsibility:** Users are solely responsible for any risks associated with using any part of the book, codes, or techniques.
- **Code Purpose:** The codes are primarily intended to demonstrate concepts and may not be optimized for efficiency or robustness. Many of the codes have not been tested by a third party.
- **Limited Use:** The book and codes are for personal use by the individual who purchased the book. Redistribution or copying to any other person without prior written permission of the author is not allowed.
- **No User Support:** No support is provided for the codes.
- **Citation:** Proper citation and acknowledgment are required when using the book, codes, ideas, or techniques.

This book is code-intensive. The examples rely on various external packages and modules, and their behavior may depend on the versions of Python and the packages used. If a code snippet does not run as expected, a version mismatch is likely the cause. Users should verify the versions of all required packages.

To check the Python environment version within a Jupyter Notebook, use the following code cell:

```
!python -V                        #! is used to execute an external command
```

```
Python 3.9.16
```

```
!jupyter notebook --version
```

```
6.1.5
```

```
import torch
print(torch.__version__)
```

```
2.5.1+cpu
```

Troubleshooting Version Issues:

If you encounter a version mismatch error, you can either adapt the code to match the version installed on your system or install the required version. It is often helpful to search the web using the exact error message to find potential solutions or guidance. Additionally, large language models can be helpful in this regard—this is the author's preferred method for resolving code execution issues.

Since our descriptions are interspersed with code cells, a given code cell may depend on earlier ones. If an error occurs when running a cell, try executing the preceding cells to ensure all dependencies have been properly initialized.

Learning Python:

This book does not cover the basics of Python, as numerous resources are available online. Interested readers may refer to Chapters 2 and 3 of Ref. [12] for a concise overview of Python for scientific computing. Since Python is relatively easy to learn, readers can also use the codes and examples in this book to gain a head start on learning Python while studying the technical content.

1.2.3 Use of External Modules or Dependencies

Module Imports in Jupyter Notebooks:

Frequently used modules in each chapter are imported in the first code cell. This cell should be executed before running any subsequent code. If you encounter a `NameError` (e.g., `NameError: name 'sp' is not defined`), return to the beginning of the chapter and re-execute the first code cell to ensure that all modules and symbolic variables are properly initialized.

```
#Often used external modules are in commonImports placed in folder grbin
#Place cursor in this cell, and press Ctrl+Enter to import dependencies.

import sys                                   # for accessing the computer system
sys.path.append('../grbin/')                   # Add in the path to your system

from commonImports import *             # Import dependencies from '../grbin/'
import grcodes as gr                          # Import the module of the author
importlib.reload(gr)                    # For in case, when grcodes is modified

from continuum_mechanics import vector
from continuum_mechanics.solids import sym_grad, strain_stress
init_printing(use_unicode=True)              # For latex-like quality printing

# Digits in print-outs
np.set_printoptions(precision=4,suppress=True,
                    formatter={'float_kind': '{:.4e}'.format})
```

The above cell includes a common import file, `commonImports.py` (provided below). This file contains all the frequently used modules in this book, helping to avoid repeated lengthy import statements and reducing redundancy. By executing this cell at the beginning of each chapter, you ensure that all necessary modules are available throughout the notebook.

```
from __future__ import print_function # Ensures print behaves like a
                                      # function in Python 2/3 compatible code

# --- For Numerical and Scientific Computation ---
import numpy as np          # Fundamental package for numerical operations
import numpy.linalg as lg             # Linear algebra routines from NumPy
import scipy.linalg as sg  # Advanced linear algebra routines from SciPy
import scipy.integrate as si          # For numerical integration~ routines
```

```
from scipy.stats import ortho_group # For generating orthogonal matrices
import autograd.numpy as anp        # NumPy-like interface with autograd
from autograd import grad        # For computing gradients automatically

# --- Symbolic Mathematics (SymPy) ---
import sympy as sp           # Core SymPy module for symbolic computation
from sympy import (
    sin, cos, exp, symbols, lambdify, init_printing, latex, # Basic funs
    pi, Matrix, sqrt, oo, integrate, diff, Derivative,       # Constants
    MatrixSymbol, simplify, nsimplify, Function, # symbolic manipulation
    factor, expand, ordered, hessian       # Algebraic, matrix operations
)
from sympy.plotting import plot as splt         # SymPy's plotting module
init_printing(use_unicode=True) # Enables LaTeX-like printing in Jupyter

# --- Data Handling and Utilities ---
import pandas as pd   # Powerful data structures and data analysis tools
import csv                             # For reading/writing CSV files
import random              # For generating random numbers and selections
import itertools    # Functions creating iterators for efficient looping
import inspect # For inspecting live objects (modules,classes,functions)
import importlib                   # For importing modules programmatically

# --- Plotting and Visualization ---
import matplotlib.pyplot as plt              # Primary plotting interface
import matplotlib as mpl                # Core Matplotlib functionalities
from matplotlib.ticker import MultipleLocator    # Fo control over ticks
from IPython.display import display, Math  # For display of mathematical
                                         # expressions in IPython/Jupyter
# --- Custom/Frequent Utilities ---
import math as ma                         # Standard Python math functions
from grcodes import drawArrow, plotfig, printM, printx # frequently used
```

Module Reloading:

Importing the same module multiple times generally does not cause issues, as Python uses a cache and skips re-importing modules that are already loaded. However, if you modify an imported module and want those changes to take effect, you need to reload the module. This can be done using the `importlib.reload()` function:

```
# grcodes was imported as gr, reload it when grcodes is modified
importlib.reload(gr)       # This is also included in the first code cell
```

1.2.4 Use of `help()`

Exploring Modules and Objects:

To access detailed information about a module or an object within a module after importing it, you can use the `help()` function. This is a useful way to explore the functionalities and attributes available in different modules.

```
help(gr.residual_energy)
```

```
Help on function residual_energy in module grcodes:

residual_energy(pinn, f_th, x, weights, x_NBC, t_NBC)
    Compute the integral of energy density for PDE Laplacian~ (u)+f=0
    using PyTorch autograd and quadrature.
    Minimizing this integral finds the solution u.
    Parameters:
    - pinn: the pinn model (PINN2D_4DBC which provides u_trial)
    - f_th: forcing/source function~ f(x, y), accepts and returns tensor
    - x: tensor of shape [N, 2], quadrature points
    - weights: tensor of shape [N, 1], quadrature weights
    - x_NBC, t_NBC: optional, not used in this function
    Returns: total energy, potential energy, work-done; scalars
```

Code Comments:

The code cells throughout the book include comments that provide additional explanations. These comments, marked by a hash symbol #, are typically placed on the right side of the code to minimize disruption while offering quick reference and guidance.

Chapter 2

PINN Formulation and Models

First, let us import the necessary external modules or dependencies for later use in this chapter.

```
# Place cursor in this cell, and press Ctrl+Enter to import dependencies.
import sys                          # for accessing the computer system
import os
sys.path.append('../grbin/')  # Add in the path to your system

from commonImports import *       # Import dependencies from '../grbin/'
import grcodes as gr                   # Import the module of the author
importlib.reload(gr)                 # When grcodes is modified, reload it
import math

init_printing(use_unicode=True)         # For latex-like quality printing
np.set_printoptions(precision=4, suppress=True)  # Digits in print-outs
```

2.1 The Framework for PINN Models

A Physics-Informed Neural Network (PINN) is a machine learning framework that integrates physical laws—typically in the form of partial differential equations (PDEs) along with boundary conditions—directly into the neural network training process. The foundational idea was introduced by Raissi, Perdikaris, and Karniadakis in 2019 [26].

Unlike traditional supervised learning approaches that rely on large amounts of labeled data, PINNs incorporate the governing equations and boundary conditions of the system as constraints in the loss function. This enables the model to learn accurate solutions to forward and inverse problems with limited or no labeled data, making them especially suitable for many scientific and engineering applications involving PDEs.

This chapter presents the key concepts, formulations, techniques, and implementations of PINN models with various configurations and implementation strategies. The primary focus is on solving two-dimensional (2D) Poisson equations under different boundary conditions. Through comprehensive examples and case studies, we will explore and analyze the behavior of these PINN models in detail.

> **Note:** The material in this chapter is densely packed, and many of the code cells are sequentially dependent . For a smooth and coherent study experience, it is highly recommended to run the code cells in order from the beginning of the chapter. Most of the PINN models presented here can be trained in tens of seconds on the author's laptop, so the total runtime for executing all the code cells should be manageable.

Consider a problem defined in the domain Ω, as shown in Fig. 2.1.

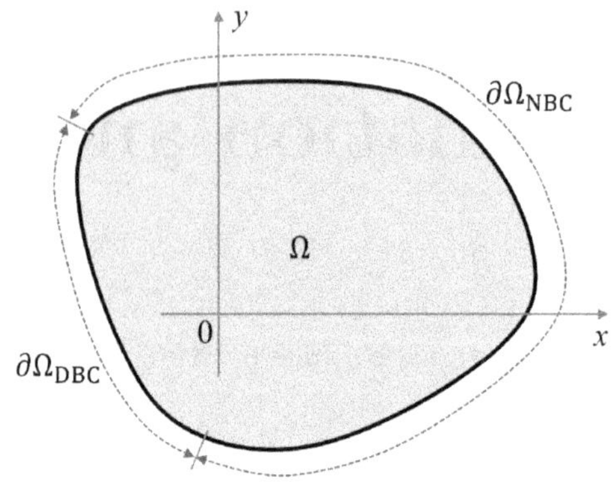

Figure 2.1: Schematic drawing of a two-dimensional (2D) domain Ω, which is bounded by its boundary composed of two parts, boundary with DBCs and boundary with NBCs: $\partial\Omega = \partial\Omega_{\mathrm{DBC}} \cup \partial\Omega_{\mathrm{NBC}}$.

A PINN augments the standard neural network loss function [12] with additional terms that penalize the network for violating known physical laws (e.g., PDE forms of conservation laws or equilibrium equations and boundary conditions), as illustrated in Fig. 2.2.

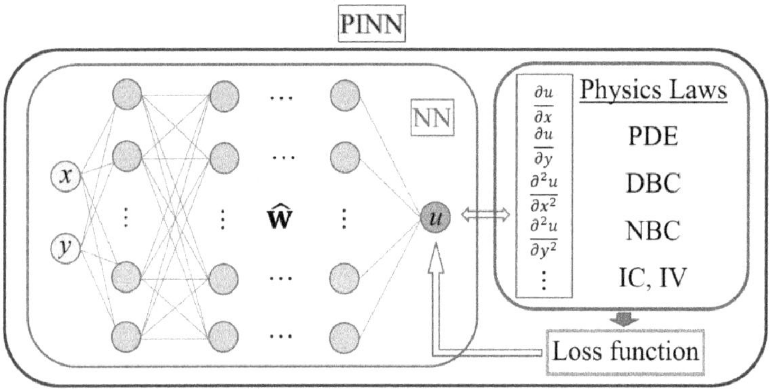

Figure 2.2: Schematic drawing of a PINN model that uses physical laws and known conditions to construct a loss function to train a neural network (NN).

For dynamic problems, known equations for initial conditions (ICs), in either differential or algebraic form, can also be included. In this way, PINNs are "informed" by the physics of the problem and its constraints, providing an alternative model for the system governed by the

PDEs, BCs, and ICs—so long as the solution is sufficiently smooth. When used on their own, they serve as alternative PDE solvers, which is the focus of this book.

2.2 General Formulation for PINNs

2.2.1 System Equations

Let the physical system be described by a PDE of the form:

$$\underbrace{\mathcal{F}(u;\mathbf{x},t,\mathbf{c})=0}_{\text{PDE}}, \quad \mathbf{x}\in\Omega,\ t\in[0,t_{\text{end}}], \tag{2.1}$$

where:

1. $\mathcal{F}$ is a differential operator that can be linear or nonlinear, and $u(\mathbf{x},t,\mathbf{c})$ is the unknown solution to the physical problem (e.g., displacement, velocity, potential, temperature, etc.). u is a function of the spatial coordinates $\mathbf{x}$, time t, and physical parameters $\mathbf{c}$, which represent coefficients in the PDE. For 2D problems, $\mathbf{x}=(x,y)$.

2. In a PINN, u is generated by a neural network (NN) denoted as $\mathcal{N}_{\hat{\mathbf{w}}}$, equipped with learning parameters $\hat{\mathbf{w}}$. Thus, we have:

$$u(\mathbf{x},t,\mathbf{c})=\mathcal{N}_{\hat{\mathbf{w}}}(\mathbf{x},t,\mathbf{c})$$

for the current parameters $\hat{\mathbf{w}}$ and inputs $(\mathbf{x},t,\mathbf{c})$. The solution u can also be written as $u_{\hat{\mathbf{w}}}(\mathbf{x},t,\mathbf{c})$.

3. The gradients or derivatives of $u_{\hat{\mathbf{w}}}(\mathbf{x},t,\mathbf{c})$ with respect to $\hat{\mathbf{w}}$, $\mathbf{x}$, t, and $\mathbf{c}$ can all be computed within $\mathcal{N}_{\hat{\mathbf{w}}}(\mathbf{x},t,\mathbf{c})$ via 'autograd', when requested. This is performed through a back-propagation process [12].

4. Ω is the spatial domain in which the problem is defined.

5. t_{end} is the final time for the problem, which is a finite positive number.

6. During training, the gradients or partial derivatives of the loss function with respect to $\hat{\mathbf{w}}$ are computed using automatic differentiation (autograd). This process is numerical but exact to machine precision [12].

7. The same autograd mechanism [12] used for training $\mathcal{N}_{\hat{\mathbf{w}}}$ is also employed to compute the derivatives of u with respect to $\mathbf{x}$ and t, as required by the PDE and by any BCs or ICs. Since autograd provides exact derivatives (within numerical precision), all required partial derivatives can be accurately evaluated.

8. The loss function is defined as the discrepancy—called the residual—between the NN's output and the PDE, BCs, and ICs. This loss is minimized using gradient-based optimization algorithms, which adjust $\hat{\mathbf{w}}$ accordingly.

Equation (2.1) (with BCs) can be expressed in either strong form or weak form. The strong form refers to the original differential equation, while the weak form involves a reformulation using test functions and integration, often reducing the differentiation order [11]. The energy method is a typical example of a weak form. Both strong and weak forms will be used in this book.

The boundary of Ω is typically decomposed as $\partial\Omega = \partial\Omega_{\text{DBC}} \cup \partial\Omega_{\text{NBC}}$, as illustrated in Fig. 2.1. Depending on the problem, the unknown function $u(\mathbf{x}, t, \mathbf{c})$ may be subject to constraints over parts of $\partial\Omega$. The boundary conditions generally fall into the following two types:

Dirichlet boundary condition (DBC):

$$u(\mathbf{x}, t) = \bar{u}(\mathbf{x}, t), \quad \forall \mathbf{x} \text{ on } \partial\Omega_{\text{DBC}}, \, t \in [0, t_{\text{end}}], \tag{2.2}$$

Here, $\bar{u}$ is the prescribed value of u for the given problem, such as fixed displacements, fixed potential, fixed temperature, etc. $\partial\Omega_{\text{DBC}}$ denotes the portion of the boundary $\partial\Omega$ where the Dirichlet boundary conditions are applied.

Neumann boundary condition (NBC):

NBCs are often given in differential form and can be generally expressed as:

$$\mathcal{B}(\mathbf{x}, t, \mathbf{c}) = 0, \quad \forall \mathbf{x} \in \partial\Omega_{\text{NBC}}, \, t \in [0, t_{\text{end}}], \tag{2.3}$$

An NBC constrains certain partial derivatives of the unknown function u with respect to $\mathbf{x}$ to prescribed values, such as strain, electric field components, or heat flux. $\partial\Omega_{\text{NBC}}$ is the portion of the boundary where these conditions apply. The differential operator $\mathcal{B}$ is implemented within the same NN using `autograd` to compute the required derivatives.

Initial condition (IC):

For time-dependent problems, initial conditions specify the state of the system at $t = 0$. These are generally expressed as:

$$\mathcal{I}(\mathbf{x}, 0, \mathbf{c}) = 0, \quad \forall \mathbf{x} \in \Omega \tag{2.4}$$

An IC constrains the function u at $t = 0$ to some prescribed value (e.g., initial displacement, temperature, etc.). The operator $\mathcal{I}$ is also implemented using the same NN.

Initial velocity condition (IV):

The initial velocity condition constrains the time derivative of u at $t = 0$, such as the velocity of displacement across the domain Ω. It takes the general form:

$$\mathcal{V}(\mathbf{x}, 0, \mathbf{c}) = 0, \quad \forall \mathbf{x} \in \Omega \tag{2.5}$$

This condition involves derivatives with respect to time, and the differential operator $\mathcal{V}$ is also implemented using the same NN and `autograd`.

2.2.2 Formulation of Loss Function

A PINN neural network $\mathcal{N}_{\hat{\mathbf{w}}}(\mathbf{x}, t, \mathbf{c})$ is equipped with a set of learnable parameters $\hat{\mathbf{w}}$, also denoted as θ in some literature. The network is trained to approximate the target solution u by minimizing the following composite loss:

$$\mathcal{L} = \mathcal{L}_{\text{data}} + w_{\text{PDE}}\mathcal{L}_{\text{PDE}} + w_{\text{DBC}}\mathcal{L}_{\text{DBC}} + w_{\text{NBC}}\mathcal{L}_{\text{NBC}} + w_{\text{IC}}\mathcal{L}_{\text{IC}} + w_{\text{IV}}\mathcal{L}_{\text{IV}} \tag{2.6}$$

where:

- $\mathcal{L}_{\text{data}}$ penalizes differences between the network outputs and any available data. In this book, we assume there are no such data, meaning that our PINN models are purely based on the governing equations, boundary conditions, and initial conditions.
- $\mathcal{L}_{\text{IV}}$ penalizes the residuals of the initial velocity conditions, where certain partial derivatives of u with respect to time are specified over the domain Ω at time $t = 0$.
- $\mathcal{L}_{\text{IC}}$ penalizes the residuals of the initial conditions, where u is prescribed within the domain Ω at time $t = 0$.
- $\mathcal{L}_{\text{NBC}}$ penalizes the residuals of the Neumann boundary conditions (NBCs), where certain partial derivatives are prescribed on $\partial\Omega_{\text{NBC}}$.
- $\mathcal{L}_{\text{DBC}}$ penalizes the residuals of the Dirichlet boundary conditions (DBCs), where u is prescribed on $\partial\Omega_{\text{DBC}}$.
- $\mathcal{L}_{\text{PDE}}$ penalizes the residuals computed from the governing equation $\mathcal{N}_{\hat{\mathbf{w}}}(\mathbf{x}, t, \mathbf{c})$ defined in Eq.(2.1) over the domain Ω.

In a PINN model, $\mathcal{L}_{\text{PDE}}$ is always included in the composite loss. The inclusion of the other terms depends on how the PINN is constructed and the specific settings of the problem.

To compute these loss terms , we must evaluate them at a set of sampled points —either inside the domain or along its boundaries—by invoking the neural network with its current set of learning parameters.

When a collocation method is used for sampling, the formula for $\mathcal{L}_{\text{PDE}}$ is given as follows:

$$\mathcal{L}_{\text{PDE}} = \frac{1}{n_{PDE}} \sum_{i=1}^{n_{PDE}} \left[\mathcal{F}_{\hat{\mathbf{w}}}\left(\mathbf{x}_i, t_i\right)\right]^2 \tag{2.7}$$

where n_{PDE} is the number of collocation points sampled (either randomly or regularly) within the problem domain using the strong-form PDE defined in Eq.(2.1). Note that the square is applied to each individual sampling point. Because of this squaring, $\mathcal{L}_{\text{PDE}}$ is always non-negative. It is a typical L2 loss, which is widely used in machine learning models.

If the strong-form PDE is linear, then the L2 loss $\mathcal{L}_{\text{PDE}}$ is quadratic in u. Clearly, to evaluate this loss effectively, a sufficient number of collocation points must be sampled.

Following the same point-wise sampling strategy , the formula for $\mathcal{L}_{\text{DBC}}$ is as follows:

$$\mathcal{L}_{\text{DBC}} = \frac{1}{n_{DBC}} \sum_{i=1}^{n_{DBC}} \left[u_{\hat{\mathbf{w}}}\left(\mathbf{x}_i, t_i\right) - \bar{u}\left(\mathbf{x}_i, t_i\right)\right]^2 \tag{2.8}$$

where n_{DBC} is the number of collocation points sampled on $\partial\Omega_{\text{DBC}}$ using the Dirichlet boundary condition defined in Eq.(2.2). Since u is produced by $\mathcal{N}_{\hat{\mathbf{w}}}$ (which is equipped with the learnable parameters $\hat{\mathbf{w}}$), we must sample a sufficient number of points to properly evaluate this loss.

Note that $\mathcal{L}_{\text{DBC}}$ is also non-negative and is quadratic in u.

Similarly, the formula for $\mathcal{L}_{\text{NBC}}$ is as follows:

$$\mathcal{L}_{\text{NBC}} = \frac{1}{n_{\text{NBC}}} \sum_{i=1}^{n_{\text{NBC}}} \left[\mathcal{B}_{\hat{\mathbf{w}}}\left(\mathbf{x}_i, t_i\right)\right]^2 \tag{2.9}$$

where n_{NBC} is the number of collocation points sampled on $\partial\Omega_{\text{NBC}}$ using the NBC defined in Eq.(2.3). We also need to sample a sufficient number of points to properly evaluate this loss. The loss term $\mathcal{L}_{\text{NBC}}$ is quadratic in u if $\mathcal{B}$ is linear.

The formula for $\mathcal{L}_{\text{IC}}$ is as follows.

$$\mathcal{L}_{\text{IC}} = \frac{1}{n_{IC}} \sum_{i=1}^{n_{IC}} \left[\mathcal{I}_{\hat{\mathbf{w}}}\left(\mathbf{x}_i, t_i\right)\right]^2 \tag{2.10}$$

where n_{IC} is the number of collocation points sampled inside the problem domain Ω using the IC defined in Eq.(2.4). Clearly, we must sample a sufficient number of points to properly evaluate this loss. The loss term $\mathcal{L}_{\text{IC}}$ is quadratic in u if $\mathcal{I}$ is linear.

The residual formula for $\mathcal{L}_{\text{IV}}$ is as follows.

$$\mathcal{L}_{\text{IV}} = \frac{1}{n_{IV}} \sum_{i=1}^{n_{IV}} \left[\dot{\mathcal{I}}_{\hat{\mathbf{w}}}\left(\mathbf{x}_i, t_i\right) - \bar{\dot{u}}\left(\mathbf{x}_i, t_i\right)\right]^2 \tag{2.11}$$

where the dots denote derivatives with respect to time, n_{IV} is the number of collocation points sampled inside the problem domain Ω using the IV defined in Eq.(2.5), which is usually the same as n_{IC}, and $\bar{\dot{u}}$ is the prescribed value for the initial velocity. The loss term $\mathcal{L}_{\text{IV}}$ is quadratic in u if $\dot{\mathcal{I}}$ is linear.

It is important to note that the required derivatives for all these loss terms are computed via automatic differentiation inside the neural network PINN. This is the fundamental reason why a PINN can be applied to solve PDEs with smooth solutions that are sufficiently differentiable.

Note that, depending on the type of problem, additional loss terms may be included in Eq.(2.6). Moreover, one may add a regularization loss term to prevent overfitting [12].

2.2.3 On the Weights of Loss Terms

In Eq.(2.6), the ws are the weights for the corresponding loss terms. Since these weights are hyperparameters and usually adjusted manually via trial and error, we assume all these weights are 1 by default in this book. We will adjust some of these weights when needed, but will not discuss in detail how to tune these parameters. Our main focus, however, will be on more automatically controllable factors through properly constructing PINN models. This is because tuning hyperparameters for a machine learning model can be very time-consuming, problem-dependent, and hence difficult to establish general rules.

There are a number of possible structures for PINN models. This chapter will detail the major basic ones. To avoid geometric complications, we confine our study in this chapter to rectangular domains only. In addition, our study in this chapter will focus solely on Poisson-type equations as examples, which are among the most fundamental PDEs governing a wide range of problems in science and engineering.

The formulation given above is quite abstract and may be difficult to comprehend. All of this should become clear as we go through the process with an example in the following section.

2.3 Preparation of PINN Models

2.3.1 Random Sampling and Seeds

In machine learning models (not limited to PINNs), random sampling is frequently used both for initialization and for selecting data points [12]. To ensure that trained PINN models and their predictive results are reproducible, it is important to set random seeds consistently across all relevant libraries. This guarantees deterministic computational behavior, allowing experiments to be reliably repeated.

The following code snippet sets the random seeds when PyTorch is used to train a machine learning model. This practice will be commonly applied in the code examples presented in this book. Changing the seed value will produce slightly different results and behavior, which readers are encouraged to explore during the example sessions.

In this book, we adopt `seed=88` as the default value for all random number generation in Python and PyTorch. Readers are welcome to modify this seed to any other integer to observe the impact on reproducibility and model outcomes.

```
def set_seed_th(seed=88):
    os.environ['PYTHONHASHSEED'] = str(seed)  # Affects hash() & python
    random.seed(seed)                                       # Python RNG
    np.random.seed(seed)                                     # NumPy RNG
    torch.manual_seed(seed)                          # CPU RNG for PyTorch
    torch.cuda.manual_seed(seed)                         # Current GPU RNG
    torch.cuda.manual_seed_all(seed)                            # All GPUs

    # Enforce deterministic algorithms in PyTorch
    torch.use_deterministic_algorithms(True, warn_only=True)

    # Make cuDNN deterministic (sometimes needed explicitly)
    torch.backends.cudnn.deterministic = True
    torch.backends.cudnn.benchmark = False

set_seed_th(seed=88)
```

The random seed setting ensures only a certain degree of consistency in the PINN models. The time logging in the case studies will vary depending on the computer's status during computation. Timing results can differ significantly when the computer is running heavy background tasks. Therefore, the analysis of computational performance should be considered only indicative.

2.3.2 Device Management

Training and inference of machine learning models require intensive computation. The use of GPUs is preferred for computational efficiency. Many computer systems may have CUDA-enabled GPUs available, but some do not (including the author's laptop). To utilize available GPUs for faster computation, we can use the following snippet to define the devices for computations.

```
# Device management: ensure CUDA is available if you plan to use GPU
device = torch.device('cuda' if torch.cuda.is_available() else 'cpu')
gr.device = device                          # inject device to gr module
torch.cuda.empty_cache()                    # Clear previous model state
print(f"Using device: {device}")
```

```
Using device: cpu
```

In our computations in this book, we use only the CPU as the device.

In the following snippet, we use `gr.device = device`. This is because the variable `device` is not defined within the `gr` module. Thus, we need to inject `device` into the `gr` module so that the relevant functions can work properly when invoked.

The following sections present the detailed procedure for training and using PINN models, focusing on Poisson's equations.

2.4 Poisson's Equations

2.4.1 The General Form

Poisson's equation is one of the most widely encountered PDEs in science and engineering. Mathematically, it takes the following general form:

$$\underbrace{\frac{\partial^2 u}{\partial x^2} + \frac{\partial^2 u}{\partial y^2} + f(x, y)}_{\mathcal{F}(u;x,y)} = 0 \tag{2.12}$$

where u is the unknown scalar field, f is a source term, and $\frac{\partial^2}{\partial x^2} + \frac{\partial^2}{\partial y^2}$ is the well-known Laplace operator and is often denoted as Δ.

Physically, several important practical problems are described by Poisson's equation, including the following:

Solid Mechanics

- *Torsion of uniform straight bars under a constant twist* [13]:
  - $u(x, y)$ (often denoted as ϕ) is the stress function distributed over the cross-sectional area of the bar, and $f(x, y)$ represents the constant twist per unit length.
- *Two-dimensional (2D) elastic solids subjected to forces* [14]:
  - $u(x, y)$ (often denoted as σ_{ii}) is the trace of the stress field over a 2D domain, and $f(x, y)$ corresponds to the body force distribution in the domain.

Steady-State Heat Transfer [20, 21]

- $u(x, y)$ (often denoted as T) is the temperature field distributed over a 2D medium, and $f(x, y)$ represents a distributed heat source or sink within the medium.

Electrostatics

- $u(x, y)$ (often denoted as E) is the electric potential distributed over a 2D domain, and $f(x, y)$ corresponds to the static charge distribution within the domain.

2.4.2 A Poisson's Equation Example

This chapter considers a two-dimensional (2D) PDE that governs a static boundary value problem defined in the rectangular domain $\Omega = (a, b) \times (c, d)$ with $x = a$ being the left edge of the rectangle. The PDE of first example of Poisson's Equation to be studied is defined as

$$\underbrace{\frac{\partial^2 u}{\partial x^2} + \frac{\partial^2 u}{\partial y^2}}_{\Delta u} + \underbrace{2\pi^2 \sin(\pi x)\sin(\pi y)}_{\text{source term } f(x,y)} = 0, \quad \text{in } \Omega \tag{2.13}$$

This Poisson's equation will be studied in great detail in this chapter, and various PINN models will be built to model it.

The next snippet defines, in both NumPy and PyTorch, the force function given in Eq.(2.13).

```
def f_np(x, y):
    '''Force or source term f in numpy'''
    return 2 * np.pi**2 * np.sin(np.pi * x) * np.sin(np.pi * y)

def f_th(x, y):
    '''Source or force term f in torch in the Poisson's equation'''
    return 2*torch.pi**2 *torch.sin(torch.pi*x) * torch.sin(torch.pi*y)
```

2.4.3 DBC for Poisson's Equation

If we assume that the unknown function $u(x, y)$ is zero on any edge of the rectangular domain, the Dirichlet boundary condition at that edge can be expressed as follows:

$$u(x, y)|_{\text{on an edge}} = \bar{u} = 0 \tag{2.14}$$

2.4.4 Exact Solution to the Example Problem

If we set $\bar{u} = 0$ for all edges of the square domain $\Omega = (0, 1) \times (0, 1)$ with $x = 0$ being the left edge of the rectangle, the **exact** solution to Eq.(2.13) can be found via simple inspection and is given as

$$u_{\text{exact}}(x, y) = \sin(\pi x)\sin(\pi y) \tag{2.15}$$

This can be confirmed easily by substituting Eq.(2.15) into Eq.(2.13).

```
# Define symbols to be used in this code cell:
x, y = sp.symbols('x, y')

# Define the surce term in Sympy:
f_sp = 2*sp.pi**2*sp.sin(sp.pi * x) * sp.sin(sp.pi * y)
# Define the exact solution in Sympy:
u_sp = sp.sin(sp.pi * x) * sp.sin(sp.pi * y)

# Compute the Poisson's equation:
PoissonEq = sp.diff(sp.diff(u_sp,x),x)+sp.diff(sp.diff(u_sp,y),y) + f_sp
```

```
display(Math(f"\\text{{Is Poisson's Eq. satisfied? }}{PoissonEq==0}"))
```

Is Poisson's Eq. satisfied? *True*

As confirmed, the exact solution satisfies Eq.(2.13). It is good practice to use a simple snippet like the one given in the preceding cell to verify equations. Let's now release the variables x and y for later use.

```
del x, y                    # release these variables for other use later
```

The following snippets define the formula in Numpy for computing the exact solution for this setting of Poisson's equation, which will be used later to assess the accuracy numerically of the predicted solution from a trained PINN.

```
# Exact Solution (in Numpy, to be used for reference later):
def exact_u(x, y, a=None, b=None, c=None, d=None):
    '''a, b, c, d are geometric parameters. Not used in this function,
    but as placefolders for possible other problems (NumPy). '''
    return np.sin(np.pi * x) * np.sin(np.pi * y)
```

2.4.5 NBC for Poisson's Equation

Since Poisson's equation is a second-order PDE, it can be set with Neumann boundary conditions (NBCs), which involve first-order derivatives. The NBC can be expressed as follows:

$$\frac{\partial u}{\partial n}\Big|_{\partial\Omega_{\text{NBC}}} = t(\mathbf{x}) \tag{2.16}$$

where $\frac{\partial u}{\partial n}$ is the directional derivative of u in the outward normal direction on the boundary $\partial\Omega_{\text{NBC}}$, and t is a prescribed value of $\frac{\partial u}{\partial n}$ on $\partial\Omega_{\text{NBC}}$. It is often set to zero, but it can be any given function, as will be shown in the example sections.

The outward normal direction can be computed using [17]:

$$\frac{\partial u}{\partial n} = u \cdot \mathbf{n} = \begin{bmatrix} \frac{\partial u}{\partial x} & \frac{\partial u}{\partial y} \end{bmatrix} \begin{bmatrix} n_x \\ n_y \end{bmatrix} = \frac{\partial u}{\partial x} n_x + \frac{\partial u}{\partial y} n_y \tag{2.17}$$

where $\mathbf{n}$ is the outward normal vector at a point on $\partial\Omega$. It has two components, n_x and n_y. For a rectangular domain of $\Omega = (a, b) \times (c, d)$, the possible NBCs can be

$$\begin{aligned}
&\text{On the left edge:} && x = a, && \mathbf{n} = \begin{bmatrix} -1 \\ 0 \end{bmatrix} && \Rightarrow && \frac{\partial u}{\partial n} = -\frac{\partial u}{\partial x} \\
&\text{On the bottom edge:} && y = c, && \mathbf{n} = \begin{bmatrix} 0 \\ -1 \end{bmatrix} && \Rightarrow && \frac{\partial u}{\partial n} = -\frac{\partial u}{\partial y} \\
&\text{On the right edge:} && x = b, && \mathbf{n} = \begin{bmatrix} 1 \\ 0 \end{bmatrix} && \Rightarrow && \frac{\partial u}{\partial n} = \frac{\partial u}{\partial x} \\
&\text{On the top edge:} && y = d, && \mathbf{n} = \begin{bmatrix} 0 \\ 1 \end{bmatrix} && \Rightarrow && \frac{\partial u}{\partial n} = \frac{\partial u}{\partial y}
\end{aligned} \tag{2.18}$$

These expressions will be used in our codes later.

For a well-posed problem, we usually need to have sufficient DBCs to ensure a unique solution. If only NBCs are given, the Poisson's equation solution is not unique.

Note that there is also a Robin boundary condition, which is a mixture of DBC and NBC. Our focus will be on DBC and NBC.

2.5 PINN Class Construction

Conventional law-based numerical methods solve PDEs like Eq.(2.13) together with DBCs and/or NBCs using analytical or numerical techniques, such as the finite element method (FEM) [21], the smoothed finite element method (S-FEM) [20], and meshfree methods [11]. These methods rely on domain discretization to approximate the solution function in a **piece-wise** manner.

PINNs use neural networks (NNs) [12] as basis functions equipped with learnable parameters to approximate the solution. These parameters are optimized using the governing PDEs, boundary conditions, and initial conditions. The resulting approximation is a **single-piece** globally smooth function, which naturally supports various orders of differentiation required by the PDEs. Our study thus begins with the construction of PINN models using multilayer perceptron (MLP) neural networks.

2.5.1 A Plain PINN Class

We begin by creating a neural network (NN) that maps (x, y) to u. A commonly used architecture is the MLP [12], which can be implemented easily. The following snippet defines a class for building such a plain NN model using PyTorch's `nn.Module`. Detailed comments are provided, and it is self-explanatory. Beginning readers are advised to study this snippet carefully, as it helps to comprehend other PINN classes to be presented later in this book (with much fewer comments).

```
class PINN2D(torch.nn.Module):
    """
    A foundational Physics-Informed Neural Network (PINN) model designed
    for solving PDEs in 2D. This class constructs a fully connected
    neural network (MLP) that maps 2D input coordinates (`x` and `y`)
    to a single scalar output, represents the predicted solution `u`
    of a partial differential equation (PDE).
    Args:
        Nn (int): number of neurons in each hidden layer~ of the MLP.
        NL (int): total number of hidden Layers in the MLP. A value of 1
           means there's one hidden layer between the input and output.
        actit (callable, optional): activation function to be applied
           after each hidden layer's affine transformation. Common choices
           include `torch.tanh` (Default), `torch.sin`, `torch.sigmoid`.
    """
    def __init__(self, Nn, NL, actit=torch.tanh):
        # Call the constructor of the parent class (nn.Module):
        super(PINN2D, self).__init__()
        self.actit = actit     # Store the chosen activation function as
```

```
                # an instance attribute for use in the forward pass

        # --- Define the Neural Network Architecture~ ---
        # `nn.ModuleList` is used to hold a list of submodules,
        #  ensuring they are registered as part of model's parameters.

        # Input Layer~: This is the first affine transformation. It maps
        # 2 input features (x, y) to 'Nn' neurons. It takes a tensor of
        # shape `[m, 2]` and outputs `[m, Nn]`.
        self.hidden = torch.nn.ModuleList([torch.nn.Linear(2, Nn)])

        # Hidden Layers: Extend the `hidden` ModuleList with `NL - 1`
        # additional linear layers. Each of these layers maps 'Nn' input
        # features to 'Nn' output features. Same width is used.
        self.hidden.extend([torch.nn.Linear(Nn,Nn) for _ in range(NL-1)])

        # Output Layer~: The final linear layer. It maps 'Nn' features
        # from the last hidden layer down to a single output feature,`u`.
        # This layer will produce a tensor of shape `[m, 1]`.
        self.out = torch.nn.Linear(Nn, 1)

        # Move all model parameters and buffers to the specified device
        # (CPU or GPU) for leveraging GPU if 'device' is set to 'cuda'.
        self.to(device)

    def forward(self, x):
        """
        Defines the forward pass (data flow) through the neural network.
        Args: x (torch.Tensor) -- The input data to the PINN. This is a
              2D tensor containing the coordinates of points of shape
              `[m, 2]`. It's crucial for `x` to have `requires_grad=True`
              if you intend to compute derivatives of the output with
              respect to the input `x` (which is typical in PINNs for
              calculating PDE and NBC residuals). `x` must also be on
              the same device as the model.
        Returns: torch.Tensor -- The network's prediction for `u` at the
              input `x`, returned as a tensor of shape `[m, 1]`.
        """
        # Initialize 'z' with input tensor 'x'. 'z' will be transformed
        # sequentially through each layer.
        z = x

        # Iterate through each layer in the `self.hidden` ModuleList.
        # For each hidden layer, apply affine transformation followed by
        # the specified activation function (`self.actit`).
        for layer in self.hidden:
            z = self.actit(layer(z))          # Apply activation function

        # Finally, pass the result through the output layer.
        # No activation function is applied here for PINNs,as the output
```

```
        # directly represents solution `u` without a bounded range.
        return self.out(z)
```

In the above construction of the PINN class, we allow the use of a user-defined activation function. Such functions can be easily defined, as illustrated in the following examples.

```
def my_actvtn(x):
    return torch.tanh(x)  # torch.sin(x), torch.relu, torch.sigmoid, ...
```

The user-defined activation function can then be passed as the `actit` argument when creating an instance of the `PINN2D` class. This flexibility allows users to experiment with various nonlinearities beyond standard choices like `Tanh`, `Sigmoid`, `sin`, `cos`, or `ReLU`, potentially improving model performance for specific PDE problems.

The following snippets include several utility functions designed to inspect the structure and complexity of the PINN model. These utilities display:

- the configuration of the neural network model layer by layer,
- the number of trainable parameters in each layer and in total, and
- the **pseudo-dimension** of the affine space defined by the model, which provides insight into the expressive capacity of the neural network. It is used later to determine the minimum number of samples needed to train the model via collocation, based on the **samples-neurons theorem** [12].

Such diagnostics are valuable for understanding the model's architecture, assessing computational cost, guiding architectural decisions, and most importantly, determining the sample numbers during the development and tuning of PINN models.

```
def count_parameters(pinn):
    '''Compute the total number of trainable parameters for an NN pinn.
    '''
    return sum(p.numel() for p in pinn.parameters() if p.requires_grad)
```

```
def print_model_parameters(pinn, print_table=True):
    '''
    Prints a detailed configuration of a PyTorch PINN model, including
    parameters for each layer, the total number of trainable parameters
    (P_total), and total pseudo-dimension of the affine space (P_star).
    Optionally, it can display this information in a formatted table.
    Args:
        pinn (torch.nn.Module): The PINN model to be inspected.
        print_table (bool, optional): If True, a detailed table of
                    parameters per layer will be printed. If False, only
                    P_star and P_total will be displayed. Defaults to True.
    Returns a tuple containing:
            - P_star (int): Total pseudo-dimension of the affine space.
            - P_total (int): Total number of trainable parameters.
    '''
    P_total = 0
    P_star = 0

    # Print table header if requested
```

```
    if print_table:
        print(f"{'Layer':<20}{'Param Shape':<15}{'# Params':<15}"
              f"{'affine_dim':<15}")
        print("-" * 61)                                    # Separator line

    # Iterate through all named parameters in the PINN model
    for name, param in pinn.named_parameters():
        # Only consider parameters that require gradients (trainable)
        if param.requires_grad:
            affin_dim = 0
            if param.dim() == 2:                     # For weight matrices
                affin_dim = param.size(1) # input dimension to the layer
            elif param.dim() == 1:                      # For bias vectors
                affin_dim = 1       # Bias adds 1 to the affine dimension

            P_star += affin_dim
            num_params = param.numel()           # Total number of elements
            P_total += num_params

            #Print detailed row for the parameter if print_table is true
            if print_table:
                print(f"{name:<20}{str(list(param.shape)):<15}"
                      f"{num_params:<15}{affin_dim}")

    if print_table:                          # Print table footer if requested
        print("-" * 61)

    # Print the summary of pseudo-dimension & total trainable parameters
    print(f"P* (Total pseudo-dimension of the affine space) = {P_star}")
    print(f"Total trainable parameters: {P_total}")

    return P_star, P_total
```

2.5.2 Example: A Simple PINN Instance

The following snippet builds a PINN instance, called `pinn`, with `NL` hidden layers, each containing `Nn` neurons:

```
p = 2              # number of "features" (for 2D PINN, it is 2: x and y)
Nn = 50                           # Number of neurons in each hidden layer
NL = 2                                          # Number of (hidden) Layers
P_star = (p + 1) + (Nn + 1) * NL          # P* number (affine space dim)
print(f"P* (Total pseudo-dimension of the affine space) = {P_star}")
```

```
P* (Total pseudo-dimension of the affine space) = 105
```

When needed, we can print the detailed configuration of the model, or just the pseudo-dimension P^* and the total number of trainable parameters P, using the following snippet — a step we will perform frequently throughout this book.

```
pinn = PINN2D(Nn=Nn, NL=NL)           # automatically to device in __init__
P_star, P_total = print_model_parameters(pinn)
#P_star, P_total = print_model_parameters(pinn, print_table=False) #P*,P
```

```
Layer               Param Shape    # Params       affine_dim
-------------------------------------------------------------
hidden.0.weight     [50, 2]        100            2
hidden.0.bias       [50]           50             1
hidden.1.weight     [50, 50]       2500           50
hidden.1.bias       [50]           50             1
out.weight          [1, 50]        50             50
out.bias            [1]            1              1
-------------------------------------------------------------
P* (Total pseudo-dimension of the affine space) = 105
Total trainable parameters: 2751
```

This foregoing table contains rich information about a 2D PINN model:

1. **Input Layer**:
   Because our PINN is 2D, the inputs or features are the two coordinates x and y. Hence, the shape of the weights for the first input layer is $[m, 2]$, where m is the number of samples used to train the neural network, which typically follows a multilayer perceptron (MLP) configuration.

2. **Output Layer**:
   Since we have only one unknown function u (for the Poisson's equation), the output consists of a single neuron. The shape of the output layer must therefore be $[m, 1]$. As such, our PINN essentially represents a single, smooth, whole-piece neural network function that performs a $2 \to 1$ mapping. This is the neural network representation of $u(x, y)$: for any given (x, y), it outputs u. Various derivatives of u with respect to x and y are computed using `autograd` within the network.

3. **Hidden Layers and Trainable Parameters**:
   This example PINN has two hidden layers, in addition to the input and output layers. Thus, it has a total of four layers, three of which contain learnable (or trainable) parameters. In this particular model, the total number of trainable parameters (weights and biases) is 2751. The formula used to compute this number in the earlier code can be found in Ref. [12].

4. **Pseudo-Dimension and Minimum Sample Requirement**:
   Most importantly, for this PINN model, the P^*-number is 105, which represents the total **pseudo-dimension** of the **affine space** of the entire `pinn`. It is also equal to the **number of neurons** used in the model. According to the **samples-neurons theorem** [12], the minimum number of total samples (m) required to train this PINN must satisfy

$$m \geq P^* \tag{2.19}$$

Condition (2.19) is important because it ensures that a sufficient number of samples are used to train the model so that the solution can be considered reliable. A detailed discussion of the samples-neurons theorem, related issues, and the derivation of the formula for computing the P^*-number can be found in Ref. [12].

Note that in standard neural network models, m may be interpreted as the mini-batch size. In our PINN context, however, m refers to the total number of sampling points (collocation points) used to train the model. The condition $m > P^*$ will be followed for all PINN models trained throughout this book.

For static problems without data, the composite loss given in Eq.(2.6) for a PINN constructed using the plain template `PINN2D` reduces to:

$$\mathcal{L} = w_{\text{PDE}}\mathcal{L}_{\text{PDE}} + w_{\text{DBC}}\mathcal{L}_{\text{DBC}} + w_{\text{NBC}}\mathcal{L}_{\text{NBC}} \tag{2.20}$$

It has three terms in general for static problems. If the problem does not have any NBC, the loss reduces to two terms:

$$\mathcal{L} = w_{\text{PDE}}\mathcal{L}_{\text{PDE}} + w_{\text{DBC}}\mathcal{L}_{\text{DBC}} \tag{2.21}$$

To obtain a unique solution, the problem must include some DBCs. Therefore, at least two terms are required if the plain template `PINN2D` is used.

2.5.3 A PINN Class with DBCs Enforced

Based on extensive studies in the standard numerical methods community, it is well known that when the approximated function satisfies sufficient Dirichlet boundary conditions (DBCs), the solution becomes stable. This observation holds generally for both strong and weak form methods [11].

A similar behavior is expected for PINN models. PDEs can be solved more effectively by enforcing DBCs exactly through the use of a trial solution when constructing the PINN.

Consider a rectangular domain $[a, b] \times [c, d]$. The DBC function, f_{DBC}, that vanishes on all four boundary edges of this rectangle can be defined as

$$f_{\text{DBC}}(x, y) = (x - a)(b - x)(c - y)(d - y) \tag{2.22}$$

This function evaluates to zero along all four edges of the rectangular domain. When all four edges are subject to Dirichlet boundary conditions with $\bar{u} = 0$, the trial solution for the unknown function $u(x, y)$ (which is zero on all edges) can be expressed using

$$\begin{aligned} u_{\text{trial}}(x, y) &= f_{\text{DBC}}(x, y)\mathcal{N}_{\hat{\mathbf{w}}}(x, y) \\ &= (x - a)(b - x)(c - y)(d - y)\,\mathcal{N}_{\hat{\mathbf{w}}}(x, y) \end{aligned} \tag{2.23}$$

where $\mathcal{N}_{\hat{\mathbf{w}}}(x, y)$ is a standard plain MLP neural network used in the PINN with trainable parameters $\hat{\mathbf{w}}$.

With $u_{\text{trial}}(x, y)$, we only need to train $\mathcal{N}_{\hat{\mathbf{w}}}(x, y)$ while ensuring that the DBCs are automatically satisfied. This approach significantly reduces the search space, as it eliminates the need to explicitly enforce DBCs through a boundary loss term. Consequently, a PINN constructed using Eq.(2.23) is usually expected to be easier to train, since the total loss includes only a single term, `loss_PDE`, which is quadratic in u.

The following snippet demonstrates how to construct a PINN with DBCs enforced at the class level, assuming that all edges of the rectangular domain are fixed with Dirichlet boundary conditions. The code is quite similar to the plain PINN2D; the only difference lies in the handling of the DBCs, and hence should be quite easy to follow.

```
class PINN2D_4DBC(nn.Module):
    """
    2D PINN model with Dirichlet boundary conditions (DBCs) (u=0)
    enforced via a trial solution on all 4 edges of a rectangular domain.
    Domain: (a, b) × (c, d), where u = 0 on all 4 edges.
    Args: Nn (int): Number of neurons per hidden layer.
          NL (int): Number of hidden layers.
          a, b, c, d (float): Domain boundaries.
          actit (callable), activation function (torch.relu, etc.)
    """
    def __init__(self, Nn, NL, actit=torch.tanh, a=0., b=1., c=0., d=1.):
        super(PINN2D_4DBC, self).__init__()
        self.actit = actit
        self.a = a
        self.b = b
        self.c = c
        self.d = d

        self.hidden = nn.ModuleList([nn.Linear(2, Nn)])     # Input layer
        # Hidden layers:
        self.hidden.extend([nn.Linear(Nn, Nn) for _ in range(NL - 1)])
        self.out = nn.Linear(Nn, 1)                        # Output layer

    def forward(self, x):
        x_ = x[:, 0:1]
        y_ = x[:, 1:2]

        # Generalized zero-DBC trial multiplier
        f_DBC = (x_-self.a) * (self.b-x_) * (y_-self.c) * (self.d-y_)

        z = x
        for layer in self.hidden:
            z = self.actit(layer(z))
        N = self.out(z)

        u_trial = f_DBC * N             # output in a trail solution form
        return u_trial
```

As shown in the snippet, the DBCs are enforced in a trial solution form to regulate the tunable component $\mathcal{N}_{\hat{\mathbf{w}}}(x, y)$. When the PINN2D_4DBC class is used, the DBCs on all four edges of the rectangular domain are automatically enforced. We will see in later examples why this is important.

The loss function is constructed based on the residuals of the PDE. Thus, we require a mechanism for computing these residuals.

2.6 Residual Evaluation via Collocation

The PDE defined in Eq.(2.13) is evaluated at a set of points denoted by $\mathbf{x}_i = (x_i, y_i)$ within the problem domain Ω. These can be collocation points used for training the PINN or evaluation points where the solution u is needed after training. In either case, the computation is performed using the neural network built in the previous section.

The input to the PINN is a 2D array containing the x and y coordinates of the evaluation points. In general, the value computed by the neural network will not satisfy the PDE exactly, resulting in a non-zero output—called the **residual**—which forms the basis of the loss function.

The required derivatives are computed using `autograd`, as provided by a Python module such as PyTorch, via a back-propagation procedure [12].

Since the PDE involves a Laplace term, we use the following function to compute it for later use.

```
def laplacian~(u, xy):
    '''
    The Laplacian of u (x,y), Δu,  using the autograd built in PyTorch.
    '''
    grads = torch.autograd.grad(u, xy, torch.ones_like(u),
                                            create_graph=True)[0]
    u_x = grads[:, 0:1]
    u_y = grads[:, 1:2]
    u_xx = torch.autograd.grad(u_x, xy, torch.ones_like(u_x),
                                    create_graph=True)[0][:, 0:1]

    u_yy = torch.autograd.grad(u_y, xy, torch.ones_like(u_y),
                                    create_graph=True)[0][:, 1:2]
    return u_xx + u_yy                                  # the Laplace Δu
```

The following snippet computes the residual for any given set of, for example, collocation points using Eq.(2.7).

```
def residual_clctn(pinn, f_th, x, weights=None, x_NBC=None, t_NBC=None):
    '''
    Compute the residual in the PDE using NN model at a set of points
    given in array x for 2D Poisson's equations.
    piin:    the NN model for PINN.
    f_th:     force/source function~ in Pytorch.
    x:        array of [m, 2] collocation points.
    weights: torch tensor of weights for integration.
              Not used for collocation schemes, but kept in signature.
    '''
    x.requires_grad = True

    x1 = x[:, 0:1]                               # get x and y coordinates
    x2 = x[:, 1:2]
    x1_device = x1.to(device)                      # do this defensively
    x2_device = x2.to(device)
```

```
    u = pinn(x)                  # u will inherit grad_u if x requires_grad

    # Residual for Poisson's equation, Δu + f = 0:
    Δu = laplacian(u, x)
    residuals = (Δu + f_th(x1_device, x2_device)) ** 2

    # Mean squared residuals~
    return torch.mean(residuals), residuals
```

Note that residual computation is used not only for training the model, but also for evaluating the trained model at a set of points that are usually differ from those used during training.

2.7 PINN Models via Collocation

Consider the example problem defined in Eqs.(2.13) and (4.2), with Dirichlet boundary conditions imposed on all four edges.

2.7.1 Generation of Collocation Points in Rectangle

To compute the residual defined in Eq.(2.7), the following snippet generates a set of **collocation** points for evaluating the PDE over a rectangular domain. This utility will be used frequently throughout the book.

```
def sampleD_clctn(N_clctn=1500, a=0.0, b=1.0, c=0.0, d=1.0):
    '''
    Sampling N_clctn collocation points inside a rectangular Domain
    defined by (a,b)×(c,d) using uniform random sampling.
    Returns: w_s_th (dummy weight tensor)
             x_Ω_th (collocation points tensor)
    '''
    # Sample N_clctn random points uniformly in (a,b)x(c,d)
    x_vals = np.random.uniform(a, b, N_clctn)
    y_vals = np.random.uniform(c, d, N_clctn)
    x_Ω = np.vstack([x_vals, y_vals]).T                # shape (N_clctn, 2)

    w_s = 1. # Dummy weight for collocation as a tensor
    w_s_th = torch.tensor(w_s,  dtype=torch.float32)

    # Convert to tensor, SET requires_grad=True, and THEN move to device
    x_Ω_th = torch.tensor(x_Ω, dtype=torch.float32,
                                requires_grad=True).to(device)

    return w_s_th.to(device), x_Ω_th              # w_s also moved to device
```

2.7.2 Generation of Uniform Points on Edges of Rectangle

To enforce DBCs, we also need to sample points on the boundary of the rectangular domain $(a, b) \times (c, d)$. Since the boundary consists of straight line segments, we can easily apply uniform sampling along each of these edges. The following code snippet samples boundary points on all four edges and assigns zero Dirichlet boundary conditions ($u = 0$) to all of them.

```
def sample4DBC_rect(N_DBC=20, a=0.0, b=1.0, c=0.0, d=1.0):
    '''
    Sample data points on all 4 edges of rectangular domain (a,b)x(c,d).
    Prepare Dirichlet boundary condition (DBC) data: u = 0 on 4 edges.
    N_DBC: number of points to sample ALONG each edge
    return: x_DBC_th, coordinates of points on 4 edges
            u_DBC_th, corresponding u values
    '''
    x_DBCs = []
    x_DBCs.extend([[x_i, c] for x_i in np.linspace(a,b, N_DBC)])# bottom
    x_DBCs.extend([[x_i, d] for x_i in np.linspace(a,b, N_DBC)])   # top
    x_DBCs.extend([[a, y_i] for y_i in np.linspace(c,d, N_DBC)])  # left
    x_DBCs.extend([[b, y_i] for y_i in np.linspace(c,d, N_DBC)]) # right
    # np.linspace includes both endpoints.

    # Convert to numpy array and remove duplicates
    x_DBC = np.unique(np.array(x_DBCs), axis=0)
    u_DBC = np.zeros((x_DBC.shape[0], 1))        # Specify u values (u=0)

    # requires_grad=True for possible plotting normal gradients for NBC)
    x_DBC_th = torch.tensor(x_DBC, dtype=torch.float32,
                                     requires_grad=True).to(device)
    u_DBC_th = torch.tensor(u_DBC, dtype=torch.float32).to(device)
                                     # given target values no need grad
    return x_DBC_th, u_DBC_th
```

2.7.3 Plot Function for Sample Points in Rectangle

To visualize the sampled points, we use the following snippet to plot them after the sampling is completed.

```
def plotSamplePointsRect(x_Ω, x_DBC, a=0.0, b=1.0, c=0.0, d=1.0,
                                        dpi=300, f_name='tmp.png'):
    '''Plot sampled collocation points.
    x_Ω:   points inside (a,b)x(c,d)
    x_DBC: points on domain edges
    '''
    x_Ω   = x_Ω.detach().numpy()
    x_DBC = x_DBC.detach().numpy()

    fig, ax = plt.subplots(figsize=(4, 3.5), dpi=dpi)
    ax.scatter(x_Ω[:, 0], x_Ω[:, 1], s=1, color='blue',label=
                                                 'Points in domain')
    ax.scatter(x_DBC[:, 0], x_DBC[:, 1], s=2, color='red',label=
                                               'Points on boundary')
    ax.set_xlabel('$x$'); ax.set_ylabel('$y$')
    ax.set_title(f'Points in ({a},{b})×({c},{d})')
    ax.legend(); ax.set_aspect('equal')
    ax.grid(True); fig.tight_layout()
```

```
    fig.savefig(f_name+'.png', dpi=500)
    plt.show(); plt.close(fig)
```

2.7.4 Example: Distribution of Collocation Points

```
set_seed_th(88)                                    # for reproducibility

# Sampling over rectangular domain (a, b) × (c, d):
a = 0.0; b = 1.0
c = 0.0; d = 1.0

# Sample collocation and boundary points
N_sample = 50
DI_rule = ''                   # use collocation no Integration rule is used
w_s, x_Ω_th = sampleD_clctn(N_clctn=N_sample,  a=a, b=b, c=c, d=d)
x_DBC_th, u_DBC_th = sample4DBC_rect(N_DBC=10, a=a, b=b, c=c, d=d)

# Plot sampled points
plotSamplePointsRect(x_Ω_th, x_DBC_th, a=a, b=b, c=c, d=d, dpi=300,
                                f_name='./imagesPN/collocation_points')
```

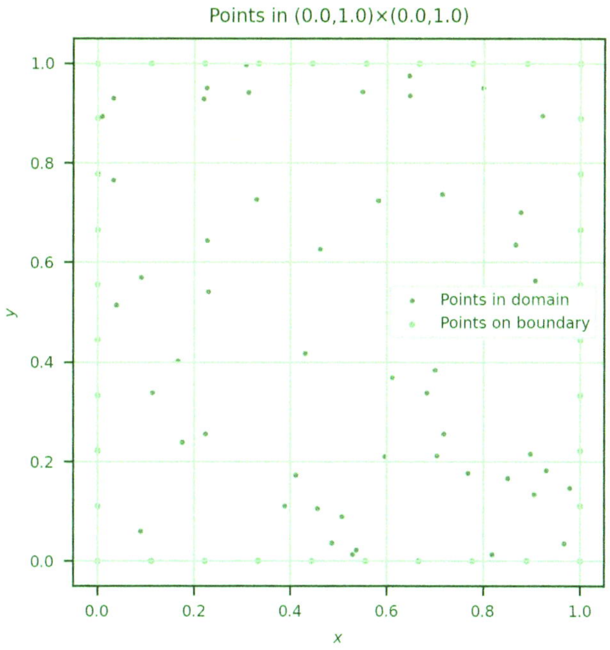

A typical case of randomly generated collocation points in the square domain $(0,1)\times(0,1)$ and uniformly sampled points along all four boundary segments is shown in the foregoing figure.

2.7.5 Trainer for PINN Models via Collocation

The residual is then minimized during training by optimizing the composite loss function defined in Eq.(2.21), which consists of two terms: the PDE loss and the DBC loss. The following snippet defines this composite loss function and computes the gradients of the loss with respect to all the trainable parameters in the constructed pinn model. Detailed comments are provided at the key steps.

```
def trainerPDE_DBC(pinn, residual, optimizer, f_th, x_Ω_th, w_s_th,
                x_DBC_th, u_DBC_th, x_NBC_th, t_NBC_th, w_NBC_th, I_NBC,
                                                             epochs=201):
    """ Train a PINN using loss_PDE + loss_DBC.
    Args: pinn: the PINN model (nn.Module)
          residual: function to compute residual loss
          optimizer: PyTorch optimizer (e.g. Adam, LBFGS)
          epochs: number of training iterations for PINN
    """
    loss_history = {"total": [], "PDE": [], "DBC": []}

    if epochs <= 301: n_prnt = 100
    else:             n_prnt = max(1, round(epochs / 3, -2))
    # Print roughly every epochs/300 steps, rounded to nearest 100

    for epoch in range(epochs):
        # Initialize loss variables for the current epoch
        loss_PDE = None; loss_DBC = None; total_loss = None

        def closure():
            nonlocal loss_PDE, loss_DBC, total_loss
            # Allow modifying variables from outer scope
            optimizer.zero_grad()
            # These calculations happen outside of torch.no_grad()
            # x and x_DBC must have requires_grad=True here
            loss_PDE, _= residual(pinn, f_th, x_Ω_th, w_s_th, x_DBC_th,
                                                              u_DBC_th)
            u_pred_DBC = pinn(x_DBC_th)
            loss_DBC   = torch.mean((u_pred_DBC-u_DBC_th)**2)
            total_loss = loss_PDE + loss_DBC                   # two terms

            total_loss.backward()                  # Perform backward pass

            return total_loss  # Return the total loss for the optimizer

        if isinstance(optimizer, torch.optim.LBFGS):
            # LBFGS optimizer step has multiple closure calls internally
            optimizer.step(closure)                # iteration in the step
            # loss_PDE, loss_DBC, total_loss were updated by the last
            total_loss_val = total_loss.item()             # closure call
            loss_PDE_val   = loss_PDE.item()
            loss_DBC_val   = loss_DBC.item()
        else: # Adam or other optimizers
            # closure performs forward pass, loss clcltn & backward pass
            total_loss_tensor = closure() #returns the total_loss tensor
            optimizer.step()
```

```
            # Get item values from the tensors updated by closure
            total_loss_val = total_loss_tensor.item()
            loss_PDE_val   = loss_PDE.item()   # It's set inside closure
            loss_DBC_val   = loss_DBC.item()

        # Logging happens with loss values obtained from optimizer step
        # This logging section is implicitly outside the graph
        # We can add torch.no_grad() here just to be explicit that no
        with torch.no_grad():              # graph is built for logging
            loss_history["total"].append(total_loss_val)
            loss_history["PDE"].append(loss_PDE_val)
            loss_history["DBC"].append(loss_DBC_val)

        if epoch % n_prnt == 0 or epoch == epochs - 1:
            print(f"Epoch {epoch:5d}, Loss: {total_loss_val:.3e}, "
                  f"PDE: {loss_PDE_val:.3e}, DBC: {loss_DBC_val:.3e}")

    return loss_history
```

To train a PINN effectively, we use both a first-order optimizer , typically adam, and a quasi-second-order optimizer , typically lbfgs. The following function schedules a two-stage training process using adam as the initial optimizer and lbfgs for fine-tuning:

```
def train_adam_lbfgs(pinn, trainer, residual, f_th, x_Ω_th, w_s_th,
            x_DBC_th, u_DBC_th, x_NBC_th, t_NBC_th, w_NBC_th, I_NBC,
                         epochs1=201, epochs2=11, use_lbfgs=True):

    '''Train a PINN model using both adam and lbfgs.
       adam trains initial epochs1, and lbfgs fine tuns epochs2 ecpochs.
    '''
    pinn.to(device) # Move model to device BEFORE creating optimizers.

    start_time = time.perf_counter()
    # 1. Train with Adam
    adam = torch.optim.Adam(pinn.parameters(), lr=1e-3)
    print("="*50)
    print(f"{epochs1} epoches training with Adam:")

    history_adam = trainer(pinn, residual, adam, f_th, x_Ω_th, w_s_th,
            x_DBC_th, u_DBC_th, x_NBC_th, t_NBC_th, w_NBC_th, I_NBC,
                                                    epochs=epochs1)
    end_time = time.perf_counter()
    time_adam = end_time - start_time
    print(f"Time taken by adam: {time_adam:.6f} seconds")

    if use_lbfgs:                               # Use the renamed flag
        start_time = time.perf_counter()
        # 2. Fine-tune with LBFGS. # Create NEW LBFGS optimizer instance
        lbfgs_optimizer = torch.optim.LBFGS(pinn.parameters(), lr=1.0,
            max_iter=500, tolerance_grad=1e-9, tolerance_change=1e-9,
```

```
                history_size=100, line_search_fn='strong_wolfe')
        print("\n" + "="*50)                        # standard
        print(f"{epochs2} epoches fine-tuning with LBFGS:")

        # Pass the SAME tensors that were used for Adam training
        history_lbfgs = trainer(pinn, residual, lbfgs_optimizer,
                                f_th, x_Ω_th, w_s_th, x_DBC_th, u_DBC_th,
                   x_NBC_th, t_NBC_th, w_NBC_th, I_NBC, epochs=epochs2)

        print(f"\nTraining complete.")
        end_time = time.perf_counter()
        time_lbfgs = end_time - start_time
        print(f"Time taken by lbfgs: {time_lbfgs:.6f} seconds")
        totalTime = time_adam + time_lbfgs
        print(f"Total time: {totalTime:.6f} seconds")
        print(f"Ratio (time_lbfgs/time_adam) of time taken per epoch:"
              f"{((time_lbfgs/epochs2)/(time_adam/epochs1)):.6f}")

        return (history_adam, history_lbfgs), totalTime
    else:
        time_lbfgs = 0.
        print(f"\nTraining complete.")
        return history_adam, time_adam      # Only Adam history and time
```

2.7.6 Case Study: Build and Train PINN via Collocation

This case study builds and trains a PINN for Poisson's equation over a unit square domain with all edges fixed at zero ($\bar{u} = 0$).

2.7.6.1 Create a Plain PINN Model and Sampling Points

The following snippet builds a PINN model and then generate sampling point for the model.

```
set_seed_th(88)                                   # for reproducibility

# PINN Model Parameters:
p  = 2                               # Input dimension (x, y) (features)
Nn = 50           # Number of neurons in a hidden layer (one may change)
NL = 2                         # Number of hidden Layers (one may change)

# Create a plain PINN model:
pinns = []                       # for logging the pinn models to be used
pinn = PINN2D(Nn=Nn, NL=NL)         # automatically to device in __init__
P_star, P_total = print_model_parameters(pinn, print_table=False)

w_PDE = 1.                                        # weight for loss_PDE
# Sampling over rectangular domain (a, b)×(c, d):
a = 0.0; b = 1.0
c = 0.0; d = 1.0
N_sample = 1*P_star     # m is N_sample, this ensures condition: m >= P*
```

```
# Using N_sample is the collocation points in the domain
DI_rule = ''            # use collocation no Domain Integration rule is used
w_s_th, x_Ω_th = sampleD_clctn(N_clctn=N_sample, a=a, b=b, c=c, d=d)

N_DBC=2*int(sqrt(N_sample))+1          # when using N_DBC (not used here)
x_DBC_th, u_DBC_th = sample4DBC_rect(N_DBC=N_DBC, a=a, b=b, c=c, d=d)

print(f"Sampled {x_Ω_th.shape[0]} collocation points in the domain.")
print(f"Is N_sample >= P* ? {N_sample >= P_star}")
print(f"Sampled {x_DBC_th.shape[0]} boundary points.")
```

```
P* (Total pseudo-dimension of the affine space) = 105
Total trainable parameters: 2751
Sampled 105 collocation points in the domain.
Is N_sample >= P* ? True
Sampled 80 boundary points.
```

This particular PINN model has two hidden layers, each with 50 neurons. The training dataset consists of 105 randomly generated sampling points inside the domain, plus 80 uniformly distributed sampling points on the four edges.

The number of samples, N_{sample} (which corresponds to m), is set equal to P^*, the total pseudo-dimension of the affine space of the entire PINN model. It also represents the total number of neurons used in the PINN model (excluding the last output neuron, which does not participate in the affine transformation). With this setting, the samples-neurons theorem in Eq.(18) is always satisfied [12], regardless of the sampling points used for Dirichlet boundary conditions (DBCs) and Neumann boundary conditions (NBCs). When a PINN model includes sampling points for DBCs and NBCs, these points are treated as extra, and Eq.(18) remains valid.

Therefore, in all PINN models we train, we consistently set $N_{\text{sample}} \geq P^*$ to ensure the samples-neurons theorem is always observed.

Note that P^* represents the minimum required number of sampling points. It guarantees solution existence for the PINN model, as long as these points are linearly independent [12]. To obtain more accurate solutions, we often use more sampling points such that

$$N_{\text{sample}} \gg P^*$$

This is particularly important when sufficiently dense sampling is needed to resolve sharply varying functions involved in the problem. For example, a sharply changing source function $f(x, y)$ over the domain.

2.7.6.2 Training of the PINN Model

The following snippet trains this PINN model.

```
Trainers = []; Residuals = []; totalTimes = []
LossTerms= []                                   # for logging loss terms
LossTerm = 'DBC'                       # loss term in addition to loss_PDE
x_NBC_th = None; t_NBC_th = None; w_NBC_th = None; I_NBC=''
```

```
Trainer  = trainerPDE_DBC
Residual = residual_clctn
losses, totalTime = train_adam_lbfgs(pinn, Trainer, Residual, f_th,
       x_Ω_th, w_s_th, x_DBC_th, u_DBC_th, x_NBC_th, t_NBC_th, w_NBC_th,
                        I_NBC, epochs1=201, epochs2=11, use_lbfgs=True)
```

```
==================================================
201 epoches training with Adam:
Epoch    0, Loss: 7.961e+01, PDE: 7.958e+01, DBC: 2.804e-02
Epoch  100, Loss: 1.029e+01, PDE: 6.998e+00, DBC: 3.296e+00
Epoch  200, Loss: 4.244e-01, PDE: 3.319e-01, DBC: 9.243e-02
Time taken by adam: 2.254071 seconds

==================================================
11 epoches fine-tuning with LBFGS:
Epoch    0, Loss: 2.599e-04, PDE: 1.311e-04, DBC: 1.288e-04
Epoch   10, Loss: 6.466e-06, PDE: 5.086e-06, DBC: 1.380e-06

Training complete.
Time taken by lbfgs: 16.077251 seconds
Total time: 18.331322 seconds
Ratio (time_lbfgs/time_adam) of time taken per epoch:130.330977
```

After training, the total loss has decreased to approximately the order of 10^{-6} from an initial order of about 10^1. This corresponds to a reduction in loss by about seven orders of magnitude, which is a strong indication that the model is well trained. The loss_DBC has decreased to roughly 10^{-6} from an initial order of about 10^{-2}, amounting to a reduction of about three orders of magnitude. Since solution accuracy typically depends on the worst case, we may expect the solution accuracy to be on the order of 10^{-3}. This provides a rough estimate for gauging the accuracy of a trained PINN model. A more precise assessment should be done by comparing against the exact solution or a reliable reference solution to the problem.

Note that a second-order optimizer, such as `lbfgs`, runs significantly slower than first-order optimizers. In this example, the time per epoch using `lbfgs` is about 100 times longer than with `adam`. This is because second-order optimizers typically require many sub-iterations within each epoch, performed inside a closure, and each sub-iteration takes more time. Therefore, it is common practice to run `adam` for a sufficient number of epochs to bring the learning parameters close to optimal, then run `lbfgs` only for a few epochs to fine-tune the solution. This strategy is used throughout this book.

2.7.6.3 Plot the Solution Contours

To quantitatively measure the error of the predicted solution from a trained PINN, we typically sample another set of m points within the problem domain. The following snippet computes and prints the relative L^2 error compared to the exact solution. The relative L^2 error is defined as follows:

$$\text{error\_L2} = \frac{\|u_{\text{pred}} - u_{\text{exact}}\|}{\|u_{\text{exact}}\|} \tag{2.24}$$

where the L^2 norm is used for both the difference between the predicted and exact solutions, as well as for the exact solution itself, evaluated at all the m sampled points for evaluation. We also compute the error variation index, errorVI, which is defined as

$$\text{errorVI} = \frac{\max_1^m |u_{\text{pred}} - u_{\text{exact}}|}{\frac{1}{m}\sum_1^m |u_{\text{pred}} - u_{\text{exact}}|} \tag{2.25}$$

It measures the largest absolute error relative to the mean absolute error. In general, a good model should have a small L^2 error, and the largest error should not be significantly larger than the average error. The following Python function computes these metrics:

```
def sln_error(u_pred, u_exact):
    '''
    Assess the error in `u_pred` against `u_exact`.
    error_L2: L2 error; errorVI: error Variation Index
    '''
    u_diff = u_pred - u_exact

    error_max = np.nanmax(np.abs(u_diff))
    print(f"Maximum Error: {error_max:.4e}")
    error_mean = np.nanmean(np.abs(u_diff))
    print(f"Mean Absolute Error: {error_mean:.4e}")

    errorVI = error_max / error_mean if error_mean != 0 else np.inf
    print(f"Error Variation Index (ratio of Max/Mean): {errorVI:.4e}")

    norm_exact = lg.norm(u_exact[~np.isnan(u_exact)], ord=2)
    if norm_exact == 0:
        error_L2 = np.inf
    else:
        error_L2 = lg.norm(u_diff[~np.isnan(u_exact)],ord=2)/norm_exact
    print(f"L2 Relative Error: {error_L2:.4e}")

    return error_L2, errorVI, u_diff
```

Since our problem is 2D, we can plot also the contours of the solutions, which can be done using the following snippet.

```
def plot_sln_rect(pinn, exact_u, f_th=None, N_plot=100, vmin=0, vmax=1,
            a=0., b=1., c=0., d=1., shrink=0.6, dpi=300, plot_it=True):
    '''
    Plot results over domain (a,b)x(c,d), pinn_pred and exact ones.
    return: error_L2, L2 error between pinn_pred and exact
            errorVI, error Variation Index.
            u_diff, difference between pinn_pred and exact
    '''
    x = np.linspace(a, b, N_plot)     # Create meshgrid over (a,b)x(c,d)
    y = np.linspace(c, d, N_plot)
    X, Y = np.meshgrid(x, y)
    XY = np.hstack((X.flatten()[:, None], Y.flatten()[:, None]))
```

```
    # Convert to PyTorch tensor and predict with model:
    XY_tensor = torch.tensor(XY, dtype=torch.float32)
    residual_pred = residual_clctn(pinn, f_th, XY_tensor)
    print(f"L2 mean residual of trained pinn: {residual_pred[0]:.4e}")

    u_pred = pinn(XY_tensor)
    u_pred = u_pred.detach().numpy().reshape(N_plot, N_plot)

    # Compute exact solution
    u_exact = exact_u(X, Y, a=a, b=b, c=c, d=d)

    error_L2, errorVI, u_diff = sln_error(u_pred, u_exact)
    if plot_it !=True:
        return error_L2, errorVI, u_diff, residual_pred

    # Plot predicted and exact solution
    fig, axs = plt.subplots(1, 3, figsize=(8, 3), dpi=dpi)

    c0 = axs[0].contourf(X, Y, u_pred, vmin=vmin, vmax=vmax, levels=15)
    axs[0].set_title('PINN Solution')
    plt.colorbar(c0, ax=axs[0], shrink=shrink)

    c1 = axs[1].contourf(X, Y,u_exact, vmin=vmin, vmax=vmax, levels=15)
    axs[1].set_title('Exact Prediction')
    plt.colorbar(c1, ax=axs[1], shrink=shrink)

    c2 = axs[2].contourf(X, Y, np.abs(u_diff), levels=100)
    axs[2].set_title('|dfference|')
    plt.colorbar(c2, ax=axs[2], shrink=shrink)

    for ax in axs:
        ax.set_xlabel('$x$')
        ax.set_ylabel('$y$')
        ax.set_aspect('equal')
        ax.grid(linewidth=0.2)

    plt.tight_layout()
    plt.show()

    return error_L2, errorVI, u_diff, residual_pred
```

The following snippet computes the solution error and plots the contours of the PINN-predicted solution, the exact solution, and their difference.

```
# Plot the contours of both PINN solution and the exact one:
error_L2s = []; errorVIs = []                    # for logging the error
error_L2, errorVI, u_diff, _ = plot_sln_rect(pinn, exact_u, f_th=f_th,
          N_plot=200, vmin=0., vmax=1., a=a,b=b,c=c,d=d, plot_it=True)
```

```
L2 mean residual of trained pinn: 2.0440e-04
Maximum Error: 3.1372e-03
```

```
Mean Absolute Error: 3.1254e-04
Error Variation Index (ratio of Max/Mean): 1.0038e+01
L2 Relative Error: 1.0084e-03
```

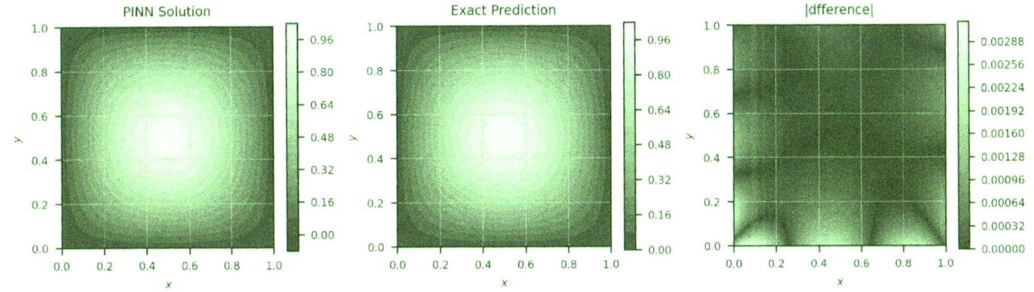

As seen, using the plain PINN2D template, a PINN model has been successfully trained, producing quite good results for this problem. The L^2 residual on the prediction (test) dataset is around the order of 10^{-4}, which aligns well with the reduction observed in the loss terms. The relative L^2 error compared to the exact solution is about the order of 10^{-3}.

In this plain PINN2D model, the loss_PDE and loss_DBC are evaluated separately and then combined to form the total loss as in Eq.(2.21). We observed that larger errors occur along the four DBC edges. Readers may consider introducing a weight for the DBC by setting w_{DBC} to a value larger or smaller than 1 to improve training effectiveness.

Let us now examine the history of these losses during training.

2.7.6.4 Plot of Loss History

The following two utility snippets plot the history of the loss terms throughout training. Since both adam and lbfgs optimizers were used consecutively, there are two separate loss histories — one for each optimizer — each containing two loss terms. The function plot3loss2 visualizes all these histories.

```
def plot3loss2(losses, optimizers, PDE='PDE', C1='DBC', C2='NBC',
                        C3='IC', C4='IV', CL='Loss', log=True, dpi=300):
    '''Plot training history of miltiple losses
       each may have various cases
    '''
    fig, axs = plt.subplots(1, 2, figsize=(7, 2), dpi=dpi)

    for i, ax in enumerate(axs):
        ax.plot(losses[i]["total"], label="Total "+CL)
        if len(losses[i]) >= 2: ax.plot(losses[i][PDE],label=PDE+' '+CL)
        if len(losses[i]) >= 3: ax.plot(losses[i][C1], label= C1+' '+CL)
        if len(losses[i]) >= 4: ax.plot(losses[i][C2], label= C2+' '+CL)
        if len(losses[i]) == 5: ax.plot(losses[i][C3], label= C3+' '+CL)

        ax.set_xlabel('Epoch')
        ax.set_ylabel('Losses')
        if log: ax.set_yscale("log")
        ax.legend()
```

```
        ax.grid(True)
        ax.set_title("PINN Training Loss History using "+optimizers[i])

    plt.tight_layout()
    plt.show()
```

Let us plot the history of various loss terms for the pinn model trained previously:

```
plot3loss2(losses, ('adam initially','lbfgs in fine tunning'),
                   PDE='PDE', C1='DBC', CL='Loss',log=False, dpi=300)
```

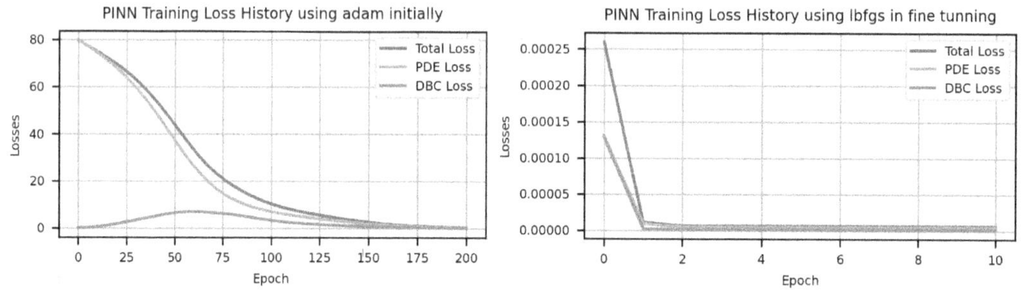

As observed in the example, both loss_{PDE} and loss_{DBC} decay nicely over the epochs during both the initial training with `adam` and the fine-tuning with `lbfgs`.

On epoch:

The term *epoch* comes from the mini-batch concept used in standard MLP models with stochastic gradient descent (SGD) algorithms such as `adam`, which is known for good performance. The training dataset is usually shuffled and divided randomly into multiple batches, each containing a small number of samples. One complete sweep over all these batches during training is called one **epoch** (see Section 9.5 of Ref. [12]).

2.7.6.5 On Stability and Efficiency

In our PINN context, however, each iteration uses *all* sampled data points in a single batch because these points represent specific locations tied to the physics of the PDE. Random reshuffling of these points is not appropriate for PINNs. Thus, each epoch in our PINN model corresponds to exactly one iteration of the gradient descent algorithm.

For training this particular model, we used 200 epochs of `adam` followed by 10 epochs of `lbfgs`. This means the PINN model is trained repeatedly on the *same* set of sampled data points. This approach differs significantly from standard direct solvers for PDEs [18], where data points are generally used only once. This explains why collocation methods in direct PDE solvers can be unstable, whereas PINNs using collocation points usually work well. On the other hand, because PINN training involves a large number of epochs, the computation time can be significantly higher compared to standard direct solvers.

We therefore note the following key insight:

> The PINN is usually stable and accurate when using collocation points because of the **repetitive use** of samples during optimization.

The PINN is usually less computationally efficient compared to standard direct solvers also because of the **repetitive use** of samples during optimization.

From the loss history curves, it is observed that `loss_PDE` decreases rapidly during the first 200 epochs with `adam`, but `loss_DBC` actually increases. This is because `loss_PDE` dominates the composite loss initially, so the `adam` optimizer focuses on balancing these terms while minimizing the total loss.

When switching to the `lbfgs` optimizer, a second-order method for fine tuning, both loss terms drop significantly and converge closer together. It takes only a few epochs (iterations) to reduce the losses to a steady low level. After a total of 210 epochs (200 with `adam` + 10 with `lbfgs`), the total loss reaches about 10^{-6} order. This two-stage training approach is typical and particularly effective in training PINNs.

2.7.6.6 Record Keeping

Since we plan to study various types of PINN models, it is useful to have a proper bookkeeping tool to track the performance of trained models on the same problem. The following snippet records the performance of this trained PINN model solving the Poisson equation over a unit square with all four edges fixed.

```
def recordSheet():
    '''
    Log the key performance indicators of PINN models for
    comparison study.
    '''
    new_error_L2_str = f"{error_L2:.3e}"

    if new_error_L2_str not in error_L2s:
        pinns.append(pinn.__class__.__name__)           # log pinn models
        Residuals.append(Residual.__name__ + DI_rule)    # log residuals
        #Trainers.append(Trainer.__name__)               # log trainers
        LossTerms.append(LossTerm) # additional terms in composite loss
        totalTimes.append(f"{totalTime:.3e}")                 # log time
        error_L2s.append(new_error_L2_str)               # log the error
        errorVIs.append(f"{errorVI:.3e}")        # error variation index
    else:
        print(f"Error_L2 value '{new_error_L2_str}'"\
              f" already exists in error_L2s. Skipping update.")

recordItems=[pinns, LossTerms, Residuals,totalTimes,error_L2s,errorVIs]
```

The following code prints out the records of all the models trained so far for the same problem.

```
def printSummary(pinns, LossTerms, Residuals, totalTimes, error_L2s,
                                                          errorVIs):
    '''Print out the summary of all trained PINN models.
    '''
    item = list(range(1, len(pinns)+1))
    print("{:<2} {:<16} {:<10} {:<11} {:<10} {:<10} {:<10}".format("#",
      "PINN model", "Loss_terms", "residual", "total_time", "error_L2s",
                                                          "errorVIs"))
```

```
    print("=" * 74)
    for i, p, t, r, tt, e, o in zip(item, pinns, LossTerms, Residuals,
                                    totalTimes, error_L2s, errorVIs):
        print("{:<2} {:<16} {:<10} {:<11} {:<10} {:<10} {:<10}".format(
                                    i, p[4:], t, r[8:], tt, e, o ))
```

Because of these logging and tracking measures, our code become a little more complicated. However, we do need these records so that we can draw proper conclusions based on data in a summarized tabular form. Let us create the record:

```
# Create record sheet (exercute this only once!):
recordSheet()
```

The following snippet is only used to empty the record or remove a record item, if it is created or updated wrongly for some reason.

```
#for item in recordItems: item.clear()       # to clear the entire record
#for item in recordItems: del item[-1]             # to delete one record
```

We can now print out the summary of the model trained:

```
printSummary(pinns, LossTerms, Residuals, totalTimes,error_L2s,errorVIs)
```

```
#  PINN model        Loss_terms residual    total_time error_L2s  errorVIs
==========================================================================
1  2D                DBC        _clctn      1.833e+01  1.008e-03  1.004e+01
```

So far, we have trained a single model, **pinn**, created using the 2D plain PINN class, with **loss_DBC** included as a loss term alongside the usual **loss_PDE** term used in all models. We used collocation sampling to compute the domain residual. The total training time (using both optimizers) is approximately 10 seconds, though this may vary. The relative L^2 error in the **pinn** predicted solution is about 0.1%, and the maximum absolute error is roughly 10 times the average absolute error.

2.7.6.7 Training with Adam Alone

Generally, achieving an accuracy on the order of 10^{-3} is quite challenging when using only the first-order optimizer **adam**. The following test demonstrates this using the same example problem:

```
set_seed_th(88)                                      # for reproducibility
# Re-create the PINN model afresh. This test will not be recorded.
pinn = PINN2D(Nn=Nn, NL=NL) # automatically moved to device in __init__

# Train it using ---adam only--- for a large number of epochs:
losses, totalTime = train_adam_lbfgs(pinn, Trainer, Residual, f_th,
        x_Ω_th, w_s_th, x_DBC_th, u_DBC_th, x_NBC_th, t_NBC_th, w_NBC_th,
                      I_NBC, epochs1=15001, epochs2=11, use_lbfgs=False)
```

```
==================================================
15001 epoches training with Adam:
Epoch      0, Loss: 7.961e+01, PDE: 7.958e+01, DBC: 2.804e-02
```

```
Epoch  5000, Loss: 2.395e-03, PDE: 4.160e-04, DBC: 1.979e-03
Epoch 10000, Loss: 3.549e-04, PDE: 1.071e-04, DBC: 2.479e-04
Epoch 15000, Loss: 4.391e-04, PDE: 3.482e-04, DBC: 9.093e-05
Time taken by adam: 125.856806 seconds

Training complete.
```

As seen, the total loss reduces by about four orders of magnitude (10^4), but the reduction in `loss_DBC` is only around two orders of magnitude (10^2), even after 15,000 epochs of training with `adam` (which is very fast per epoch). This indicates that the true minimum has not yet been reached. Also, the entire training process can take a couple of minutes, which is much slower than the two-optimizer strategy used previously. The following snippets plot the history of these loss terms from a single stage training.

```
def plot3loss(loss_h, PDE='PDE', C1='DBC', C2='NBC', C3='IC', CL='Loss',
                                                        log=True, dpi=300):
    '''
    Plot loss history of an optimizer with various loss terms
    '''
    plt.figure(figsize=(6, 2), dpi=dpi)                   # inches
    plt.plot(loss_h["total"], label="Total "+CL)
    if len(loss_h) >= 2: plt.plot(loss_h[PDE],label=PDE+' '+CL)
    if len(loss_h) >= 3: plt.plot(loss_h[C1], label= C1+' '+CL)
    if len(loss_h) >= 4: plt.plot(loss_h[C2], label= C2+' '+CL)
    if len(loss_h) == 5: plt.plot(loss_h[C3], label= C3+' '+CL)
    if log: plt.yscale("log")
    plt.xlabel("Epoch")
    plt.ylabel("Loss")
    plt.legend()
    plt.title("PINN Training Loss History using adam")
    plt.grid(True)
    plt.show()
```

```
plot3loss(losses, PDE='PDE', C1='DBC', CL='Loss', log=True)
```

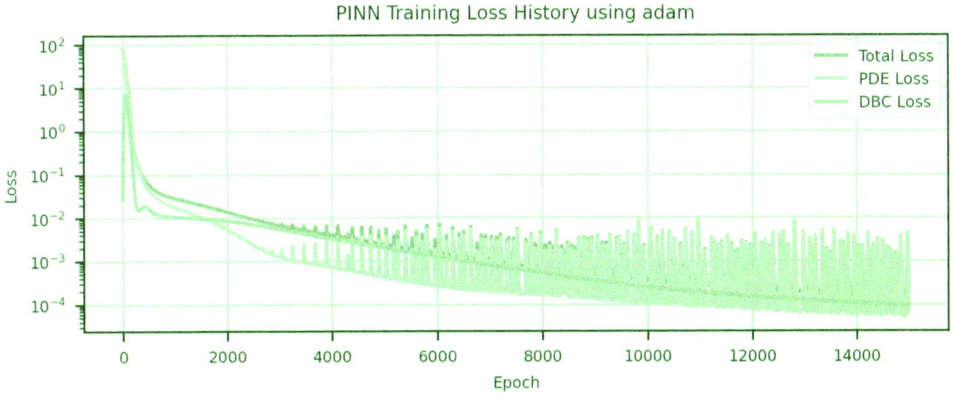

When using only a first-order optimizer (such as `adam`), the total loss decay roughly follows the typical gradient descent behavior (see Section 9.4 of Ref. [12]). More importantly, severe oscillations occur once the number of epochs reaches around 2500. This happens because, near the minimum, the gradients become very small; if the learning rate is not reduced accordingly, an overshooting phenomenon can arise, resulting in oscillations. Even with a learning rate schedule, effectively preventing such oscillations can be difficult, as finding a suitable schedule often requires multiple trial-and-error attempts. Moreover, the oscillation behavior is problem-dependent. To fully address this oscillation, one needs access to the second derivatives of the loss function near the stationary point [12]. Many studies have been conducted on overcoming such oscillations [3].

In the author's opinion, the most practical approach is to switch to a second-order optimizer (which involves the Hessian matrix, a second-order symmetric tensor of the loss function), or a quasi second-order optimizer like `lbfgs` (which approximates the Hessian matrix [17]). This approach can reduce the loss (for collocation-based problems like ours) to the order of 10^{-5}. If we want to reduce the loss even further, a full second-order optimizer or even higher-order derivatives (such as third derivatives) may be necessary. However, this becomes extremely challenging because it involves handling third-order tensors, which are computationally prohibitively expensive at present.

The following snippet plots the results and the error distribution of the trained model. The L^2 error is on the order of 10^{-2}. Readers may uncomment these lines to view the plots.

```
# # Plot the contours of both PINN solution and the exact one:
# plot_sln_rect(pinn, exact_u, N_plot=100, vmin=0., vmax=1.,
#                                            a=a,b=b,c=c,d=d)
```

2.7.7 Derivatives Using FDM

Alternatively, we can compute the spatial derivatives of various orders for u with respect to the coordinates x and y using the finite difference method (FDM). This approach is feasible because, with a trained neural network model, we can evaluate $u(x, y)$ at any location by simply querying the model using the current set of learned parameters.

First-order Derivatives via Finite Difference

Given that we can evaluate $u(x, y)$ at arbitrary points, we can estimate its partial derivatives to any desired order using standard FDM schemes. For partial derivatives, we hold the other coordinate constant while varying the one of interest:

- To compute $\frac{\partial u}{\partial x}$, we fix y and vary x.
- To compute $\frac{\partial u}{\partial y}$, we fix x and vary y.

The same principle applies to higher-order derivatives. Using the central finite difference approximations for the first-order partial derivatives in 2D are given by:

$$\begin{aligned} u_{,x}(x, y) &\equiv \frac{\partial u}{\partial x}(x, y) \approx \frac{u(x+h, y) - u(x-h, y)}{2h} \\ u_{,y}(x, y) &\equiv \frac{\partial u}{\partial y}(x, y) \approx \frac{u(x, y+h) - u(x, y-h)}{2h} \end{aligned} \tag{2.26}$$

where h is a small step size away from x or y, and it should be much smaller than the size of the domain to ensure good accuracy.

Second-Order Derivatives via Finite Difference

The second-order partial derivatives for 2D cases can be computed using central finite difference formulas twice, which gives

$$
\begin{aligned}
u_{,xx}(x, y) &\equiv \frac{\partial^2 u}{\partial x^2} \approx \frac{u(x-h, y) - 2u(x, y) + u(x+h, y)}{h^2} \\
u_{,yy}(x, y) &\equiv \frac{\partial^2 u}{\partial y^2} \approx \frac{u(x, y-h) - 2u(x, y) + u(x, y+h)}{h^2}
\end{aligned} \tag{2.27}
$$

These finite difference formulas provide a practical way to estimate derivatives numerically without requiring autograd tools.

The following snippet define the PDE residual computed using the FDM for an NN model for the Poisson's equation:

```python
def residual_FDM(pinn, f_th, x, x_NBC, g_NBC, weights=1.0, h=0.02):
    '''
    Compute the residual using the finite difference method (FDM)
    Parameters:
    - pinn: a PINN model
    - f_th: forcing or source function f(x, y), returns torch.Tensor
    - x: interior points as a tensor of shape [m, 2]
    - weights: optional (not used in collocation-based residuals)
    - h: finite difference step size
    - x_NBC, g_NBC: optional Neumann boundary inputs (not used here)
    returns: mean squared PDE residual, pointwise squared residuals
    '''
    x = x.detach()  # Make sure we don't track gradient during shifting
    u = pinn(x)

    uW = pinn(x + torch.tensor([[-h, 0.]], dtype=torch.float32))
    uE = pinn(x + torch.tensor([[ h, 0.]], dtype=torch.float32))
    uN = pinn(x + torch.tensor([[0.,  h]], dtype=torch.float32))
    uS = pinn(x + torch.tensor([[0., -h]], dtype=torch.float32))

    u_xx = (uW - 2.0 * u + uE) / h**2                    # central difference
    u_yy = (uN - 2.0 * u + uS) / h**2

    f_val = f_th(x[:, 0:1], x[:, 1:2])
    residuals = (u_xx + u_yy + f_val)**2

    return torch.mean(residuals), residuals
```

Let us now use `residual_FDM` to train a PINN model and obtain the solution using the same dataset. All we need to do is replace the function used for residual computation with the one based on finite difference method (FDM), as follows:

```python
set_seed_th(88)                                          # for reproducibility
```

```
# Use FDM to compute the residual, residual_FDM()
pinn = PINN2D(Nn=Nn, NL=NL)           # Re-create the PINN model a fresh
DI_rule = ''          # use collocation no Domain Integration rule is used
Trainer  = trainerPDE_DBC
Residual = residual_FDM
LossTerm = 'DBC'                      # loss term in addition to loss_PDE

losses, totalTime = train_adam_lbfgs(pinn, Trainer, Residual, f_th,
       x_Ω_th, w_s_th, x_DBC_th, u_DBC_th, x_NBC_th, t_NBC_th, w_NBC_th,
                        I_NBC, epochs1=201, epochs2=11, use_lbfgs=True)
```

```
==================================================
201 epoches training with Adam:
Epoch     0, Loss: 7.961e+01, PDE: 7.958e+01, DBC: 2.804e-02
Epoch   100, Loss: 1.030e+01, PDE: 7.001e+00, DBC: 3.297e+00
Epoch   200, Loss: 4.248e-01, PDE: 3.324e-01, DBC: 9.243e-02
Time taken by adam: 0.854242 seconds

==================================================
11 epoches fine-tuning with LBFGS:
Epoch     0, Loss: 5.828e-03, PDE: 2.148e-03, DBC: 3.680e-03
Epoch    10, Loss: 5.828e-03, PDE: 2.148e-03, DBC: 3.680e-03

Training complete.
Time taken by lbfgs: 2.565262 seconds
Total time: 3.419504 seconds
Ratio (time_lbfgs/time_adam) of time taken per epoch:54.872430
```

```
# Evaluation of the accuracy of the PINN solution against the exact one:
error_L2, errorVI, u_diff, _ = plot_sln_rect(pinn, exact_u, f_th=f_th,
          N_plot=200, vmin=0., vmax=1., a=a,b=b,c=c,d=d, plot_it=False)

# plot3loss2(losses, ('adam initially','lbfgs in fine tunning'),
#          PDE='PDE', C1='DBC', CL='Loss',log=False) # plot loss curves
```

```
L2 mean residual of trained pinn: 1.7176e-02
Maximum Error: 1.8495e-01
Mean Absolute Error: 3.0408e-02
Error Variation Index (ratio of Max/Mean): 6.0822e+00
L2 Relative Error: 7.1294e-02
```

```
# Update record sheet (exercute this only once!):
recordSheet()
```

```
printSummary(pinns, LossTerms, Residuals, totalTimes,error_L2s,errorVIs)
```

```
#  PINN model        Loss_terms residual     total_time error_L2s  errorVIs
=========================================================================
1  2D                DBC        _clctn       1.833e+01  1.008e-03  1.004e+01
2  2D                DBC        _FDM         3.420e+00  7.129e-02  6.082e+00
```

The use of `residual_FDM` to compute the residual produced a less accurate solution compared to using autograd. This is due to the approximation errors inherent in finite difference methods (FDM) for estimating derivatives. However, one benefit of using FDM is that the computation within the `lbfgs` optimizer becomes much faster — approximately 5 times as fast — resulting in a substantial reduction in overall training time. This may be attributed to the simplified computation of approximate derivatives, which in turn accelerates the estimation of the Hessian matrix of the loss function with respect to the learning parameters. This insight suggests that incorporating FDM-based derivatives could inspire the development of new, efficient second-order optimizers. More in-depth studies may be needed on the use of FDM in PINN models for efficient training.

2.7.8 Integral Residual Evaluation

Random sampling can lead to biased collocation point distributions. As an alternative, we can evaluate the loss using its integral form. Numerical integration can be performed using either Gauss quadrature or the trapezoidal rule. Gauss quadrature is more efficient for smooth integrands, while the trapezoidal rule is often more robust and versatile, especially when using uniformly distributed sample points.

Since the neural network approximation is smooth over the domain, the residual—and therefore the loss function—is also smooth. We can thus treat it as a continuous function over space and compute the loss via integration. The values of the residual at sample points inside and on the boundary of the domain can be used in this integral evaluation. The formula for computing the PDE loss $\mathcal{L}_{\text{PDE}}$ is expressed as:

$$\mathcal{L}_{\text{PDE}} = \frac{1}{A}\int_{\Omega} |\mathcal{R}(x,y)|^2 \approx \frac{1}{A}\sum_{i=1}^{n_{int}} w_i\, |\mathcal{R}(x_i,y_i)|^2 \tag{2.28}$$

where $\mathcal{R}$ stands for a residual, (x_i, y_i) are the collocation points in the domain, w_i are the associated quadrature weights, A is the area of the integral domain, and n_{int} is the number of integration (or quadrature) points—sampled either uniformly (if using the trapezoidal rule) or placed at the Gaussian quadrature points (if using Gaussian quadrature) [20]. The values w_i are the corresponding integration weights . The loss defined in Eq.(2.28) is an **integrated squared residual**, called the *integral residual*.

The division by A is optional; it is related to the use of the loss weight w_{PDE}. In our examples, this normalization is either omitted or simply set as $A = 1$. We include it in the formula to be consistent with the collocation scheme.

The computation of the other loss terms, $\mathcal{L}_{\text{DBC}}$ and/or $\mathcal{L}_{\text{NBC}}$, follows a similar structure as Eq.(2.28). The only difference is that these become **line integrals** evaluated along the respective boundaries and are typically normalized by the total length of the boundary. We will not repeat those formulas here.

2.7.8.1 Quadrature Points and Weights Generation

To numerically evaluate Eq.(2.28), we follow the steps below:

- **Step 1**: Generate the 1D integration points and their associated weights using either the **Gaussian quadrature** or the **trapezoidal rule** for the interval $[0, 1]$ (or any subinterval of the spatial domain).

- **Step 2**: Generate the 2D grid of integration points using `meshgrid()` and then `torch.stack()` (or `np.meshgrid()` and `np.stack()` if using NumPy). We denote the 1D grids in x and y directions as:

$$\mathbf{x} = [x_1, x_2, \dots, x_{n_x}] \tag{2.29}$$

$$\mathbf{y} = [y_1, y_2, \dots, y_{n_y}] \tag{2.30}$$

`and the corresponding weights as`

$$\mathbf{w}_x = [w_{x(1)}, w_{x(2)}, \dots, w_{x(n_x)}] \tag{2.31}$$

$$\mathbf{w}_y = [w_{y(1)}, w_{y(2)}, \dots, w_{y(n_y)}] \tag{2.32}$$

The following functions compute the integral points and weights for 1D domains.

```
def Gauss_points_weights(n, a, b):
    '''
    Generate Gauss quadrature points & weights on [a, b]
    Parameters: n - int, Number of quadrature points
                a - float, Start of the interval
                b - float, End of the interval
    Returns: points - torch.Tensor, Gauss quadrature points on [a, b]
             weights - torch.Tensor, Corresponding weights
    '''
    points, weights = np.polynomial.legendre.leggauss(n)  # for [-1, 1]
    # Map from [-1, 1] to [a, b]
    points = 0.5 * (points + 1) * (b - a) + a
    weights = 0.5 * (b - a) * weights
    return torch.tensor(points,  dtype=torch.float32), \
           torch.tensor(weights, dtype=torch.float32)

def trapz_points_weights(n, a=0., b=1.):
    '''Trapezoidal Quadrature Points and Weights
    '''
    x = torch.linspace(a, b, n)
    h = (b - a) / (n - 1)
    w = torch.full_like(x, h)
    w[ 0] *= 0.5
    w[-1] *= 0.5
    return x, w
```

- **Step 3**: Compute the 2D grid points using

$$\mathbf{X}, \mathbf{Y} = \text{meshgrid}(\mathbf{x}, \mathbf{y}) \tag{2.33}$$

$$\mathbf{XY} = \text{stack}([\mathbf{X}, \mathbf{Y}], \text{axis} = 1) \tag{2.34}$$

The **weights** for the 2D integral by taking the **outer product** of the 1D weights:

$$\mathbf{W} = \mathbf{w}_x \otimes \mathbf{w}_y \tag{2.35}$$

where $\mathbf{w}_x$ and $\mathbf{w}_y$ are the 1D weights in x and y directions, respectively, given in Eq.(2.31). This results in a 2D array of weights $\mathbf{W}$ with shape (n_x, n_y).

- **Step 4**: Evaluate the residual at each integration point, compute its square, multiply by the corresponding weight, and sum all contributions (squared residuals) to obtain $\mathcal{L}_{\mathrm{PDE}}$.

The following snippet computes the sampling points and weights in 2D domain, when an integration or quadrature is used.

```
def sampleD_int(quadratureRule, N_x=9, N_y=9, a=0., b=1., c=0., d=1.):
    '''2D Domain Quadrature Assembly'''
    points_x, weights_x = quadratureRule(N_x, a, b)
    points_y, weights_y = quadratureRule(N_y, c, d)

    X, Y = torch.meshgrid(points_x, points_y, indexing='ij')
    x_Omega = torch.stack([X.reshape(-1), Y.reshape(-1)], dim=1)

    weights_2d = torch.ger(weights_x, weights_y).reshape(-1, 1)
    return weights_2d, x_Omega
```

2.7.8.2 Code for Integral Residual Computation

The computation of the PDE residual via integration is similar to that in the collocation method. The only difference is that **weights are applied** to the squared residuals at the integration (or quadrature) points. The weighted sum of these squared residuals then defines the total squared residual.

```
def residual_int(pinn, f_th, x, weights, DBCfun=None, x_NBC=None,
                                                          g_NBC=None):
    '''Compute integrated squared residual of 2D PDE via torch autograd.
       x: (m, 2) tensor of integration points over the domain.
       weights: (m, 1) tensor of quadrature weights.
    '''
    x.requires_grad = True
    u = pinn(x)                                                # u: (m, 1)

    # First-order gradients
    grad_u = torch.autograd.grad(u, x, grad_outputs=torch.ones_like(u),
                                 create_graph=True)[0]  # shape: (m, 2)
    u_x = grad_u[:, 0:1]
    u_y = grad_u[:, 1:2]

    # Second-order gradients
    u_xx = torch.autograd.grad(u_x, x, grad_outputs=torch.ones_like(u_x),
                                       create_graph=True)[0][:, 0:1]
    u_yy = torch.autograd.grad(u_y, x, grad_outputs=torch.ones_like(u_y),
                                       create_graph=True)[0][:, 1:2]
```

```
    # PDE residual: u_xx + u_yy + f(x, y)
    res = u_xx + u_yy + f_th(x[:,0:1],x[:,1:2])     # returns shape (m,1)

    residuals = res**2                                      # square first
    integral = torch.sum(weights * residuals)               # weighted sum
    return integral, residuals
```

2.7.8.3 Example: PINN Using Gauss Quadrature

When using a numerical integration scheme, it is important to assess its accuracy to estimate how much error may be introduced by the integration method. This can be done by integrating a known function with an exact integral value. In our example, we integrate the exact solution function given in Eq.(2.15) using SymPy. The formula is as follows:

$$\int_0^1 \int_0^1 \sin(\pi x)\sin(\pi y)\,dx\,dy = \left(\int_0^1 \sin(\pi x)dx\right)^2 = \frac{4}{\pi^2} \approx 0.405285 \tag{2.36}$$

The following Sympy code confirms this.

```
import sympy as sp
x, y = sp.symbols('x, y', real=True)              # Define symbolic symbols
f = sp.sin(sp.pi*x) * sp.sin(sp.pi*y)       # Define the symbolic function

# Compute the double integral over [0, 1] x [0, 1]
integral = sp.integrate(f, (x, 0, 1), (y, 0, 1))
display(Math(r"\int_0^1 \bigg( \int_0^1" f"{latex(f)}" r" dx \bigg)dy "\
             f"={latex(integral)}={integral.evalf(6)}"))
```

$$\int_0^1 \left(\int_0^1 \sin(\pi x)\sin(\pi y)dx\right)dy = \frac{4}{\pi^2} = 0.405285$$

We can then compute the numerical integral value using Gaussian quadrature and compare it with the exact value. The snippet is as follows:

```
set_seed_th(88)                                          # for reproducibility

# a=0.; b=1.; c=0.; d=1.                               # the same geometric data
# p = 2; Nn=50; NL=2     # the same NN: NL hidden layers, nn neurons each

# Generate dataset for integration:
DI_rule = 'GQ'                                          # use Gaussian Qudrature
w_s_th, x_Ω_th=sampleD_int(Gauss_points_weights, N_x=7, N_y=7,
                                              a=a, b=b, c=c, d=d)
x_DBC_th, u_DBC_th = sample4DBC_rect(N_DBC=20, a=a, b=b, c=c, d=d)
print(f"Sampled {x_Ω_th.shape[0]} collocation points in the domain.")

# Check integration accuracy using the exact u:
points = x_Ω_th.detach().numpy()
f_vals = exact_u(points[:, 0], points[:, 1]).reshape(-1,1)  # f at nodes
weights = w_s_th.detach().numpy()
```

```
I_approx = np.sum(f_vals * weights)                # numerical integral value

I_exact = 4 / np.pi**2                              # integral of the exact u

I_error_L2 = lg.norm(I_exact - I_approx) / lg.norm(I_exact)
print(f"L2 relative error for integrating exact u: {I_error_L2:.4e}")
```

```
Sampled 49 collocation points in the domain.
L2 relative error for integrating exact u: 4.9081e-09
```

It is observed that the accuracy of integrating the exact solution using a 7×7 grid with Gaussian quadrature reaches the order of 10^{-9}, which is highly accurate. We shall use this approach in our PINN model.

```
# Create a new PINN model:
pinn = PINN2D(Nn=Nn, NL=NL)
P_star, P_total = print_model_parameters(pinn, print_table=False)
print(f"Sampled {x_Ω_th.shape[0]} collocation points in the domain.")
print(f"Sampled {x_DBC_th.shape[0]} boundary points.")

# Use Gauss quadrature to compute the squared residual:
Trainer  = trainerPDE_DBC
Residual = residual_int
LossTerm = 'DBC'                      # loss term in addition to loss_PDE
losses, totalTime = train_adam_lbfgs(pinn, Trainer, Residual, f_th,
       x_Ω_th, w_s_th, x_DBC_th, u_DBC_th, x_NBC_th, t_NBC_th, w_NBC_th,
                             I_NBC, epochs1=201, epochs2=11, use_lbfgs=True)
```

```
P* (Total pseudo-dimension of the affine space) = 105
Total trainable parameters: 2751
Sampled 49 collocation points in the domain.
Sampled 76 boundary points.
==================================================
201 epoches training with Adam:
Epoch     0, Loss: 9.553e+01, PDE: 9.550e+01, DBC: 2.804e-02
Epoch   100, Loss: 1.100e+01, PDE: 6.576e+00, DBC: 4.428e+00
Epoch   200, Loss: 9.411e-01, PDE: 5.908e-01, DBC: 3.503e-01
Time taken by adam: 1.034791 seconds

==================================================
11 epoches fine-tuning with LBFGS:
Epoch     0, Loss: 6.196e-04, PDE: 3.115e-04, DBC: 3.081e-04
Epoch    10, Loss: 5.348e-06, PDE: 9.651e-07, DBC: 4.383e-06

Training complete.
Time taken by lbfgs: 17.730296 seconds
Total time: 18.765088 seconds
Ratio (time_lbfgs/time_adam) of time taken per epoch:313.088128
```

```
# Plot the contours of both PINN solution and the exact one:
error_L2, errorVI, u_diff, _ = plot_sln_rect(pinn, exact_u, f_th=f_th,
```

```
            N_plot=200, vmin=0., vmax=1., a=a,b=b,c=c,d=d, plot_it=True)
plot3loss2(losses, ('adam initially','lbfgs in fine tunning'),
                     PDE='PDE', C1='DBC', C2='NBC', C3='Loss',log=False)
```

```
L2 mean residual of trained pinn: 1.1640e-03
Maximum Error: 4.1122e-03
Mean Absolute Error: 6.7528e-04
Error Variation Index (ratio of Max/Mean): 6.0896e+00
L2 Relative Error: 1.8546e-03
```

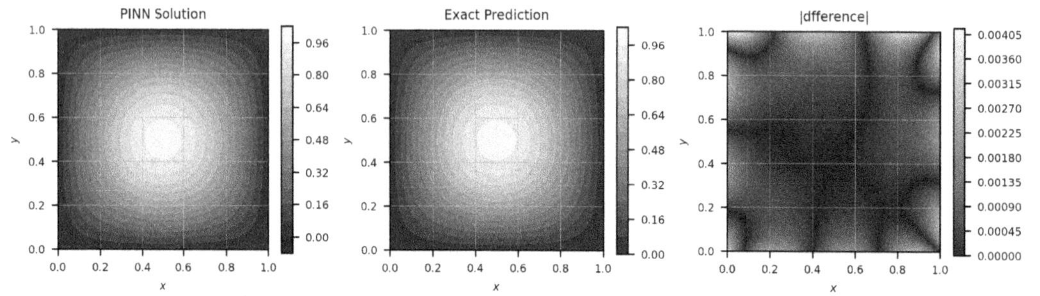

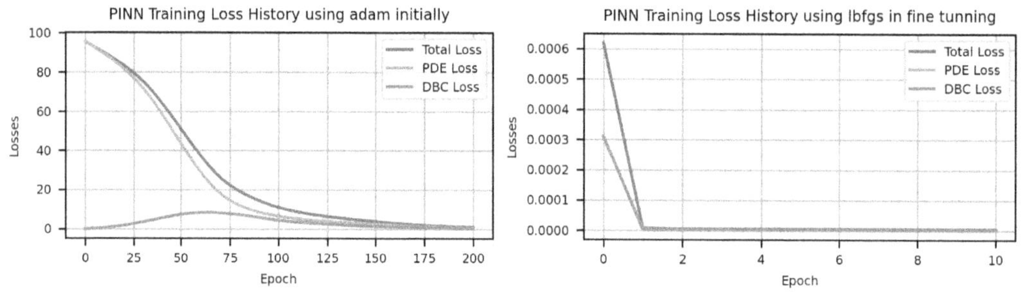


```
# Update record sheet (exercute this only once!):
recordSheet()
```

```
printSummary(pinns, LossTerms, Residuals, totalTimes,error_L2s,errorVIs)
```

#	PINN model	Loss_terms	residual	total_time	error_L2s	errorVIs
1	2D	DBC	_clctn	1.833e+01	1.008e-03	1.004e+01
2	2D	DBC	_FDM	3.420e+00	7.129e-02	6.082e+00
3	2D	DBC	_intGQ	1.877e+01	1.855e-03	6.090e+00

The relative error of the solution using the Gaussian quadrature is on the order of $\mathcal{O}(-3)$, which is less accurate compared to that obtained using the collocation method (model-1). The error variation index is lower than that from random collocation. Note that in this case, the number of sampling points inside the domain is smaller than the P^* value, because when Gaussian quadrature is used, we have implicitly imposed additional constraints—namely, that the squared residual varies in a higher-order polynomial form.

2.7.8.4 Example: PINN Using Trapezoidal Rule

The following snippet performs the same computation but uses the trapezoidal rule. As usual, let us first check the accuracy of the integration scheme using the following snippet:

```
set_seed_th(88)                                        # for reproducibility

# Generate the training dateset:
DI_rule = 'TR'                                         # use Trapezoidal Rule
w_s_th, x_Ω_th=sampleD_int(trapz_points_weights, N_x=11, N_y=11,
                                                    a=a, b=b, c=c, d=d)
x_DBC_th, u_DBC_th = sample4DBC_rect(N_DBC=20, a=a, b=b, c=c, d=d)
print(f"Sampled {x_Ω_th.shape[0]} collocation points in the domain.")
print(f"Sampled {x_DBC_th.shape[0]} boundary points.")

# Check integration accuracy using the exact u:
points = x_Ω_th.detach().numpy()
f_vals = exact_u(points[:, 0], points[:, 1]).reshape(-1,1)  # f at nodes
weights = w_s_th.detach().numpy()
I_approx = np.sum(f_vals * weights)                # numerical integral values
I_exact = 4 / np.pi**2                              # integral of the exact u
error_L2 = lg.norm(I_exact - I_approx) / lg.norm(I_exact)
print(f"L2 relative error for integrating the exact u: {error_L2:.4e}")
```

```
Sampled 121 collocation points in the domain.
Sampled 76 boundary points.
L2 relative error for integrating the exact u: 1.6408e-02
```

The accuracy of the trapezoidal rule for integrating the exact solution using an 11×11 grid is on the order of 10^{-2}, which is significantly less accurate compared to the Gaussian quadrature, as expected. Let us now apply the trapezoidal rule for residual evaluation in our PINN model for the same problem.

```
# Use trapezoidal integration to compute the residual, residual_int
pinn = PINN2D(Nn=Nn, NL=NL)                                      # new model
losses, totalTime = train_adam_lbfgs(pinn, Trainer, Residual, f_th,
       x_Ω_th, w_s_th, x_DBC_th, u_DBC_th, x_NBC_th, t_NBC_th, w_NBC_th,
                          I_NBC, epochs1=201, epochs2=11, use_lbfgs=True)
```

```
==================================================
201 epoches training with Adam:
Epoch     0, Loss: 9.556e+01, PDE: 9.553e+01, DBC: 2.804e-02
Epoch   100, Loss: 1.139e+01, PDE: 7.095e+00, DBC: 4.292e+00
Epoch   200, Loss: 1.028e+00, PDE: 7.334e-01, DBC: 2.943e-01
Time taken by adam: 0.873569 seconds

==================================================
11 epoches fine-tuning with LBFGS:
Epoch     0, Loss: 1.942e-04, PDE: 1.277e-04, DBC: 6.647e-05
Epoch    10, Loss: 6.316e-06, PDE: 5.446e-06, DBC: 8.702e-07

Training complete.
```

```
Time taken by lbfgs: 10.209444 seconds
Total time: 11.083013 seconds
Ratio (time_lbfgs/time_adam) of time taken per epoch:213.554269
```

```
# Plot the contours of both PINN solution and the exact one:
error_L2, errorVI, u_diff, _ = plot_sln_rect(pinn, exact_u, f_th=f_th,
          N_plot=200, vmin=0., vmax=1., a=a,b=b,c=c,d=d, plot_it=False)

recordSheet()                                   # update record sheet
# plot3loss2(losses, ('adam initially','lbfgs in fine tunning'),
#   PDE='PDE', C1='DBC', C2='NBC', C3='Loss',log=False) # plot loss terms
```

```
L2 mean residual of trained pinn: 4.9947e-06
Maximum Error: 2.4526e-03
Mean Absolute Error: 2.6791e-04
Error Variation Index (ratio of Max/Mean): 9.1545e+00
L2 Relative Error: 7.9229e-04
```

```
printSummary(pinns, LossTerms, Residuals, totalTimes,error_L2s,errorVIs)
```

```
#  PINN model       Loss_terms residual    total_time error_L2s  errorVIs
=========================================================================
1  2D               DBC        _clctn      1.833e+01  1.008e-03  1.004e+01
2  2D               DBC        _FDM        3.420e+00  7.129e-02  6.082e+00
3  2D               DBC        _intGQ      1.877e+01  1.855e-03  6.090e+00
4  2D               DBC        _intTR      1.108e+01  7.923e-04  9.155e+00
```

The relative error of the solution using the trapezoidal rule of integration is also on the order of 10^{-3}. The accuracy is higher than that of the collocation model-1 and also higher than that obtained using the Gaussian quadrature (model-3) in this case. In terms of computation time, however, the trapezoidal rule is faster for this case.

2.8 Formulation for Energy Method (Weak Form)

When the weak form of the PDE is used, the energy of the system can be evaluated. It includes:

- the potential energy within the domain, usually denoted by U
- the work done by the force inside the domain (i.e., the source term in the PDE), denoted as W_f
- the work done by the traction on the Neumann boundary, denoted as W_t

The total potential energy is then defined as [21]:

$$U_T = U - W_f - W_t \tag{2.37}$$

This formulation leads to a **single quadratic functional** U_T. The detailed formulation leading to Eq.(2.37) for Poisson's equation is given below.

2.8.1 Classical PDE (Strong Form)

Let $\Omega \subset \mathbb{R}^d$ be a domain in d-dimensional space. For our 2D problems, $d = 2$. The boundary of the domain is denoted by $\partial\Omega = \partial\Omega_{\text{DBC}} \cup \partial\Omega_{\text{NBC}}$, where $\partial\Omega_{\text{DBC}}$ is the Dirichlet boundary (on which function values are prescribed), and $\partial\Omega_{\text{NBC}}$ is the Neumann boundary (on which the normal derivative or traction is specified).

The **strong form** of Poisson's equation, as given in Eq.(2.12), is rewritten as:

$$-\Delta u(\mathbf{x}) = f(\mathbf{x}), \quad \mathbf{x} \in \Omega \tag{2.38}$$

where $\Delta u = \nabla \cdot \nabla u$ is the Laplacian of u, ∇u is the gradient of u, and $f(\mathbf{x})$ is a given force (or source) function.

The Dirichlet boundary condition (DBC) is restated as:

$$u(\mathbf{x}) = 0, \mathbf{x} \in \partial\Omega_{\text{DBC}} \tag{2.39}$$

2.8.2 Introduction of Test Function

Introduce a smooth (at least piecewise differentiable) test function $v(\mathbf{x})$ that vanishes on $\partial\Omega_{\text{DBC}}$. Multiply both sides of the strong form equation Eq.(2.38) by v:

$$-\Delta u \cdot v = f \cdot v \tag{2.40}$$

Then integrate both sides of the above equation over the domain Ω:

$$\int_\Omega -\Delta u\, v\, d\mathbf{x} = \int_\Omega f v\, d\mathbf{x} \tag{2.41}$$

2.8.3 Integrate by Parts

Now, apply integration by parts (or Green's Theorem [17]):

$$\int_\Omega \nabla u \cdot \nabla v\, d\mathbf{x} - \int_{\partial\Omega} v \frac{\partial u}{\partial n}\, ds = \int_\Omega f v\, d\mathbf{x} \tag{2.42}$$

or

$$\int_\Omega \nabla u \cdot \nabla v\, d\mathbf{x} - \int_{\partial\Omega_{\text{DBC}}} \underbrace{v}_{0} \frac{\partial u}{\partial n}\, ds - \int_{\partial\Omega_{\text{NBC}}} v \underbrace{\frac{\partial u}{\partial n}}_{t}\, ds = \int_\Omega f v\, d\mathbf{x} \tag{2.43}$$

where $\frac{\partial u}{\partial n}$ is the normal directional derivative on the boundary and ds is the differential boundary measure (the domain boundary must be at least piecewise differentiable), and the Neumann boundary condition, $\dfrac{\partial u}{\partial n} = t(\mathbf{x})$ on $\partial\Omega_{\text{NBC}}$, is used.

Since $v(\mathbf{x})$ is zero on $\partial\Omega_{\text{DBC}}$, the second integral in Eq.(2.43) becomes zero. Thus, the problem of solving Eqs.(2.38) with (2.39) can now be cast as follows: find $u \in \mathbb{V}$ (where $\mathbb{V}$ is an appropriate function space, such as $H_0^1(\Omega)$), such that:

$$\int_\Omega \nabla u \cdot \nabla v \, d\mathbf{x} = \int_\Omega v f \, d\mathbf{x} + \int_{\partial\Omega_{\text{NBC}}} v t \, ds \quad \forall v \in \mathbb{V} \tag{2.44}$$

This is the weak form corresponding to the strong form defined by Eqs.(2.38) and (2.39), and is widely used in various numerical methods [11].

2.8.4 Total Potential Energy for Poisson's Equation

We now define the following **total potential energy** of the system governed by the strong form of Poisson's equation. This is an energy **functional** of the function $u(x, y)$:

$$U_T(u) = \underbrace{\frac{1}{2}\int_\Omega \nabla u \cdot \nabla u \, d\mathbf{x}}_{U} - \underbrace{\int_\Omega u f \, d\mathbf{x}}_{W_f} - \underbrace{\int_{\partial\Omega_{\text{NBC}}} u \, t \, ds}_{W_t} \tag{2.45}$$

where the coefficient $\frac{1}{2}$ in front of the first term is required to correctly represent the potential energy within the domain. The second term represents the work done by the force function $f(x, y)$ on $u(x, y)$ over the domain, and the third term accounts for the work done by the traction t along the Neumann boundary $\partial\Omega_{\text{NBC}}$. It is clear that the energy functional $U_T(u)$ is a function of the function $u(x, y)$. We require that the function $u(x, y)$ be *admissible*, meaning it must belong to a Hilbert space $H_0^1(\Omega)$, in which the squared first derivatives of any function are still bounded. The subscript 0 is a shorthand indicating that the function must satisfy the DBCs throughout the entire process.

2.8.5 Principle of Minimum Total Potential Energy

The principle of minimum total potential energy states that

$$\delta[U_T(u)] = \delta[U - W_f - W_t] = 0 \tag{2.46}$$

where δ denotes the variation. This is a linear operation applied to a functional with respect to its function argument—analogous to differentiation, but for functionals. The key difference is that differentiation applies to functions with respect to their independent variables, whereas variation applies to functionals with respect to their input functions. It evaluates the change in the functional when the input function undergoes a small perturbation.

One important requirement is that the perturbed function must still belong to the Hilbert space $H_0^1(\Omega)$. Therefore, when we apply the principle of minimum total potential energy, the Dirichlet boundary conditions (DBCs) must be satisfied *a priori*. For this reason, DBCs are also called *essential boundary conditions* , because they are essential for the principle of minimum total potential energy to hold [20, 21].

Equation (2.46) states that the energy functional $U_T(u)$ must be stationary with respect to variations in the function u. The variation of $U_T(u)$ is written as

$$\delta U_T = \delta \left[\frac{1}{2}\int_\Omega \nabla u \cdot \nabla u \, d\mathbf{x} - \int_\Omega u f \, d\mathbf{x} - \int_{\partial\Omega_{\text{NBC}}} u \, t \, ds \right] \tag{2.47}$$

It is clear that the method of minimum total potential energy seeks a function $u \in H_0^1(\Omega)$ that minimizes the total potential energy $U_T(u)$ [11, 20, 21].

2.8.6 Equivalence of Energy Principle and Weak Form

Next, we demonstrate that the minimization problem for the total potential energy is equivalent to the weak form stated in Eq.(2.44).

To do so, we perform the variation of $U_T(u)$ with respect to u—analogous to taking the derivative using the chain rule—in Eq.(2.47):

$$\delta U_T = \int_\Omega \nabla u \cdot \delta[\nabla(u)]\, d\mathbf{x} - \int_\Omega \delta u f \, d\mathbf{x} - \int_{\partial\Omega_{\mathrm{NBC}}} \delta u\, t\, ds \tag{2.48}$$

Since the integration and variation operators are both linear and are with respect to different quantities, we can move δ inside the integrals. The order of these operations is immaterial as long as both are executed. Similarly, because variation and differentiation are also linear and with respect to different objects, we can interchange their order as well—performing δ first, followed by ∇, which gives:

$$\delta U_T = \int_\Omega \nabla u \cdot \nabla[\delta u]\, d\mathbf{x} - \int_\Omega \delta u f \, d\mathbf{x} - \int_{\partial\Omega_{\mathrm{NBC}}} \delta u\, t\, ds \tag{2.49}$$

Since the foregoing equation must hold for arbitrary variations, δu is independent of u and still belongs to $H_0^1(\Omega)$. We can therefore denote δu as v.

With this substitution and using the stationary condition in Eq.(2.46), we obtain:

$$\int_\Omega \nabla u \cdot \nabla v \, d\mathbf{x} - \int_\Omega v f \, d\mathbf{x} - \int_{\partial\Omega_{\mathrm{NBC}}} v\, t\, ds = 0 \tag{2.50}$$

which holds for all $v \in H_0^1(\Omega)$, and is therefore exactly the weak form given in Eq.(2.44). This confirms that the stationary point of the functional $U_T(u)$ corresponds to the solution u obtained from the weak formulation in Eq.(2.44).

In conclusion, the energy method seeks the function $u^*(x, y)$ within an admissible space $\mathbb{V}$ such that:

$$u^*(x, y) = \min_{u\in\mathbb{V}} U_T = \min_{u\in\mathbb{V}} \left[\underbrace{\frac{1}{2}\int_\Omega \nabla u \cdot \nabla u \, d\mathbf{x}}_{U} - \underbrace{\int_\Omega u f \, d\mathbf{x}}_{W_f} - \underbrace{\int_{\partial\Omega_{\mathrm{NBC}}} u\, t\, ds}_{W_t} \right] \tag{2.51}$$

The solution $u^*(x, y)$ minimizes the total potential energy U_T defined in Eq.(2.37), ensuring that the system is at a stationary state (in equilibrium). We can now use optimizers in a machine learning model to find $u^*(x, y)$, which leads to a PINN model with an energy loss function.

2.8.7 Proof of Minimization

To prove that the stationary point is indeed a minimum, we need to perform the second-order variation of the total potential energy with respect to u. The expression is as follows:

$$
\begin{aligned}
\delta^2[U_T] &= \delta[\delta U_T] \\
&= \delta\left[\int_\Omega \nabla u \cdot \nabla(\delta u)\, d\mathbf{x} - \int_\Omega \delta u f\, d\mathbf{x} - \int_{\partial\Omega_{\text{NBC}}} \delta u\, t\, ds\right] \\
&= \int_\Omega \delta(\nabla u) \cdot \nabla(\delta u)\, d\mathbf{x} - \underbrace{\delta\left[\int_\Omega \delta u f\, d\mathbf{x} + \int_{\partial\Omega_{\text{NBC}}} \delta u\, t\, ds\right]}_{=0} \\
&= \int_\Omega \nabla(\delta u) \cdot \nabla(\delta u) \geq 0
\end{aligned}
\tag{2.52}
$$

in which we used Eq.(2.49). When moving the variation operator into the square brackets, we used the following two facts:

1. The function ∇u (a vector) is independent of the scalar function u—similar to how dx is treated in differentiation. Therefore,
   $\delta[\nabla u \cdot \nabla(\delta u)] = \nabla(\delta u) \cdot \nabla(\delta u)$,
   which is the dot product of a vector with itself and is thus always nonnegative.

2. Both $f(x, y)$ and $t(x, y)$ are given functions, so
   $\delta[\delta u f] = 0$ and $\delta[\delta u, t] = 0$,
   which causes the last two terms within the square brackets to vanish.

We note the following:

1. Equation (2.52) shows that the second variation of U_T is nonnegative. Therefore, the stationary point must be a minimum. It is also unique because the energy functional U_T is quadratic in u.

2. The nonnegativity of the second variation of the total potential energy implies that the energy functional is convex, which is a highly desirable property when used in machine learning models [12]. We will leverage this convexity in training PINN models by minimizing the energy-based residual .

2.8.8 Applications

- The weak formulation reduces the regularity requirement: you only need $u \in H^1(\Omega)$, meaning that the function u needs to be only first-order differentiable. The second-order differentiability required in the original PDE is no longer necessary.

- It forms the foundation of Finite Element Methods [21], Smoothed Finite Element Methods [20], and meshfree methods [11].

- We will also adopt it in our PINN models. In this context, the function space $\mathbb{V}$ is approximated by a neural network $\mathcal{N}_{\hat{\mathbf{w}}}(x, y)$ that employs continuous activation and loss functions. This ensures sufficient smoothness, placing the network output in $H^1(\Omega)$. To enforce that the trial functions lie in $H_0^1(\Omega)$ (i.e., satisfy the DBCs), we can construct the trial solution to moderate $\mathcal{N}_{\hat{\mathbf{w}}}(x, y)$ accordingly.

The energy level at any set of learning parameters in a PINN model can then be computed using the integration approach defined in Eq.(2.45), with $u(x, y)$ and its first derivatives generated by $\mathcal{N}_{\hat{\mathbf{w}}}(x, y)$. The same minimization procedure used in training PINNs can now be applied to a loss function defined by the total potential energy.

Note that the total potential energy U_T is an integral expression and is a single-piece quadratic functional in u. In linear FEM, minimizing U_T leads to a system of linear equations that can be solved deterministically [21]. In the case of PINNs, training is expected to be more stable and faster, as will be demonstrated in the example section.

2.8.9 Energy Computation in PINN

In a PINN implementation, the total potential energy can be computed using Eq.(2.51) and integration points:

$$U_T(u) = \sum_{i=1}^{N_\Omega} w_i \left(\frac{1}{2} \|\nabla u(\mathbf{x}_i)\|^2 - f(\mathbf{x}_i) u(\mathbf{x}_i) \right) - \sum_{i=1}^{N_{\text{NBC}}} w_i t(\mathbf{x}_i) u(\mathbf{x}_i) \tag{2.53}$$

where u is the current output from the neural network (NN) of the PINN, N_Ω is the number of integration points used within the domain Ω, and N_{NBC} is the number of integration points on the Neumann boundary. The symbol w_i denotes the weight associated with the ith integration point, either in the interior of Ω (for the first summation) or on the Neumann boundary (for the second summation).

Therefore, when the energy method is used, the loss function for the PINN consists of only a single term:

$$\mathcal{L} = \mathcal{L}_{\text{PDE}} = U_T \tag{2.54}$$

This is a quadratic functional in u to be minimized. The $\mathcal{L}_{\text{NBC}}$ term is **naturally** embedded as a linear component of the quadratic energy expression. Hence, the name of 'natural boundary condition'. The Dirichlet boundary condition (DBC) term $\mathcal{L}_{\text{DBC}}$ can be exactly enforced through the construction of the trial solution at the class level. Consequently, a PINN model based on the energy method is generally easier to train due to the smooth and convex nature of the loss landscape.

Unlike the traditional $\mathcal{L}^2$ loss used in strong form-based PINNs, the loss $\mathcal{L}$ derived from the energy method is not guaranteed to be positive. This is because U_T includes subtracted work terms and may take negative values depending on the configuration of $f(x, y)$ and $t(x, y)$.

2.8.10 On Sampling Points for the Energy Method

It is important to note that when the energy method is used, the loss function becomes the total potential energy U_T. The minimum number of required sampling points still follows the **samples-neurons theorem**, $m \geq P^*$. However, the **accuracy** of the resulting PINN model depends heavily on how accurately the total potential energy U_T is integrated over the domain.

In general, more sampling points may be needed to ensure accurate integration of U_T. If the integration is not accurate, increasing the number of training epochs may not necessarily improve the model's accuracy. This behavior contrasts with that of the collocation-based method , where fewer collocation points may be used (still satisfying $m \geq P^*$), and training accuracy can still improve by increasing the number of epochs.

We therefore emphasize the following:

> Accurate integration of the total potential energy is critical to improving the accuracy of the trained PINN model when using the energy method.

Additionally, the **uniqueness** of the solution obtained by a PINN model trained using the energy method can be enhanced. This topic will be discussed further in a later section on solution uniqueness .

The following snippet illustrates how to compute the energy level $U_T(u)$ for a 2D problem using integration:

```
def residual_energy(pinn, f_th, x, weights, x_NBC, t_NBC):
    '''
    Compute the integral of energy density for PDE Laplacian(u)+f=0
    using PyTorch autograd and quadrature.
    Minimizing this integral finds the solution u.
    Parameters:
    - pinn: the pinn model (PINN2D_4DBC which provides u_trial)
    - f_th: forcing/source function f(x, y), accepts and returns tensor
    - x: tensor of shape [N, 2], quadrature points
    - weights: tensor of shape [N, 1], quadrature weights
    - x_NBC, t_NBC: optional, not used in this function
    Returns: total energy, potential energy, work-done; scalars
    '''
    x.requires_grad = True
    u = pinn(x)                    # Get the trial solution u_trial(x, y)

    # Compute gradients of u with respect to x and y
    grad = torch.autograd.grad(outputs=u, inputs=x,
      grad_outputs=torch.ones_like(u), # Gradient of scalar wrt. tensor
      create_graph=True, retain_graph=True)[0]    # Get gradients wrt. x

    u_x = grad[:, 0:1]                           # du/dx (Shape: m, 1)
    u_y = grad[:, 1:2]                           # du/dy (Shape: m, 1)

    # Get the source term value at each point
    f_val = f_th(x[:, 0:1], x[:, 1:2])                  # Shape (m, 1)

    # Compute the potential energy:
    U = 0.5*torch.sum(weights * (u_x**2 + u_y**2))         # Scalar, U
    # Compute the work done by the force/source:
    Wf = torch.sum(weights * u * f_val)                    # Scalar Wf
    # In energy minimization, we minimize energy itself, not its square.

    return U-Wf, U, Wf
```

2.8.11 Energy Computation via FDM

The total potential energy can also be computed using the finite difference method (FDM). In this approach, automatic differentiation (autograd) is not required for evaluating the derivatives of u with respect to x and y, and therefore backward propagation is not needed during energy evaluation.

Instead, the spatial derivatives in the energy functional are computed using the central difference scheme directly through the neural network outputs. This can offer computational efficiency in certain implementations. The following snippet illustrates how to implement this approach:

```
def residual_energyFDM(pinn, f_th, x, weights, x_NBC, t_NBC, h=0.01):
    '''
    Compute PDE energy residual via finite difference method (FDM)
    for 2D at quadrature points. No Neumann boundary.
    Weak PDE and NBCs are unified in a single energy (scalar).
    Parameters: - pinn: the PINN model
                - f_th: forcing function f(x, y), returns a tensor
                - x: tensor of shape [N, 2], quadrature points
                - weights: tensor of shape [N, 1], quadrature weights
                - x_NBC, t_NBC: optional, not used here
    Returns: total_energy, e_tensor
    '''
    u  = pinn(x)
    uW = pinn(x + torch.tensor([[-h,0.]],dtype=x.dtype,device=x.device))
    uE = pinn(x + torch.tensor([[ h,0.]],dtype=x.dtype,device=x.device))
    uN = pinn(x + torch.tensor([[0., h]],dtype=x.dtype,device=x.device))
    uS = pinn(x + torch.tensor([[0.,-h]],dtype=x.dtype,device=x.device))

    u_x = (uE - uW) / (2.0 * h)
    u_y = (uN - uS) / (2.0 * h)

    f_val = f_th(x[:, 0:1], x[:, 1:2])

    # Compute the potential energy:
    U = 0.5*torch.sum(weights * (u_x**2 + u_y**2))          # Scalar, U
    # Compute the work done by the force/source:
    Wf = torch.sum(weights * u * f_val)                     # Scalar Wf
    # In energy minimization, we minimize energy itself, not its square.

    return U-Wf, U, Wf
```

2.9 Training of PINN with Trial Solution

We use the previously defined PINN class with the trial solution form, namely the PINN2D_DBC class, to build our model. In this setup, the Dirichlet boundary condition (DBC) is already enforced by construction. Therefore, the loss function consists solely of the PDE residual term. This class can also work seamlessly with the energy method. The trainer function is accordingly modified as follows.

```
def trainer_int_PDE(pinn, residual, optimizer, f_th, x_Ω_th, w_s_th,
            x_DBC_th, u_DBC_th, x_NBC_th, t_NBC_th, w_NBC_th, I_NBC,
                                                        epochs=201):
    """For training a PINN using loss_PDE
          (e.g. for energy method or integrated residual)
    Args: pinn: the PINN model (nn.Module), expected on device.
          residual: function to compute energy or residual loss.
```

```
        optimizer: PyTorch optimizer (e.g. Adam, LBFGS).
        epochs: number of training iterations for PINN.
        f_th: the force/source function f(x, y)
        w_s_th: weights, a tensor on device [N, 1]
        x_Ω_th: integral or collocation points,
                tensor on device [N, 2] (has requires_grad=True)
    x_DBC_th, u_DBC_th, x_NBC_th, t_NBC_th, w_NBC_th, I_NBC: dummy
    """
    pinn.train()                        # Ensure model is in training mode

    # Determine if the optimizer is LBFGS to print internal loss
    is_lbfgs = isinstance(optimizer, torch.optim.LBFGS)
    lbfgs_closure_calls = 0 #  Counter for internal LBFGS closure calls

    loss_history = {"total": [], "PDE": []}     # for logging the losses

    if epochs <= 301: n_prnt = 100              # Adjust print frequency
    else:             n_prnt = max(1, round(epochs/3, -2))
    #print_lbfgs_internal = True #print loss within LBFGS closure calls

    for epoch in range(epochs):
        loss_PDE_tensor = None; total_loss_tensor = None    # Initialize

        def closure():
            nonlocal loss_PDE_tensor, total_loss_tensor
            nonlocal lbfgs_closure_calls
            optimizer.zero_grad()                    # Clear old gradients
            # Calculate the losses:
            loss_PDE_tensor, *_ = residual(pinn, f_th, x_Ω_th, w_s_th,
                                                 x_NBC_th, t_NBC_th)
            total_loss_tensor = loss_PDE_tensor           # only one term

            #Check if the loss tensor is NaN or Inf BEFORE backward pass
            if not torch.isfinite(total_loss_tensor).all():
                print(f"Warning: Loss tensor is not finite (NaN or Inf)"
                      f" at epoch {epoch}, closure call "
                      f"{lbfgs_closure_calls if is_lbfgs else 'N/A'}."
                      f" Stopping training.")
                return total_loss_tensor

            total_loss_tensor.backward()
            return total_loss_tensor

        # --- Optimizer Step ---
        if is_lbfgs:
            lbfgs_closure_calls = 0          # Reset counter at the start
            #The optimizer.step() for LBFGS calls closure multiple times
            optimizer.step(closure)
            #  total_loss_tensor is updated at last closure call.
            total_loss_val = total_loss_tensor.item()
```

```
            loss_PDE_val = loss_PDE_tensor.item()

        else: # Adam or other optimizers
            # optimizer.step() for Adam typically calls the closure once.
            total_loss_tensor_returned = closure()
            optimizer.step()
            # The nonlocal tensors loss_PDE_tensor and total_loss_tensor
            # were also updated by closure.
            total_loss_val = total_loss_tensor_returned.item()
            loss_PDE_val = loss_PDE_tensor.item()

        # Check if loss is NaN or Inf AFTER the optimizer step
        if not math.isfinite(total_loss_val):
            print(f"Warning: Loss value is not finite (NaN or Inf) "
                  f"after step at epoch {epoch}. Stopping training.")
            break

        with torch.no_grad():
            loss_history["total"].append(total_loss_val)
            loss_history["PDE"  ].append(loss_PDE_val)

        # Print epoch summary less frequently if epochs is large
        if epoch % n_prnt == 0 or epoch == epochs - 1:
             print(f"Epoch {epoch+1:5d}, Loss: {total_loss_val:.3e}, "
                  f"PDE: {loss_PDE_val:.3e} ")

    return loss_history
```

2.9.1 Example: Trapezoidal Rule for Energy Integral

In this example, we train a PINN by minimizing the energy residual integrated over the problem domain for a Poisson equation. The model is constructed using the `PINN2D_4DBC` class, which enforces Dirichlet boundary conditions (DBCs) on all four edges at the class level. The loss function is defined as the total potential energy. All other settings remain unchanged.

```
set_seed_th(88)                                         # for reproducibility
# a=0.; b=1.; c=0.; d=1.                                     # the same setup
# p = 2; Nn=50; NL=2

# Generate dataset: Energy method, trapezoidal rule
DI_rule = 'TR'
w_s_th, x_Ω_th = sampleD_int(trapz_points_weights,N_x=100,N_y=100,
                                                  a=a, b=b, c=c, d=d)

# Create new PINN model with DBC enforced
pinn = PINN2D_4DBC(Nn, NL=NL, a=a, b=b, c=c, d=d)
print_model_parameters(pinn, print_table=False)         # print only P*, P

x_DBC_th=None; u_DBC_th=None                          # no need for loss DBC
x_NBC_th=None; t_NBC_th=None                          # no need for loss NBC
```

```
# The residual_energy calculates the energy functional:
Trainer  = trainer_int_PDE
Residual = residual_energy
LossTerm = 'PDE_only'
losses, totalTime = train_adam_lbfgs(pinn, Trainer, Residual, f_th,
       x_Ω_th, w_s_th, x_DBC_th, u_DBC_th, x_NBC_th, t_NBC_th, w_NBC_th,
                          I_NBC, epochs1=201, epochs2=11, use_lbfgs=True)
```

```
P* (Total pseudo-dimension of the affine space) = 105
Total trainable parameters: 2751
==================================================
201 epoches training with Adam:
Epoch     1, Loss: -5.192e-02, PDE: -5.192e-02
Epoch   101, Loss: -1.935e+00, PDE: -1.935e+00
Epoch   201, Loss: -2.349e+00, PDE: -2.349e+00
Time taken by adam: 2.201927 seconds

==================================================
11 epoches fine-tuning with LBFGS:
Epoch     1, Loss: -2.467e+00, PDE: -2.467e+00
Epoch    11, Loss: -2.467e+00, PDE: -2.467e+00

Training complete.
Time taken by lbfgs: 2.811018 seconds
Total time: 5.012945 seconds
Ratio (time_lbfgs/time_adam) of time taken per epoch:23.327276
```

Readers may notice that the loss values can become negative. Training tends to be faster when using the energy residual, which is the integral of a single-piece quadratic functional in u. This behavior is especially pronounced when employing the quasi-second-order optimizer `lbfgs`.

Let's plot the PINN solution, the exact solution, and the corresponding error.

```
# Plot the contours of both PINN solution and the exact one:
error_L2, errorVI, u_diff, _ = plot_sln_rect(pinn, exact_u, f_th=f_th,
           N_plot=200, vmin=0., vmax=1., a=a,b=b,c=c,d=d, plot_it=True)

plot3loss2(losses, ('adam initially','lbfgs in fine tunning'),
                    PDE='PDE', C1='DBC', C2='NBC', C3='Loss',log=False)

# Create record sheet (exercute this only once!):
recordSheet()
```

```
L2 mean residual of trained pinn: 3.6953e-02
Maximum Error: 9.1853e-04
Mean Absolute Error: 2.4724e-04
Error Variation Index (ratio of Max/Mean): 3.7151e+00
L2 Relative Error: 6.2977e-04
```

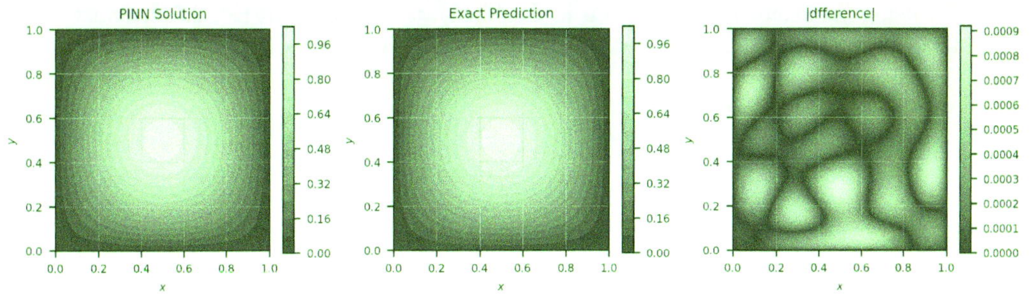

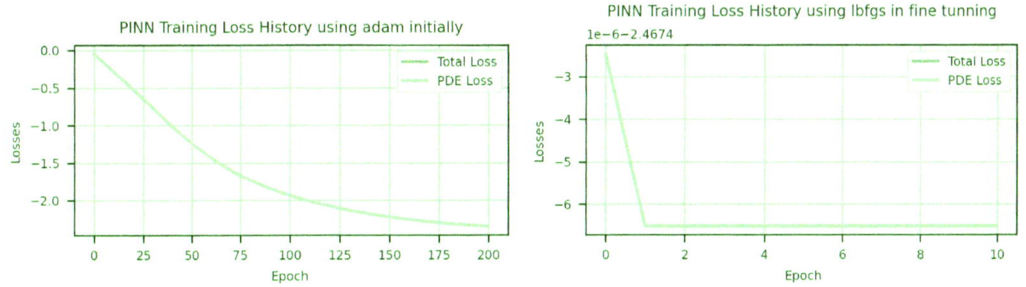

As observed, the solution accuracy is quite high, with the relative error on the order of 10^{-4}. The error is now concentrated within the domain, while the Dirichlet boundary shows no error. This is due to the use of the `PINN2D_4DBC` class, where the DBCs are strictly enforced during model construction.

Once the PINN model is trained, we can easily compute the energy components in the system using the same `residual_energy()` function. The corresponding code snippet is given below.

```
x_NBC, t_NBC = None, None
UnWf, U_PINN, Wf_PINN = residual_energy(pinn, f_th, x_Ω_th, w_s_th,
                                                          x_NBC, t_NBC)
print(f"U_PINN  = {U_PINN:.7f}, 2U_PINN = {2*U_PINN:.7f}, "
      f"Wf_PINN = {Wf_PINN:.7f}")
```

```
U_PINN  = 2.4672654, 2U_PINN = 4.9345307, Wf_PINN = 4.9346719
```

It is found that $2U \approx W_f$ (matching up to 4 significant digits), meaning that the potential energy in the system is approximately half of the work done by the applied force when the total potential energy is minimized (i.e., at equilibrium). This result is a direct confirmation of the well-known **Clapeyron's theorem** (see Section 13.2.8 in Ref. [13]). Thus, the relation $2U \approx W_f$ serves as strong evidence that the computed solution is a good approximation.

According to the theory of the weak formulation, if the approximate function u lies in $H_0^1(\Omega)$—as is true in our example—and the integration is exact, then the approximate potential energy should provide a **lower bound** for the exact potential energy [11]:

$$U_{\text{exact}} \geq U_{\text{PINN}} \tag{2.55}$$

Here, U_{PINN} is the potential energy computed from the trained PINN model (calculated in the previous code cell via integration), and U_{exact} is the exact potential energy, which can be evaluated using the exact solution given in Eq.(2.15), the formula in Eq.(2.51), and symbolic computation via Sympy. The snippet to compute this is as follows.

```
# Exact Solution (Sympy, for energy computation):
x, y = sp.symbols('x, y', real=True)          # define sympolic variables
exact_u_sp = sp.sin(sp.pi * x) * sp.sin(sp.pi * y)     # exact solution

u_x_sp = diff(exact_u_sp, x)                          # first derivatives
u_y_sp = diff(exact_u_sp, y)

U_sp = sp.integrate(integrate(u_x_sp**2 + u_y_sp**2,(x,0,1)),(y,0,1))/2
U_sp = U_sp.evalf()                                    # potential energy
print(f"U_sp = {U_sp}, U_difference = {U_sp-U_PINN}")
```

```
U_sp = 2.46740110027234, U_difference = 0.000135732764404928
```

It is observed that the exact potential energy is higher than the potential energy computed from the PINN model, confirming that Eq.(2.55) holds true in this case.

```
printSummary(pinns, LossTerms, Residuals, totalTimes,error_L2s,errorVIs)
```

```
#  PINN model      Loss_terms residual    total_time error_L2s  errorVIs
========================================================================
1  2D              DBC        _clctn      1.833e+01  1.008e-03  1.004e+01
2  2D              DBC        _FDM        3.420e+00  7.129e-02  6.082e+00
3  2D              DBC        _intGQ      1.877e+01  1.855e-03  6.090e+00
4  2D              DBC        _intTR      1.108e+01  7.923e-04  9.155e+00
5  2D_4DBC         PDE_only   _energyTR   5.013e+00  6.298e-04  3.715e+00
```

We note the following for this Poisson problem:

1. The relative error of the solution from model-5 that uses `PINN2D_4DBC` with the `residual_energy` loss is on the order of 10^{-4}, which is better than all other models trained so far.
2. The computation speed ranks as the second fastest among the tested methods.
3. The error variation index (EVI) over the domain is the lowest, indicating more uniform accuracy.

2.9.2 Example: Gaussian Quadrature for Energy Integral

The following snippet trains the PINN model using Gaussian quadrature for energy computation with the `PINN2D_4DBC` class. The code remains the same as in the previous section, except for the change in the integration scheme.

```
set_seed_th(88)                                      # for reproducibility

# Generate dataset using a small number of Gaussian points:
DI_rule = 'GQ'
w_s_th, x_Ω_th=sampleD_int(Gauss_points_weights, N_x=12, N_y=12,
                                                a=a, b=b, c=c, d=d)
# Create new PINN model with DBC enforced in class, and train it:
pinn = PINN2D_4DBC(Nn, NL=NL, a=a, b=b, c=c, d=d)
```

```
Trainer  = trainer_int_PDE
Residual = residual_energy
losses, totalTime = train_adam_lbfgs(pinn, Trainer, Residual, f_th,
       x_Ω_th, w_s_th, x_DBC_th, u_DBC_th, x_NBC_th, t_NBC_th, w_NBC_th,
                           I_NBC, epochs1=201, epochs2=11, use_lbfgs=True)
```

```
==================================================
201 epoches training with Adam:
Epoch     1, Loss: -5.192e-02, PDE: -5.192e-02
Epoch   101, Loss: -1.935e+00, PDE: -1.935e+00
Epoch   201, Loss: -2.350e+00, PDE: -2.350e+00
Time taken by adam: 0.545584 seconds

==================================================
11 epoches fine-tuning with LBFGS:
Epoch     1, Loss: -2.467e+00, PDE: -2.467e+00
Epoch    11, Loss: -2.467e+00, PDE: -2.467e+00

Training complete.
Time taken by lbfgs: 0.685681 seconds
Total time: 1.231265 seconds
Ratio (time_lbfgs/time_adam) of time taken per epoch:22.964878
```

```
# Compute the solution error:
error_L2, errorVI, u_diff, _ = plot_sln_rect(pinn, exact_u, f_th=f_th,
          N_plot=200, vmin=0., vmax=1., a=a,b=b,c=c,d=d, plot_it=False)

# Update the record (exercute this only once!):
recordSheet()

# plot3loss2(losses, ('adam initially','lbfgs in fine tunning'),
#          PDE='PDE', C1='DBC', C2='NBC', C3='Loss',log=False) # optional
```

```
L2 mean residual of trained pinn: 3.5393e-02
Maximum Error: 1.2495e-03
Mean Absolute Error: 3.0631e-04
Error Variation Index (ratio of Max/Mean): 4.0793e+00
L2 Relative Error: 7.7919e-04
```

The following snippet computes the energy components using the trained PINN model: the potential energy U, the work done by the force over the domain W_f, and the total potential energy $U_T = U + W_f$. It also computes the exact potential energy for comparison:

```
UnWf_PINN, U_PINN, Wf_PINN = residual_energy(pinn, f_th, x_Ω_th, w_s_th,
                                                          x_NBC, t_NBC)
print(f"U_PINN={U_PINN:.7f},  2U_PINN={2*U_PINN:.7f},  "
      f"Wf_PINN={Wf_PINN:.7f}")
print(f"U_sp={U_sp:.7f},     U_difference={U_sp-U_PINN:.7f}")
```

```
U_PINN=2.4680064,  2U_PINN=4.9360127,  Wf_PINN=4.9353819
```

```
U_sp=2.4674011,     U_difference=-0.0006053
```

This time, the lower bound property of the PINN solution given by Eq.(2.55) does not hold, but the difference is very small. This indicates that the model may not be fully trained, or there could be inaccuracies in the numerical integration.

```
printSummary(pinns, LossTerms, Residuals, totalTimes,error_L2s,errorVIs)
```

```
#  PINN model        Loss_terms residual     total_time error_L2s  errorVIs
===========================================================================
1  2D                DBC        _clctn       1.833e+01  1.008e-03  1.004e+01
2  2D                DBC        _FDM         3.420e+00  7.129e-02  6.082e+00
3  2D                DBC        _intGQ       1.877e+01  1.855e-03  6.090e+00
4  2D                DBC        _intTR       1.108e+01  7.923e-04  9.155e+00
5  2D_4DBC           PDE_only   _energyTR    5.013e+00  6.298e-04  3.715e+00
6  2D_4DBC           PDE_only   _energyGQ    1.231e+00  7.792e-04  4.079e+00
```

As seen, using Gaussian quadrature yields accuracy only slightly lower than that of model-5, which uses the trapezoidal rule for integration. However, the computation is significantly faster. The EVI is roughly the same as model-5's and much lower than all other models. Overall, model-6 demonstrates the good performance so far in terms of accuracy, efficiency, and EVI.

2.9.3 Example: Use of Collocation Residual

Next, we employ the strong form PDE and sample data via collocation points, computing the residual by integrating with a dummy weight of 1. We then train a PINN using the PINN2D_4DBC template. Essentially, this is a collocation-based PINN with a single PDE loss term for the same problem. The snippet is as follows.

```
set_seed_th(88)                                   # for reproducibility

# Generate training dateset, random collocation points over (a,b)×(c,d):
DI_rule = ''                    # collocation, no integration rule is used
w_s_th, x_Ω_th=sampleD_clctn(N_clctn=N_sample, a=a, b=b, c=c, d=d)
#x_DBC_th, u_DBC_th = sample4DBC_rect(N_DBC=20,a=a, b=b, c=c, d=d)
                                                               # no need
print(f"{w_s_th}, # of sample points: {len(x_Ω_th)}")
print(f"Is N_sample >= P* ? {len(x_Ω_th) >= P_star}")

# Create new PINN model with DBC enforced, and train it:
pinn = PINN2D_4DBC(Nn, NL=NL, a=a, b=b, c=c, d=d)

Trainer  = trainer_int_PDE
Residual = residual_clctn
losses, totalTime = train_adam_lbfgs(pinn, Trainer, Residual, f_th,
       x_Ω_th, w_s_th, x_DBC_th, u_DBC_th, x_NBC_th, t_NBC_th, w_NBC_th,
                        I_NBC, epochs1=201, epochs2=11, use_lbfgs=True)

# executed in 12.9s, finished 10:04:04 2025-05-16
```

```
1.0, # of sample points: 105
Is N_sample >= P* ? True
==================================================
201 epoches training with Adam:
Epoch      1, Loss: 7.950e+01, PDE: 7.950e+01
Epoch    101, Loss: 2.065e+01, PDE: 2.065e+01
Epoch    201, Loss: 1.175e+01, PDE: 1.175e+01
Time taken by adam: 1.417443 seconds

==================================================
11 epoches fine-tuning with LBFGS:
Epoch      1, Loss: 5.218e-06, PDE: 5.218e-06
Epoch     11, Loss: 5.218e-06, PDE: 5.218e-06

Training complete.
Time taken by lbfgs: 3.148743 seconds
Total time: 4.566187 seconds
Ratio (time_lbfgs/time_adam) of time taken per epoch:40.591491
```

```
# Plot the contours of both PINN solution and the exact one:
error_L2, errorVI, u_diff, _ = plot_sln_rect(pinn, exact_u, f_th=f_th,
           N_plot=200, vmin=0., vmax=1., a=a,b=b,c=c,d=d, plot_it=True)

plot3loss2(losses, ('adam initially','lbfgs in fine tunning'),
                    PDE='PDE', C1='DBC', C2='NBC', C3='Loss',log=False)

# Update the record (exercute this only once!):
recordSheet()
```

```
L2 mean residual of trained pinn: 1.7678e-05
Maximum Error: 7.8609e-05
Mean Absolute Error: 1.7498e-05
Error Variation Index (ratio of Max/Mean): 4.4925e+00
L2 Relative Error: 5.1956e-05
```

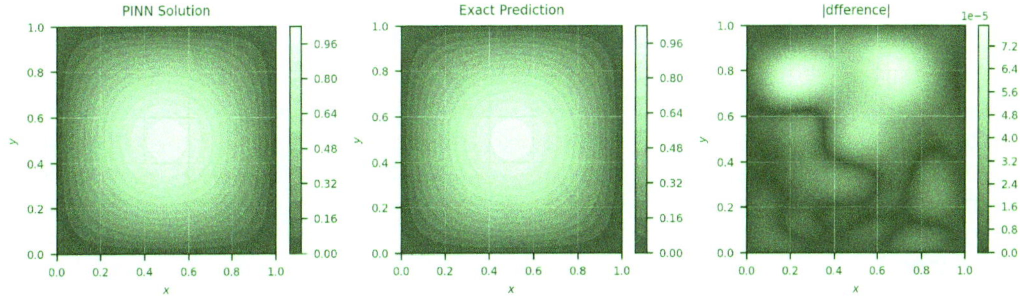

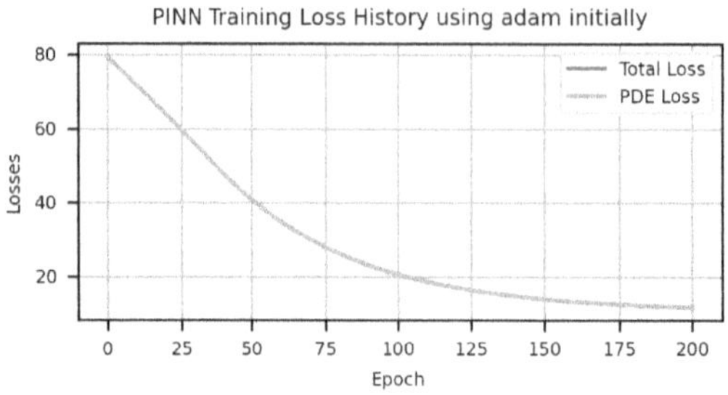

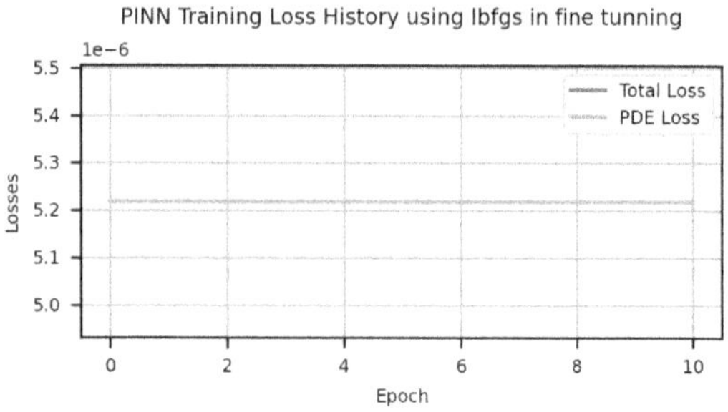

As seen, the solution accuracy is also very high, with the relative error on the order of 10^{-5}.

Although the strong form does not guarantee the lower bound property for the energy, let us nevertheless examine the energy bound property of the PINN model trained via collocation. For this purpose, we still use Gaussian quadrature to perform the integration needed to compute the energies. The snippet is as follows.

```
w_s_th, x_Ω_th=sampleD_int(Gauss_points_weights, N_x=20, N_y=20,
                                                  a=a, b=b, c=c, d=d)

UnWf_PINN, U_PINN, Wf_PINN = residual_energy(pinn, f_th, x_Ω_th, w_s_th,
                                                       x_NBC, t_NBC)
print(f"U_PINN = {U_PINN:.7f},  2U_PINN = {2*U_PINN:.7f},  "
     f"Wf_PINN = {Wf_PINN:.7f}")
print(f"U_sp = {U_sp:.7f},    U_difference = {U_sp-U_PINN:.7f}")
```

```
U_PINN = 2.4672701,  2U_PINN = 4.9345403,  Wf_PINN = 4.9346714
U_sp = 2.4674011,    U_difference = 0.0001310
```

```
printSummary(pinns, LossTerms, Residuals, totalTimes,error_L2s,errorVIs)
```

#	PINN model	Loss_terms	residual	total_time	error_L2s	errorVIs
1	2D	DBC	_clctn	1.833e+01	1.008e-03	1.004e+01
2	2D	DBC	_FDM	3.420e+00	7.129e-02	6.082e+00
3	2D	DBC	_intGQ	1.877e+01	1.855e-03	6.090e+00
4	2D	DBC	_intTR	1.108e+01	7.923e-04	9.155e+00
5	2D_4DBC	PDE_only	_energyTR	5.013e+00	6.298e-04	3.715e+00
6	2D_4DBC	PDE_only	_energyGQ	1.231e+00	7.792e-04	4.079e+00
7	2D_4DBC	PDE_only	_clctn	4.566e+00	5.196e-05	4.493e+00

```
printSummary(pinns, LossTerms, Residuals, totalTimes,error_L2s,errorVIs)
```

#	PINN model	Loss_terms	residual	total_time	error_L2s	errorVIs
1	2D	DBC	_clctn	9.615e+00	1.008e-03	1.004e+01
2	2D	DBC	_FDM	2.137e+00	7.129e-02	6.082e+00
3	2D	DBC	_intGQ	8.096e+00	1.855e-03	6.090e+00
4	2D	DBC	_intTR	1.061e+01	7.923e-04	9.155e+00
5	2D_4DBC	PDE_only	_energyTR	4.729e+00	6.298e-04	3.715e+00
6	2D_4DBC	PDE_only	_energyGQ	1.264e+00	7.792e-04	4.079e+00

```
7  2D_4DBC           PDE_only   _clctn       3.742e+00  5.196e-05  4.493e+00
```

As shown, training a PINN with the collocation method using the `PINN2D_4DBC` template drastically improves the solution accuracy by about one order of magnitude compared to the second-best model—the energy method with Gaussian quadrature. However, the computation time can be longer.

We reiterate that when using `PINN2D_4DBC` to create a PINN model, the error occurs only inside the domain, with zero error on the Dirichlet boundary conditions (DBC) by construction. Consequently, the error variation index is generally smaller, as summarized in the preceding table. This behavior is further illustrated in the next example.

2.9.4 Example: Trapezoidal Rule and Integral Residual

The following snippet trains the PINN model using the trapezoidal integration rule and the `PINN2D_DBC` class with a trial solution for the boundary conditions. Hence, only the PDE loss needs to be minimized:

```
set_seed_th(88)                                       # for reproducibility

# Generate dataset using the trapezoidal rule:
DI_rule = 'TR'
w_s_th, x_Ω_th=sampleD_int(trapz_points_weights, N_x=20, N_y=20,
                                                      a=a, b=b, c=c, d=d)
print(f"# of weights: {len(w_s_th)}, # of sample points: {len(x_Ω_th)}")

# Create new PINN model with DBC enforced, and integral residual:
pinn = PINN2D_4DBC(Nn, NL=NL, a=a, b=b, c=c, d=d)

Trainer  = trainer_int_PDE
Residual = residual_int
losses, totalTime = train_adam_lbfgs(pinn, Trainer, Residual, f_th,
       x_Ω_th, w_s_th, x_DBC_th, u_DBC_th, x_NBC_th, t_NBC_th, w_NBC_th,
                            I_NBC, epochs1=201, epochs2=11, use_lbfgs=True)
```

```
# of weights: 400, # of sample points: 400
==================================================
201 epoches training with Adam:
Epoch     1, Loss: 9.537e+01, PDE: 9.537e+01
Epoch   101, Loss: 2.460e+01, PDE: 2.460e+01
Epoch   201, Loss: 1.227e+01, PDE: 1.227e+01
Time taken by adam: 1.422916 seconds

==================================================
11 epoches fine-tuning with LBFGS:
Epoch     1, Loss: 8.193e-06, PDE: 8.193e-06
Epoch    11, Loss: 3.884e-06, PDE: 3.884e-06

Training complete.
Time taken by lbfgs: 4.115916 seconds
Total time: 5.538832 seconds
Ratio (time_lbfgs/time_adam) of time taken per epoch:52.855561
```

```
# Plot the contours of both PINN solution and the exact one:
error_L2, errorVI, u_diff, _ = plot_sln_rect(pinn, exact_u, f_th=f_th,
            N_plot=200, vmin=0., vmax=1., a=a,b=b,c=c,d=d, plot_it=True)

plot3loss2(losses, ('adam initially','lbfgs in fine tunning'),
                    PDE='PDE', C1='DBC', C2='NBC', C3='Loss',log=False)

# Update the record (exercute this only once!):
recordSheet()
```

```
L2 mean residual of trained pinn: 3.3613e-06
Maximum Error: 1.2315e-05
Mean Absolute Error: 2.7222e-06
Error Variation Index (ratio of Max/Mean): 4.5240e+00
L2 Relative Error: 7.2979e-06
```

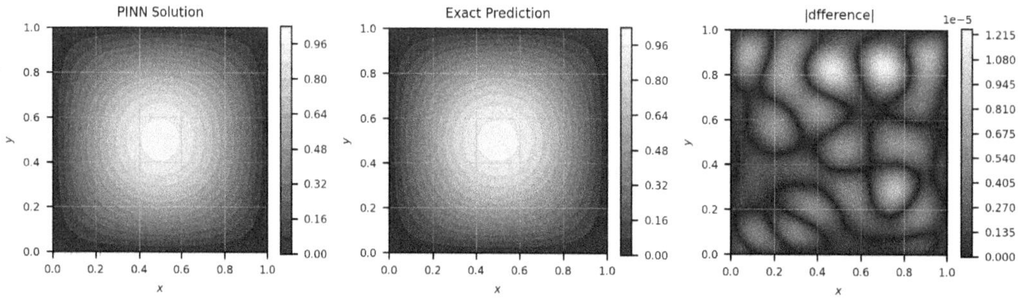

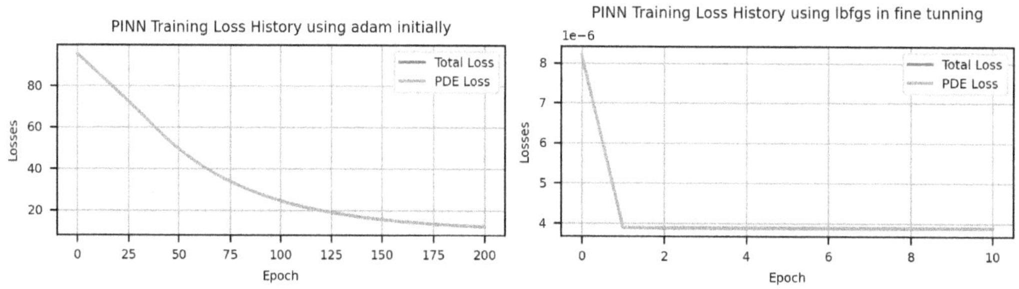

As found, the solution accuracy is very high, with the relative error on the order of 10^{-6}. Let us take a look at the performance of this model:

```
printSummary(pinns, LossTerms, Residuals, totalTimes,error_L2s,errorVIs)
```

```
#  PINN model      Loss_terms residual    total_time error_L2s  errorVIs
==========================================================================
1  2D              DBC        _clctn      1.833e+01  1.008e-03  1.004e+01
2  2D              DBC        _FDM        3.420e+00  7.129e-02  6.082e+00
3  2D              DBC        _intGQ      1.877e+01  1.855e-03  6.090e+00
4  2D              DBC        _intTR      1.108e+01  7.923e-04  9.155e+00
5  2D_4DBC         PDE_only   _energyTR   5.013e+00  6.298e-04  3.715e+00
6  2D_4DBC         PDE_only   _energyGQ   1.231e+00  7.792e-04  4.079e+00
7  2D_4DBC         PDE_only   _clctn      4.566e+00  5.196e-05  4.493e+00
```

```
8  2D_4DBC          PDE_only   _intTR      5.539e+00  7.298e-06  4.524e+00
```

From the record table built so far, we note the following:

1. Using `PINN2D_4DBC`, where all Dirichlet boundary conditions (DBCs) are enforced at the class level (the "hard way"), yields very accurate solutions. This can be clearly seen by comparing Model-1 with Model-7, and Model-4 with Model-8. Moreover, training is faster, and the error variation index (EVI) is lower.

2. The proper treatment of DBCs is key to achieving accurate solutions. Employing a trial solution that automatically enforces DBCs is an effective approach. This holds true for both strong form methods (collocation type) and weak form methods (energy type) used in training PINN models. This insight aligns with traditional computational methods such as FEM, smoothed FEM, and meshfree methods, where DBCs are known as essential boundary conditions. The current results from the PINN models support this understanding.

3. When using `PINN2D_4DBC`, the loss function contains only one term, which simplifies the optimization landscape and helps optimizers converge more efficiently. This factor contributes significantly to the improved accuracy and efficiency observed.

4. Interestingly, PINN models trained using the energy method are generally less accurate compared to those trained using collocation and integration of the strong form PDE. This observation is largely consistent with traditional computational methods, where weak form approaches are often more stable and versatile, while strong form methods (collocation) can yield higher accuracy—especially at collocation points—provided the method is formulated properly and stable [18]. We note the following:

   The PINN solution using the strong form PDE with collocation points benefits from the **repetitive use** of samples during optimization, which helps overcome stability issues and better enforces the PDE constraints at all the collocation points. The computation is generally slower.

   On the other hand, the energy method minimizes the total potential energy at the global level via integration. The primary variable (u) evaluated at points in the problem domain may not be as accurate as that from the collocation method. However, the computation is generally faster.

2.10 Energy or Collocation

Careful study of various PINN models leads to the following points:

1. A neural network with smooth activation functions essentially represents a very smooth function. The use of `autograd` provides exact derivatives of all required orders in the PDE. This ensures that derivatives are bounded and accurately evaluated.

2. The collocation approach enforces the satisfaction of the PDE (and DBCs) exactly at **each and every** sampled point in an L^2 sense. This is a strict requirement, so if the global minimum of the loss is found, the solution is guaranteed to be accurate. While this approach may require more training time compared to the energy method, as demonstrated in the previous results, it tends to yield higher accuracy.

3. In traditional collocation methods (e.g., Chapter 6 of Ref. [18]), the main challenge is solving the resulting discretized linear system that comes from minimizing the L^2 loss.

Such systems may suffer from instability and ill-conditioning, heavily depending on the location of collocation points, which are used only once in deriving the solution.

4. Neural networks alleviate this instability issue because they can be trained with an arbitrarily large number of collocation points, repeatedly fed through multiple epochs. This abundance of data helps the optimizer overcome potential instability and improves the chance of finding a global minimum of the L^2 loss, even if it requires more computation time.

5. In contrast, the energy method (weak form) evaluates the total energy of the **entire** system as a single scalar quantity. Loss functions based on this scalar energy are naturally easier to minimize, leading to faster training times, as observed in previous experiments. However, the energy method generally yields less accurate results due to several factors:

   (a) the neural network optimizer may fail to find the true minimum of the total potential energy;

   (b) the loss does not directly measure the errors at points in the domain; and

   (c) numerical errors in the energy integration can degrade accuracy.

In summary, while the energy method offers a computationally efficient and stable approach for training PINNs by minimizing a scalar functional over the entire domain, it often yields less accurate solutions compared to collocation-based methods using PINNs of the same capacity. The collocation approach, despite requiring more training time, enforces PDE satisfaction pointwise and benefits from the neural network's smoothness and automatic differentiation, resulting in higher accuracy.

Therefore, the choice between energy and collocation methods in PINNs involves a trade-off between computational efficiency and solution accuracy, with collocation generally preferred when accuracy is paramount.

2.11 PINN for Problems with Neumann Boundary

To investigate the impact of Neumann boundary conditions (NBCs), we consider the same Poisson problem but posed different as follows:

1. The Dirichlet boundary condition (DBC) is enforced only on the left edge of the unit square domain ($x = 0$).

2. The remaining three edges (bottom, top, and right) are prescribed with NBCs. The target NBC values are computed using the exact solution given in Eq.(2.15).

Since this is essentially the same problem formulated differently, a properly trained PINN should still approximate the exact solution accurately. This setup allows us to rigorously evaluate the PINN's handling of NBCs.

2.11.1 PINN with Trial Solution for One Edge

Because the DBC is imposed only on the left edge ($x = 0$), the trial solution can be constructed using the factor $(x - 0) = x$ to automatically satisfy the DBC on this edge:

$$u_{\text{trial}}(x, y) = x \, \mathcal{N}_{\hat{\mathbf{w}}}(x, y)$$

where $\mathcal{N}_{\hat{\mathbf{w}}}(x, y)$ is a standard multilayer perceptron (MLP) neural network with learning parameters $\hat{\mathbf{w}}$.

The following snippet defines a PINN class that enforces the DBC automatically on the left edge using this trial solution formulation.

```
class PINN2D_LeftDBC(nn.Module):
    """
    2D PINN model with Dirichlet boundary conditions (DBCs) (u=0)
    enforced via a trial solution on LEFT edge of a rectangular domain.
    Domain: (a, b) × (c, d), where u = 0 on the left edge.
    Args: Nn (int): Number of neurons per hidden layer.
          NL (int): Number of hidden layers.
          a, b, c, d (float): Domain boundaries. Only a is used here.
          activation (nn.Module): e.g., nn.Tanh(), nn.ReLU().
    """
    def __init__(self, Nn, NL, actit=nn.Tanh(), a=0., b=1., c=0., d=1.):
        super().__init__()
        self.a = a
        self.actit = actit

        layers = [nn.Linear(2, Nn)]
        for _ in range(NL - 1):
            layers.append(self.actit)
            layers.append(nn.Linear(Nn, Nn))
        self.hidden = nn.Sequential(*layers)
        self.out = nn.Linear(Nn, 1)

    def forward(self, x):
        x_ = x[:, 0:1]
        NBC_term = x_ - self.a                          # DBC on the left edge
        z = self.hidden(x)
        N = self.out(z)
        return NBC_term * N
```

In this case, when a collocation-based method is used the composite loss for a PINN created using the template PINN2D_LeftDBC reduces to:

$$\mathcal{L} = w_{\text{PDE}}\mathcal{L}_{\text{PDE}} + w_{\text{NBC}}\mathcal{L}_{\text{NBC}} \tag{2.56}$$

It has two terms, with the $\mathcal{L}_{\text{NBC}}$ term accounting for the Neumann boundary conditions imposed along the bottom, top, and right edges of the rectangular domain.

We will use this class to build PINN models where the Dirichlet boundary condition (DBC) is enforced only on the left edge, while Neumann boundary conditions (NBCs) are applied on the other three edges. In this scenario, if the strong form PDE is used, the total loss function must include an additional NBC loss term. However, when using the energy method, the loss function remains a single quadratic term in u.

2.11.2 Setting of NBCs

Consider a rectangular domain with the left edge fixed (DBC). The target NBC functions on the bottom, top, and right edges are derived from the exact solution given in Eq.(2.15). The following code snippet computes these NBC target functions for use during training:

```
x, y = sp.symbols('x, y', real=True)          # Define symbolic symbols
sp_u = sp.sin(sp.pi*x) * sp.sin(sp.pi*y)    # function over [0,1]x[0,1]

dudy_y0 = sp.diff(sp_u, y).subs(y,0)                    # bottom edge: y =0
display(Math(f" \\frac{{\\partial u}}{{\\partial n}}\\big|_{{y=0}} = "
             f"-\\frac{{\\partial u}}{{\\partial y}}\\big|_{{y=0}} = "
             f"{latex(-dudy_y0)}"))

dudy_y1 = sp.diff(sp_u, y).subs(y,1)                       # top edge: y =1
display(Math(f" \\frac{{\\partial u}}{{\\partial n}}\\big|_{{y=1}} = "
             f" \\frac{{\\partial u}}{{\\partial y}}\\big|_{{y=1}} = "
             f"{latex(dudy_y1)}"))

dudx_x1 = sp.diff(sp_u, x).subs(x,1)                     # right edge: x =1
display(Math(f" \\frac{{\\partial u}}{{\\partial n}}\\big|_{{x=1}} = "
             f" \\frac{{\\partial u}}{{\\partial x}}\\big|_{{x=1}} = "
             f"{latex(dudx_x1)}"))
```

$$\frac{\partial u}{\partial n}\big|_{y=0} = -\frac{\partial u}{\partial y}\big|_{y=0} = -\pi \sin(\pi x)$$

$$\frac{\partial u}{\partial n}\big|_{y=1} = \frac{\partial u}{\partial y}\big|_{y=1} = -\pi \sin(\pi x)$$

$$\frac{\partial u}{\partial n}\big|_{x=1} = \frac{\partial u}{\partial x}\big|_{x=1} = -\pi \sin(\pi y)$$

These NBC functions are defined in the following snippet in both numpy and torch for later use.

```
def t_np(x):
    ''' traction function for Neumann BC, numpy'''
    return -np.pi * np.sin(np.pi * x)

def t_th(x):
    ''' traction function for Neumann BC, torch'''
    return - torch.pi*torch.sin(torch.pi*x)
```

The detailed problem setup for implementation in code is as follows:

- **Problem:** Solve the Poisson equation (2.13) over the rectangular domain $(a, b) \times (c, d)$.
- **DBCs:** Applied on the left edge of the rectangle ($x = a$).
- **NBCs:** Applied on the remaining three edges — bottom ($y = c$), top ($y = d$), and right ($x = b$). Each of these edges has its own outward normal vector:
    - Bottom edge ($y = c$): $\mathbf{n} = (0, -1)$

– Top edge ($y = d$): $\mathbf{n} = (0, 1)$

– Right edge ($x = b$): $\mathbf{n} = (1, 0)$

2.11.3 Computation of NBC Loss

To compute the NBC loss, we first evaluate the normal directional derivatives of the PINN output u on these Neumann boundaries. The NBC loss is then calculated as the discrepancy between these directional derivatives and the known NBC values, using either:

1. **Collocation points** along the boundaries to compute the mean squared error (residual) of the NBCs.
2. **Numerical integration** along these edges via Gaussian quadrature or the trapezoidal rule to accurately evaluate the NBC residual contribution to the loss.

The following code snippet provides an implementation of these computations.

```
def loss_NBC_fun(pinn, x_NBC_th, t_NBC_th, w_NBC_th, I_NBC,
                                                        a=0,b=1,c=0,d=1):
    '''Compute NBC loss for `pinn` for
       bottom, top, and right edges of a rectangle.
       x_NBC_th, Dirichlet boundary points
       (m, 2) tensor requires_grad=True, on device)
       t_NBC_th, target values on NBC,(m,1) tensor, on device
       w_NBC_th, weights on NBC, (m,1) tensor, on device.
       I_NBC: integration rule.
             'TR'-- trapezolidal rule; 'GQ'-- Gaussian quadrature
             'mse'-- mean squared pointwise residual
    return: loss_NBC_th, loss on NBC
    '''
    x_NBC_th.requires_grad = True
    u_pred = pinn(x_NBC_th)

    # Compute gradient of u w.r.t. x_NBC_th, initialized as 1s.
    grads = torch.autograd.grad(u_pred, x_NBC_th,    # [du/dx, du/dy]
            grad_outputs=torch.ones_like(u_pred), create_graph=True)[0]

    x_ = x_NBC_th[:, 0:1]
    y_ = x_NBC_th[:, 1:2]
    du_dn_pred = torch.zeros_like(u_pred)# normal directional derivative

    # Bottom edge (y = c): normal = (0, -1) -> use -du/dy
    bottom_mask = torch.abs(y_ - c) < 1e-5
    du_dn_pred += torch.where(bottom_mask, -grads[:, 1:2],
                                            torch.zeros_like(u_pred))
    # Top edge    (y = d): normal = (0, +1) -> use +du/dy
    top_mask = torch.abs(y_ - d) < 1e-5
    du_dn_pred += torch.where(top_mask, grads[:, 1:2],
                                            torch.zeros_like(u_pred))
    # Right edge  (x = b): normal = (+1, 0) -> use +du/dx
    right_mask = torch.abs(x_ - b) < 1e-5
```

```
    du_dn_pred += torch.where(right_mask, grads[:, 0:1],
                                       torch.zeros_like(u_pred))
    r2 = (du_dn_pred - t_NBC_th) ** 2
    if I_NBC == 'mse':
        # MSE loss between predicted and target Neumann BC values
        loss_NBC_th = torch.mean(r2)
    elif I_NBC == 'TR' or I_NBC == 'GQ':
        loss_NBC_th = torch.sum(w_NBC_th*r2)
    else:
        print("Please specify integration rule as 'mse', 'TR', or 'GQ'")

    return loss_NBC_th      # loss in torch
```

2.11.4 Generation of NBC Points and Weights

The following code snippet generates sample points on the Neumann boundaries—specifically the bottom, top, and right edges—of the rectangular domain. These sampled points are used to evaluate the Neumann boundary loss based on Eq.(2.9). The integration over each edge is performed using the trapezoidal rule.

```
def sampleNBCs_rect_TR(t_np, N_NBC=20, a=0.0, b=1.0, c=0.0, d=1.0):
    '''
    Sample uniformly boundary points on 3 edges of domain [a,b] x [c,d]
    for Neumann boundary condition (NBC) on bottom, top, right edges.
    Also computes trapezoidal weights for NBC points.
    '''
    x_NBCs = []; t_NBCs = []; w_s_NBCs = []                   # initialize

    x_horz = np.linspace(a, b, N_NBC, dtype=np.float32)
    y_vert = np.linspace(c, d, N_NBC, dtype=np.float32)

    delta_x = (b - a) / (N_NBC - 1)          # Trapezoidal rule spacings
    delta_y = (d - c) / (N_NBC - 1)

    for i in range(N_NBC):                            # Bottom edge (y = c)
        xi = x_horz[i]
        x_NBCs.append([xi, c])
        t_NBCs.append([t_np(xi)])
        weight = delta_x * (0.5 if (i == 0 or i == N_NBC - 1) else 1.0)
        w_s_NBCs.append(weight)

    for i in range(N_NBC):                               # Top edge (y = d)
        xi = x_horz[i]
        x_NBCs.append([xi, d])
        t_NBCs.append([t_np(xi)])
        weight = delta_x * (0.5 if (i == 0 or i == N_NBC - 1) else 1.0)
        w_s_NBCs.append(weight)

    for i in range(N_NBC):                             # Right edge (x = b)
        yi = y_vert[i]
```

```
        x_NBCs.append([b, yi])
        t_NBCs.append([t_np(yi)])
        weight = delta_y * (0.5 if (i == 0 or i == N_NBC - 1) else 1.0)
        w_s_NBCs.append(weight)

    return (torch.tensor(x_NBCs), torch.tensor(t_NBCs),
            torch.tensor(w_s_NBCs).unsqueeze(1))
```

Alternatively, the integration along each edge can be carried out using the **Gaussian quadrature** method, which offers higher accuracy for smooth (polynomial) integrands.

```
def sampleNBCs_rect_GQ(t_np, N_NBC=20, a=0.0, b=1.0, c=0.0, d=1.0):
    '''
    Sample boundary points on 3 edges (bottom, top, right) of domain
    [a,b] x [c,d] using Gaussian quadrature.
    return: coordinates on 3 edges,
            corresponding integration weights.
    '''
    x_NBCs = []; t_NBCs = []; w_s_NBCs = []

    # Get Gauss-Legendre nodes and weights on [-1, 1]
    xi_ref, wi_ref = np.polynomial.legendre.leggauss(N_NBC)

    # Map nodes and weights to [a, b]
    x_nodes = 0.5 * (b - a) * xi_ref + 0.5 * (b + a)
    w_x     = 0.5 * (b - a) * wi_ref

    for xi, wi in zip(x_nodes, w_x):                # Bottom edge (y = c)
        x_NBCs.append([xi, c])
        t_NBCs.append([t_np(xi)])
        w_s_NBCs.append(wi)

    for xi, wi in zip(x_nodes, w_x):                   # Top edge (y = d)
        x_NBCs.append([xi, d])
        t_NBCs.append([t_np(xi)])
        w_s_NBCs.append(wi)

    # Map nodes and weights to [c, d] for vertical edge
    y_nodes = 0.5 * (d - c) * xi_ref + 0.5 * (d + c)
    w_y = 0.5 * (d - c) * wi_ref

    for yi, wi in zip(y_nodes, w_y):                 # Right edge (x = b)
        x_NBCs.append([b, yi])
        t_NBCs.append([t_np(yi)])
        w_s_NBCs.append(wi)

    return (torch.tensor(x_NBCs, dtype=torch.float32),
            torch.tensor(t_NBCs, dtype=torch.float32),
            torch.tensor(w_s_NBCs, dtype=torch.float32).unsqueeze(1))
```

2.11.5 Trainer of PINN Using PDE and NBC Loss

The following function defines a trainer for a PINN model where the total loss is composed of two terms: `loss_PDE` and `loss_NBC`.

This function supports flexible configuration, allowing users to choose between different forms of PDE residuals (e.g., strong form or energy form) and various optimizers (e.g., Adam or L-BFGS).

Detailed explanations of the input arguments and usage are provided in the doc-string and comments of the function below.

```
def trainerPDE_NBC(pinn, residual, optimizer, f_th, x_Ω_th, w_s_th,
                x_DBC_th, u_DBC_th, x_NBC_th, t_NBC_th, w_NBC_th, I_NBC,
                                             a=a,b=b,c=c,d=d, epochs=8):
    """
    For training a PINN with loss_PDE+loss_NBC.
    Args: pinn: the PINN model (nn.Module)
          residual: function to compute residual loss
          optimizer: PyTorch optimizer (e.g. Adam, LBFGS)
          epochs: number of training iterations for PINN
          x_DBC_th and u_DBC_th are not used here.
    """
    loss_history = {"total": [], "PDE": [], "NBC": []}

    if epochs <= 301: n_prnt = 100
    else:             n_prnt = max(1, round(epochs / 3, -2))
    # Print roughly every epochs/300 steps, rounded to nearest 100

    for epoch in range(epochs):
        # Initialize loss variables for the current epoch
        loss_PDE = None; loss_NBC = None; total_loss = None

        def closure():
            nonlocal loss_PDE, loss_NBC, total_loss
            # Allow modifying variables from outer scope
            optimizer.zero_grad()
            # These calculations happen outside of torch.no_grad()
            # x and x_NBC must have requires_grad=True here
            loss_PDE, _= residual(pinn, f_th, x_Ω_th, w_s_th, x_NBC_th,
                                                               t_NBC_th)
            loss_NBC = loss_NBC_fun(pinn, x_NBC_th, t_NBC_th, w_NBC_th,
                                               I_NBC, a=a,b=b,c=c,d=d)
            total_loss = loss_PDE + loss_NBC

            total_loss.backward()               # Perform backward pass

            return total_loss # Return the total loss for the optimizer

        if isinstance(optimizer, torch.optim.LBFGS):
            # LBFGS optimizer step has multiple closure calls internally
```

```
            # The loss value returned by step is the loss of the last
            optimizer.step(closure)          # iteration in the step
            # loss_PDE, loss_NBC, total_loss were updated by the last
            total_loss_val = total_loss.item()         # closure call
            loss_PDE_val = loss_PDE.item()
            loss_NBC_val = loss_NBC.item()

        else: # Adam or other optimizers
            # closure performs forward pass, loss clcltn & backward pass
            total_loss_tensor = closure() #returns the total_loss tensor
            optimizer.step()
            # Get item values from the tensors updated by closure
            total_loss_val = total_loss_tensor.item()
            loss_PDE_val = loss_PDE.item() #These were set inside closure
            loss_NBC_val = loss_NBC.item() #These were set inside closure

        # Logging happens with loss values obtained from optimizer step
        # This logging section is implicitly outside the graph
        # We can add torch.no_grad() here just to be explicit that no
        with torch.no_grad():            # graph is built for logging
            loss_history["total"].append(total_loss_val)
            loss_history["PDE"].append(loss_PDE_val)
            loss_history["NBC"].append(loss_NBC_val)

        if epoch % n_prnt == 0 or epoch == epochs - 1:
             print(f"Epoch {epoch:5d}, Loss: {total_loss_val:.3e}, "
                   f"PDE: {loss_PDE_val:.3e}, NBC: {loss_NBC_val:.3e}")

    return loss_history
```

2.11.6 Plot Function of Solution Surface

The following utility function generates 3D surface plots for visualizing the solution predicted by the PINN model, the exact analytical solution, and the pointwise error between them. These plots provide intuitive insights into the accuracy and distribution of the model error over the computational domain.

```
from mpl_toolkits.mplot3d import Axes3D

# ----- Model evaluation and surface plotting -----
def plot_sln_Surface(pinn, exact_u, f_th=None, N_plot=50,
             a=0., b=1., c=0., d=1., shrink=0.7, dpi=300, plot_it=True):
    '''Evaluate model.
      Plot surfaces for the PINN and exact solutions, and error.
    '''
    x_plot = np.linspace(a, b, N_plot)
    y_plot = np.linspace(c, d, N_plot)
    X_plot, Y_plot = np.meshgrid(x_plot, y_plot)
    x_eval_np = np.hstack((X_plot.flatten()[:, None],
                           Y_plot.flatten()[:, None]))
```

```
    x_eval_th = torch.tensor(x_eval_np, dtype=torch.float32)
    residual_pred = residual_clctn(pinn, f_th, x_eval_th)
    print(f"L2 mean residual of trained pinn: {residual_pred[0]:.4e}")

    u_pred = pinn(x_eval_th).detach().cpu().numpy().reshape(X_plot.shape)
    u_exact= exact_u(X_plot, Y_plot)

    error_L2, errorVI, u_diff = sln_error(u_pred, u_exact)

    if plot_it != True:
        return error_L2, errorVI, u_diff, residual_pred

    width = 9; height = width/3.4
    fig = plt.figure(figsize=(width, height), dpi=dpi)
    ax1 = fig.add_subplot(131, projection='3d')
    surf = ax1.plot_surface(X_plot, Y_plot, u_pred, cmap='viridis')
    #fig.colorbar(surf, ax=ax1, shrink=shrink, aspect=10)
    ax1.set_title('PINN Solution $u_{PINN}(x,y)$')
    ax1.set_xlabel('x'); ax1.set_ylabel('y')

    ax2 = fig.add_subplot(132, projection='3d')
    surf = ax2.plot_surface(X_plot, Y_plot, u_exact, cmap='viridis')
    #fig.colorbar(surf, ax=ax2, shrink=shrink, aspect=10)
    ax2.set_title('Exact Solution $u_{exact}(x,y)$')
    ax2.set_xlabel('x'); ax2.set_ylabel('y')

    ax3 = fig.add_subplot(133, projection='3d')
    surf= ax3.plot_surface(X_plot, Y_plot, u_diff, cmap='plasma')
    #fig.colorbar(surf, ax=ax3, shrink=shrink, aspect=10)
    ax3.set_title('Error $(u_{pred} - u_{exact})$')
    ax3.set_xlabel('x'); ax3.set_ylabel('y')
    fig.subplots_adjust(wspace=0.)
    plt.show()

    return error_L2, errorVI, u_diff, residual_pred
```

2.11.7 Example: Solution to Poisson's Equation with NBCs

We now build a PINN model to solve the Poisson equation defined over a square domain, where the left edge is subjected to DBC, and the remaining three edges (bottom, top, and right) are subjected to NBCs.

2.11.7.1 Use of Collocation Method

In this example, we use collocation points to compute the PDE residuals. The total loss function consists of two components: `loss_PDE`, accounting for the strong-form PDE residuals, and `loss_NBC`, enforcing the Neumann conditions along the boundaries.

```
set_seed_th(88)                                        # for reproducibility
```

```
# --- collocation, use `loss_PDE` and `loss_NBC`, same setup ---
# a=0.; b =1.; c =0.; d=1.                  # same setup for (a,b)×(c,d)
# p=2 ; Nn=50; NL=2        # same setup for NN, NL layers, nn neurons each

# Sample NBC boundary -- trapezoidal rule -- use N_DBC defined earlier:
I_NBC = 'TR'  # Integration rule for computing NBC losse in loss_NBC_fun
x_NBC_th, t_NBC_th, w_NBC_th = sampleNBCs_rect_TR(t_np, N_NBC=N_DBC,
                                                   a=a, b=b, c=c, d=d)
# Domain data -- random collocation --,(a,b)×(c,d)
DI_rule = ''                             # no integration for the domain
w_s_th, x_Ω_th = sampleD_clctn(N_clctn=N_sample, a=a, b=b, c=c, d=d)

# Create a new PINN model with DBC enforced on the left edge
pinn = PINN2D_LeftDBC(Nn, NL, a=a, b=b, c=c, d=d)
print_model_parameters(pinn, print_table=False)        # print only P*, P

x_DBC_th = None; u_DBC_th = None
LossTerm = 'NBC'                         # loss term in addition to loss_PDE
Trainer  = trainerPDE_NBC
Residual = residual_clctn
losses, totalTime = train_adam_lbfgs(pinn, Trainer, Residual, f_th,
       x_Ω_th, w_s_th, x_DBC_th, u_DBC_th, x_NBC_th, t_NBC_th, w_NBC_th,
                         I_NBC, epochs1=201, epochs2=11, use_lbfgs=True)

print(f"Sampled {x_Ω_th.shape[0]} collocation points in the domain.")
print(f"Is N_sample >= P* ? {x_Ω_th.shape[0] >= P_star}")
print(f"Sampled {x_NBC_th.shape[0]} NBC boundary points.")
```

```
P* (Total pseudo-dimension of the affine space) = 105
Total trainable parameters: 2751
==================================================
201 epoches training with Adam:
Epoch     0, Loss: 1.019e+02, PDE: 8.548e+01, NBC: 1.641e+01
Epoch   100, Loss: 4.068e+01, PDE: 3.045e+01, NBC: 1.023e+01
Epoch   200, Loss: 3.712e+01, PDE: 2.858e+01, NBC: 8.539e+00
Time taken by adam: 1.062020 seconds

==================================================
11 epoches fine-tuning with LBFGS:
Epoch     0, Loss: 7.045e-04, PDE: 6.051e-04, NBC: 9.942e-05
Epoch    10, Loss: 5.158e-05, PDE: 4.480e-05, NBC: 6.778e-06

Training complete.
Time taken by lbfgs: 13.481560 seconds
Total time: 14.543580 seconds
Ratio (time_lbfgs/time_adam) of time taken per epoch:231.958698
Sampled 105 collocation points in the domain.
Is N_sample >= P* ? True
Sampled 63 NBC boundary points.
```

```
error_L2, errorVI, u_diff, _= plot_sln_Surface(pinn, exact_u, f_th=f_th,
                    N_plot=50, a=a, b=b, c=c, d=d, dpi=300, plot_it=True)

plot3loss2(losses, ('adam initially','lbfgs in fine tunning'),
                                      PDE='PDE', C1='NBC', log=False)
```

```
L2 mean residual of trained pinn: 2.3877e-04
Maximum Error: 6.7268e-04
Mean Absolute Error: 1.9870e-04
Error Variation Index (ratio of Max/Mean): 3.3855e+00
L2 Relative Error: 5.1402e-04
```

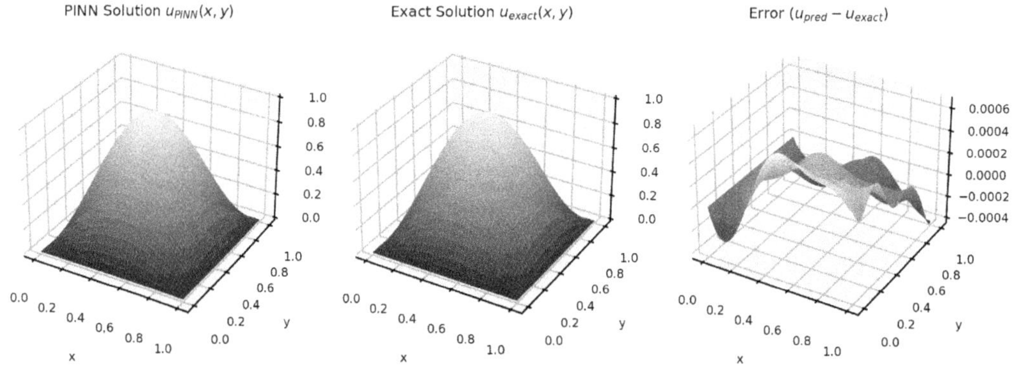

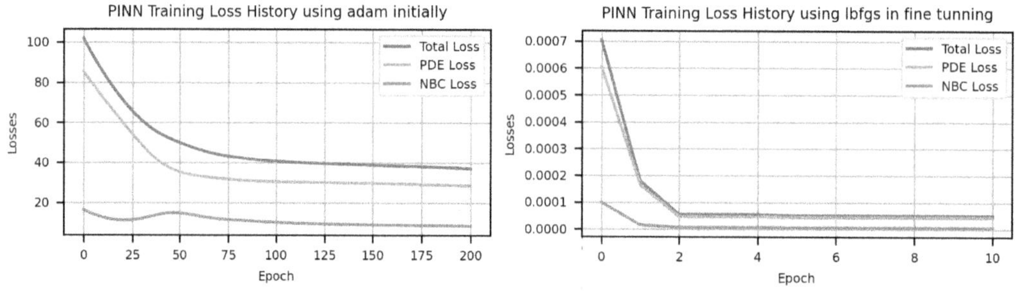


```
recordSheet()                                    # Update the record
```

We obtained a highly accurate solution, and the relative error is on the order of 10^{-4}.

```
printSummary(pinns, LossTerms, Residuals, totalTimes,error_L2s,errorVIs)
```

#	PINN model	Loss_terms	residual	total_time	error_L2s	errorVIs
1	2D	DBC	_clctn	1.833e+01	1.008e-03	1.004e+01
2	2D	DBC	_FDM	3.420e+00	7.129e-02	6.082e+00
3	2D	DBC	_intGQ	1.877e+01	1.855e-03	6.090e+00
4	2D	DBC	_intTR	1.108e+01	7.923e-04	9.155e+00
5	2D_4DBC	PDE_only	_energyTR	5.013e+00	6.298e-04	3.715e+00
6	2D_4DBC	PDE_only	_energyGQ	1.231e+00	7.792e-04	4.079e+00
7	2D_4DBC	PDE_only	_clctn	4.566e+00	5.196e-05	4.493e+00

```
8  2D_4DBC          PDE_only   _intTR       5.539e+00  7.298e-06  4.524e+00
9  2D_LeftDBC       NBC        _clctn       1.454e+01  5.140e-04  3.385e+00
```

We note the following observations:

1. When a portion of the boundary is changed to a Neumann boundary condition (NBC), the solution accuracy decreases. The accuracy of Model-9 is significantly lower than that of Model-7, despite the fact that the Dirichlet boundary condition (DBC) on the left edge is still enforced at the class level.
2. Both the accuracy and total training time of Model-9 are similar to those of Model-1, which also uses the `trainerPDE_DBC` function. This is expected, as both models involve composite loss functions with two terms.
3. This example reinforces the general trend that composite loss functions—those containing multiple distinct loss terms—are typically more challenging for optimizers to minimize effectively.

2.11.7.2 Use of Integral Residual via Gaussian Quadrature

```
set_seed_th(88)                                        # for reproducibility

# Sample NBC boundary using --- Gauss points ----:
I_NBC = 'GQ'   # Integration rule for computing NBC loss in loss_NBC_fun
x_NBC_th, t_NBC_th, w_NBC_th = sampleNBCs_rect_GQ(t_np, N_NBC=5,
                                                    a=a, b=b, c=c, d=d)

# Generate domain data with ---Gauss_points---, (a,b)×(c,d)
DI_rule = 'GQ'                  # Integration rule for computing domain loss
w_s_th, x_Ω_th = sampleD_int(Gauss_points_weights, N_x=10, N_y=10,
                                                    a=a, b=b, c=c, d=d)
# new PINN model, DBC enforced on the left edge:
pinn = PINN2D_LeftDBC(Nn, NL, a=a, b=b, c=c, d=d)

Trainer  = trainerPDE_NBC
Residual = residual_int
losses, totalTime = train_adam_lbfgs(pinn, Trainer, Residual, f_th,
       x_Ω_th, w_s_th, x_DBC_th, u_DBC_th, x_NBC_th, t_NBC_th, w_NBC_th,
                           I_NBC, epochs1=201, epochs2=11, use_lbfgs=True)
```

```
==================================================
201 epoches training with Adam:
Epoch     0, Loss: 1.187e+02, PDE: 1.023e+02, NBC: 1.640e+01
Epoch   100, Loss: 4.075e+01, PDE: 3.055e+01, NBC: 1.020e+01
Epoch   200, Loss: 3.690e+01, PDE: 2.898e+01, NBC: 7.913e+00
Time taken by adam: 1.082172 seconds

==================================================
11 epoches fine-tuning with LBFGS:
Epoch     0, Loss: 7.083e-04, PDE: 5.937e-04, NBC: 1.147e-04
Epoch    10, Loss: 6.547e-05, PDE: 6.386e-05, NBC: 1.609e-06
```

```
Training complete.
Time taken by lbfgs: 18.532069 seconds
Total time: 19.614240 seconds
Ratio (time_lbfgs/time_adam) of time taken per epoch:312.918368
```

```
error_L2, errorVI, u_diff, _= plot_sln_Surface(pinn, exact_u, f_th=f_th,
                    N_plot=50, a=a, b=b, c=c, d=d, dpi=300, plot_it=True)
plot3loss2(losses, ('adam initially','lbfgs in fine tunning'),
                                          PDE='PDE', C1='NBC', log=False)
# Update the record (exercute this only once!):
recordSheet()
```

```
L2 mean residual of trained pinn: 8.7045e-05
Maximum Error: 1.8807e-04
Mean Absolute Error: 4.4789e-05
Error Variation Index (ratio of Max/Mean): 4.1990e+00
L2 Relative Error: 1.1123e-04
```

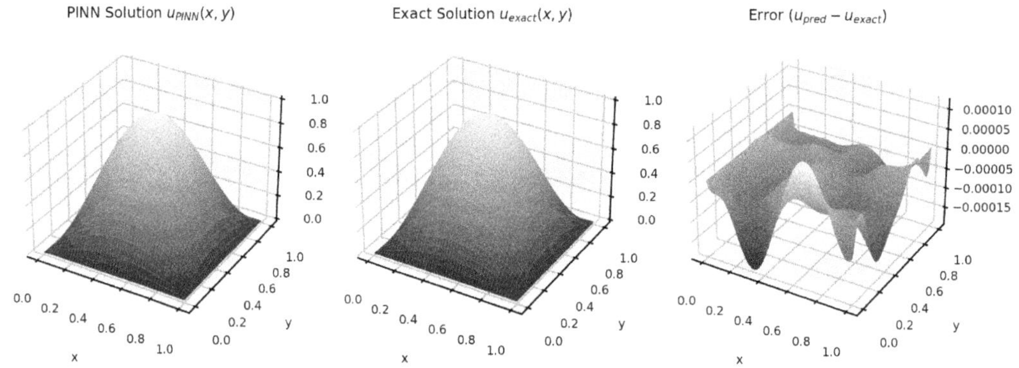

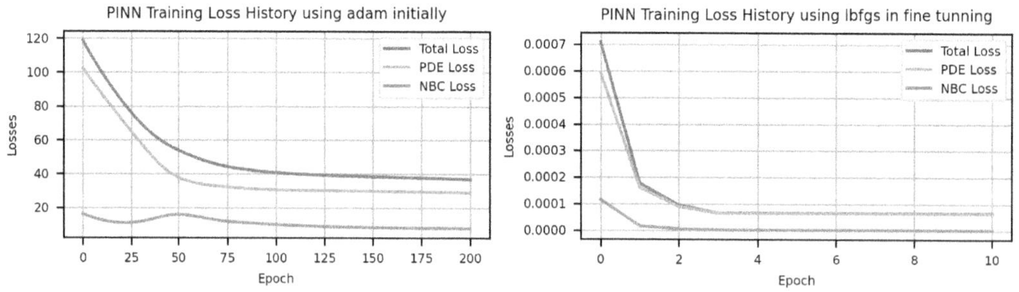

We obtained a reasonably accurate solution using only 100 Gaussian quadrature points, with the relative error reaching the order of 10^{-4}.

The following snippet tests whether the Adam optimizer alone—without further refinement by a second-stage optimizer like L-BFGS—is capable of finding a good solution to the same problem.

```
printSummary(pinns, LossTerms, Residuals, totalTimes,error_L2s,errorVIs)
```

```
#  PINN model        Loss_terms residual    total_time error_L2s  errorVIs
==========================================================================
```

```
1  2D          DBC       _clctn     1.833e+01  1.008e-03  1.004e+01
2  2D          DBC       _FDM       3.420e+00  7.129e-02  6.082e+00
3  2D          DBC       _intGQ     1.877e+01  1.855e-03  6.090e+00
4  2D          DBC       _intTR     1.108e+01  7.923e-04  9.155e+00
5  2D_4DBC     PDE_only  _energyTR  5.013e+00  6.298e-04  3.715e+00
6  2D_4DBC     PDE_only  _energyGQ  1.231e+00  7.792e-04  4.079e+00
7  2D_4DBC     PDE_only  _clctn     4.566e+00  5.196e-05  4.493e+00
8  2D_4DBC     PDE_only  _intTR     5.539e+00  7.298e-06  4.524e+00
9  2D_LeftDBC  NBC       _clctn     1.454e+01  5.140e-04  3.385e+00
10 2D_LeftDBC  NBC       _intGQ     1.961e+01  1.112e-04  4.199e+00
```

For this case, the use of Gaussian quadrature of the collocated residual yields significantly higher accuracy than the collocation approach, though it is slightly slower in training.

The following snippet trains a PINN model using only the `adam` optimizer.

```
set_seed_th(88)                                    # for reproducibility

# Create a new PINN model with DBC enforced on left edge, not recorded
pinn = PINN2D_LeftDBC(Nn, NL, a=a, b=b, c=c, d=d)

losses, totalTime = train_adam_lbfgs(pinn, Trainer, Residual, f_th,
       x_Ω_th, w_s_th, x_DBC_th, u_DBC_th, x_NBC_th, t_NBC_th, w_NBC_th,
                      I_NBC, epochs1=15001, epochs2=11, use_lbfgs=False)
```

```
==================================================
15001 epoches training with Adam:
Epoch     0, Loss: 1.187e+02, PDE: 1.023e+02, NBC: 1.640e+01
Epoch  5000, Loss: 1.370e-02, PDE: 1.204e-02, NBC: 1.661e-03
Epoch 10000, Loss: 2.568e-03, PDE: 2.308e-03, NBC: 2.593e-04
Epoch 15000, Loss: 1.868e-03, PDE: 1.701e-03, NBC: 1.672e-04
Time taken by adam: 82.727740 seconds

Training complete.
```

```
plot3loss(losses, PDE='PDE', C1='NBC', log=True)
```

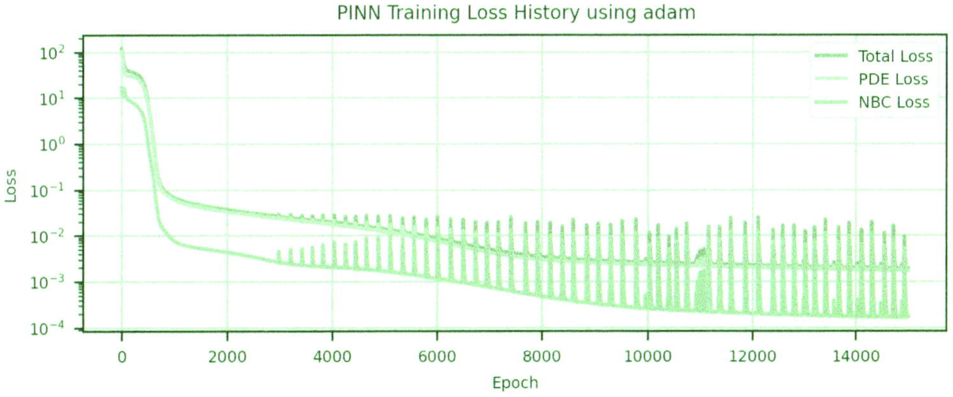

Using `adam` alone reduces the loss only to the 10^{-3} order, indicating that incorporating `LBFGS`

is necessary for higher accuracy. Once again, we observe the characteristic oscillations when adam attempts to minimize the loss aggressively, relying solely on first-order gradients.

2.11.7.3 Use of Integral Residual via Trapezoidal Rule

```
set_seed_th(88)                                    # for reproducibility

# on NBC boundary, use trapezoidal rule:
I_NBC = 'TR'  # Integration rule for computing NBC loss in loss_NBC_fun
x_NBC_th, t_NBC_th, w_NBC_th = sampleNBCs_rect_TR(t_np, N_NBC=N_DBC,
                                                   a=a, b=b, c=c, d=d)
# --- trapz_points in the Domain ---
DI_rule = 'TR'          # Domain Integration rule for computing loss_PDE
N_x = int(sqrt(N_sample))+1
N_y = N_x
w_s_th, x_Ω_th = sampleD_int(trapz_points_weights, N_x, N_y,
                                                   a=a, b=b, c=c, d=d)
print(f"Sampled {x_Ω_th.shape[0]} collocation points in the domain.")
print(f"Sampled {x_NBC_th.shape[0]} Neumann boundary points.")

# Initialize model:
pinn = PINN2D_LeftDBC(Nn, NL, a=a, b=b, c=c, d=d)
LossTerm = 'NBC'                          # loss term in addition to loss_PDE
# Train model with combined PDE, NBC loss, integrated squared residual:
Trainer  = trainerPDE_NBC
Residual = residual_int
losses, totalTime = train_adam_lbfgs(pinn, Trainer, Residual, f_th,
       x_Ω_th, w_s_th, x_DBC_th, u_DBC_th, x_NBC_th, t_NBC_th, w_NBC_th,
                        I_NBC, epochs1=201, epochs2=11, use_lbfgs=True)
```

```
Sampled 121 collocation points in the domain.
Sampled 63 Neumann boundary points.
==================================================
201 epoches training with Adam:
Epoch     0, Loss: 1.186e+02, PDE: 1.022e+02, NBC: 1.641e+01
Epoch   100, Loss: 4.292e+01, PDE: 3.243e+01, NBC: 1.049e+01
Epoch   200, Loss: 3.890e+01, PDE: 3.050e+01, NBC: 8.397e+00
Time taken by adam: 0.981748 seconds

==================================================
11 epoches fine-tuning with LBFGS:
Epoch     0, Loss: 9.098e-04, PDE: 6.949e-04, NBC: 2.149e-04
Epoch    10, Loss: 6.420e-05, PDE: 5.581e-05, NBC: 8.389e-06

Training complete.
Time taken by lbfgs: 12.852707 seconds
Total time: 13.834455 seconds
Ratio (time_lbfgs/time_adam) of time taken per epoch:239.220181
```

```
error_L2, errorVI, u_diff, _= plot_sln_Surface(pinn, exact_u, f_th=f_th,
                N_plot=50, a=a, b=b, c=c, d=d, dpi=300, plot_it=True)

plot3loss2(losses, ('adam initially','lbfgs in fine tunning'),
                                        PDE='PDE', C1='NBC', log=False)
recordSheet()     # Update the record
```

```
L2 mean residual of trained pinn: 4.5310e-05
Maximum Error: 4.4346e-04
Mean Absolute Error: 9.4771e-05
Error Variation Index (ratio of Max/Mean): 4.6792e+00
L2 Relative Error: 2.7968e-04
```

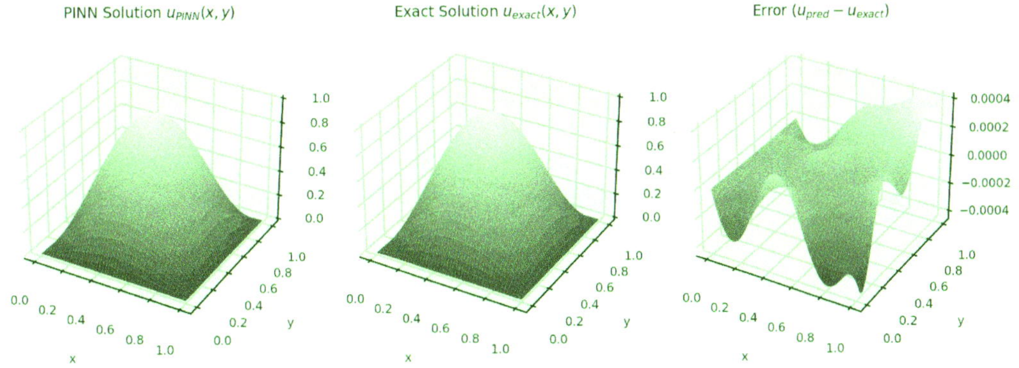

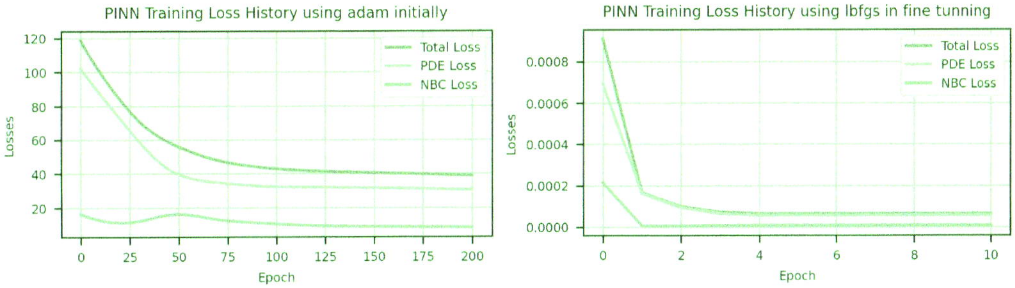

We obtained a quite accurate solution with the trapezoidal rule, achieving a relative error on the order of 10^{-4}.

```
printSummary(pinns, LossTerms, Residuals, totalTimes,error_L2s,errorVIs)
```

```
#  PINN model        Loss_terms residual     total_time error_L2s  errorVIs
===========================================================================
1  2D                DBC        _clctn       1.833e+01  1.008e-03  1.004e+01
2  2D                DBC        _FDM         3.420e+00  7.129e-02  6.082e+00
3  2D                DBC        _intGQ       1.877e+01  1.855e-03  6.090e+00
4  2D                DBC        _intTR       1.108e+01  7.923e-04  9.155e+00
5  2D_4DBC           PDE_only   _energyTR    5.013e+00  6.298e-04  3.715e+00
6  2D_4DBC           PDE_only   _energyGQ    1.231e+00  7.792e-04  4.079e+00
7  2D_4DBC           PDE_only   _clctn       4.566e+00  5.196e-05  4.493e+00
8  2D_4DBC           PDE_only   _intTR       5.539e+00  7.298e-06  4.524e+00
```

9	2D_LeftDBC	NBC	_clctn	1.454e+01	5.140e-04	3.385e+00
10	2D_LeftDBC	NBC	_intGQ	1.961e+01	1.112e-04	4.199e+00
11	2D_LeftDBC	NBC	_intTR	1.383e+01	2.797e-04	4.679e+00

It is found that using integration for residual computation via the trapezoidal rule produces a less accurate solution compared to the Gaussian quadrature rule.

2.11.7.4 Use of Energy Loss

When using the energy method, the loss function consists of a single term—the total potential energy—even when the problem includes NBCs. The total potential energy functional incorporates both the work done by the force/source term over the domain and the work done by the tractions on the three Neumann edges.

2.11.7.4.1 Code for Energy Level Computation The following snippet computes the energy components: the work done by the force term, the work done by the traction on the NBC boundary, and the total potential energy:

```
def residual_E_NBC(pinn, f_th, x_Omega, w_Omega, x_NBC, t_NBC, w_NBC):
    '''
    Compute the three energy terms for the energy method.
    '''
    x_Omega.requires_grad = True
    u_Omega = pinn(x_Omega)
    grads = torch.autograd.grad(u_Omega, x_Omega,
            grad_outputs=torch.ones_like(u_Omega), create_graph=True)[0]
    u_x = grads[:, 0:1]
    u_y = grads[:, 1:2]

    Wf = f_th(x_Omega[:, 0:1], x_Omega[:, 1:2])*u_Omega
    U = 0.5 * (u_x**2 + u_y**2)                   # potential energy density
    Wf = torch.sum(w_Omega * Wf)                   # work done by force in Ω
    U = torch.sum(w_Omega*U)                         # potential energy in Ω

    u_NBC = pinn(x_NBC)
    integrand_NBC = t_NBC * u_NBC
    Wt = torch.sum(w_NBC * integrand_NBC)        # energy on NB by traction

    return U, Wf, Wt          # Potential energy, work done f, and by t.
```

2.11.7.4.2 Trainer Function for Energy Method The following snippet defines the trainer function where the loss is the total potential energy, composed of three energy components.

```
def trainerE(pinn, residual, optimizer, f_th, x_Ω_th, w_s_th, x_DBC_th,
            u_DBC_th, x_NBC_th, t_NBC_th, w_NBC_th, I_NBC, epochs=201):
    """
    Trainer for a PINN that uses energy loss function.
    Args: pinn:        the PINN model (nn.Module)
          residual:  function to compute residual loss
```

```
        optimizer: PyTorch optimizer (e.g. Adam, LBFGS)
        epochs:    number of training iterations for PINN
        x_DBC_th, u_DBC_th, and I_NBC: not used here.
    """
    loss_history = {"total": [], "U": [], "Wf": [], "Wt": []}

    if epochs <= 301: n_prnt = 100
    else:             n_prnt = max(1, round(epochs / 3, -2))

    for epoch in range(epochs):
        # Initialize loss variables for the current epoch
        U = None; Wf = None; Wt = None; total_loss = None

        def closure():
            nonlocal U, Wf, Wt, total_loss
            # Allow modifying variables from outer scope
            optimizer.zero_grad()
            U, Wf, Wt = residual(pinn, f_th, x_Ω_th, w_s_th, x_NBC_th,
                                            t_NBC_th, w_NBC_th)
            total_loss = U - Wf - Wt
            total_loss.backward()              # Perform backward pass

            return total_loss  # Return the total loss for the optimizer

        if isinstance(optimizer, torch.optim.LBFGS):
            # LBFGS optimizer step has multiple closure calls internally
            optimizer.step(closure)
            total_loss_val = total_loss.item()
            loss_U_val = U.item()
            loss_Wf_val = Wf.item()
            loss_Wt_val = Wt.item()

        else: # Adam or other optimizers
            total_loss_tensor = closure()
            optimizer.step()
            total_loss_val = total_loss_tensor.item()
            loss_U_val = U.item()
            loss_Wf_val = Wf.item()
            loss_Wt_val = Wt.item()

        # Logging:
        with torch.no_grad():          # no graph is built for logging
            loss_history["total"].append(total_loss_val)
            loss_history["U"].append(loss_U_val)
            loss_history["Wf"].append(loss_Wf_val)
            loss_history["Wt"].append(loss_Wt_val)

        if epoch % n_prnt == 0 or epoch == epochs - 1:
            print(f"Epoch {epoch:5d}, Loss: {total_loss_val:.3e}, "
                  f"U: {loss_U_val:.3e}, Wf: {loss_Wf_val:.3e}, "
```

```
                    f"Wt: {loss_Wt_val:.3e}")

    return loss_history
```

2.11.7.4.3 Training a PINN via Energy Method We now apply the energy method to train the PINN for the same problem with Neumann boundary conditions (NBCs).

```
set_seed_th(88)                                      # for reproducibility

#            --- Energy, trapz_points in Domain ---
DI_rule = "TR"
w_s_th, x_Ω_th = sampleD_int(trapz_points_weights, N_x=100,N_y=100,
                                                    a=a, b=b, c=c, d=d)
#            --- trapz_points on NBC boundary ---
I_NBC = "TR"
x_NBC_th, t_NBC_th, w_NBC_th = sampleNBCs_rect_TR(t_np, N_NBC=100,
                                                    a=a, b=b, c=c, d=d)
print(f"Sampled {x_Ω_th.shape[0]} collocation points in the domain.")
print(f"Sampled {x_NBC_th.shape[0]} Neumann boundary points.")

# --- Create new model  ---
pinn = PINN2D_LeftDBC(Nn, NL=NL, a=a, b=b, c=c, d=d)

# Train the model with energy loss
Trainer  = trainerE
Residual = residual_E_NBC
LossTerm = 'PDE_only'
losses, totalTime = train_adam_lbfgs(pinn, Trainer, Residual, f_th,
       x_Ω_th, w_s_th, x_DBC_th, u_DBC_th, x_NBC_th, t_NBC_th, w_NBC_th,
                         I_NBC, epochs1=201, epochs2=11, use_lbfgs=True)
```

```
Sampled 10000 collocation points in the domain.
Sampled 300 Neumann boundary points.
==================================================
201 epoches training with Adam:
Epoch     0, Loss: 1.554e-01, U: 3.408e-02, Wf: 7.121e-01, Wt: -8.334e-01
Epoch   100, Loss: -1.167e+00, U: 1.070e+00, Wf: 4.320e+00, Wt: -2.084e+00
Epoch   200, Loss: -1.690e+00, U: 1.480e+00, Wf: 4.550e+00, Wt: -1.380e+00
Time taken by adam: 1.842218 seconds

==================================================
11 epoches fine-tuning with LBFGS:
Epoch     0, Loss: -2.467e+00, U: 2.468e+00, Wf: 4.933e+00, Wt: 1.705e-03
Epoch    10, Loss: -2.467e+00, U: 2.468e+00, Wf: 4.933e+00, Wt: 1.705e-03

Training complete.
Time taken by lbfgs: 2.959575 seconds
Total time: 4.801792 seconds
Ratio (time_lbfgs/time_adam) of time taken per epoch:29.355648
```

```
# --- Plot Results ---
error_L2, errorVI, u_diff, _= plot_sln_Surface(pinn, exact_u, f_th=f_th,
                    N_plot=50, a=a, b=b, c=c, d=d, dpi=300, plot_it=True)

plot3loss2(losses, ('adam initially','lbfgs in fine tunning'),
                    PDE="U", C1='Wf', C2='Wt', C3='energy', log=False)

recordSheet()                                          # Update the record
```

```
L2 mean residual of trained pinn: 1.5918e-01
Maximum Error: 3.8468e-03
Mean Absolute Error: 1.0475e-03
Error Variation Index (ratio of Max/Mean): 3.6723e+00
L2 Relative Error: 2.6693e-03
```

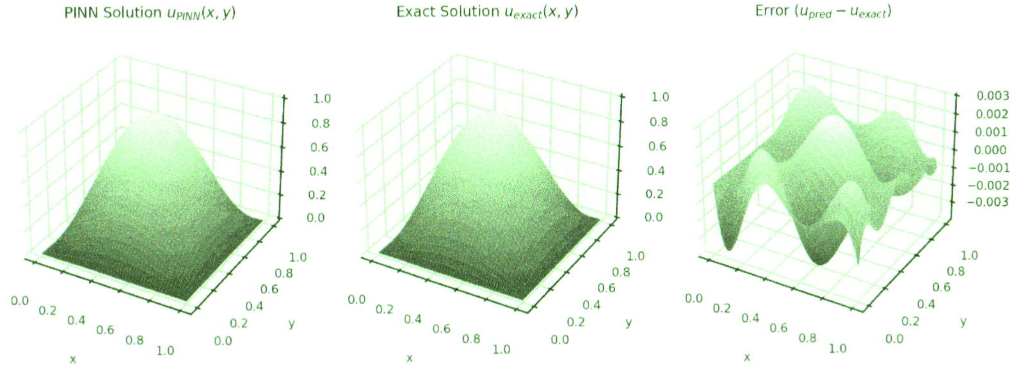

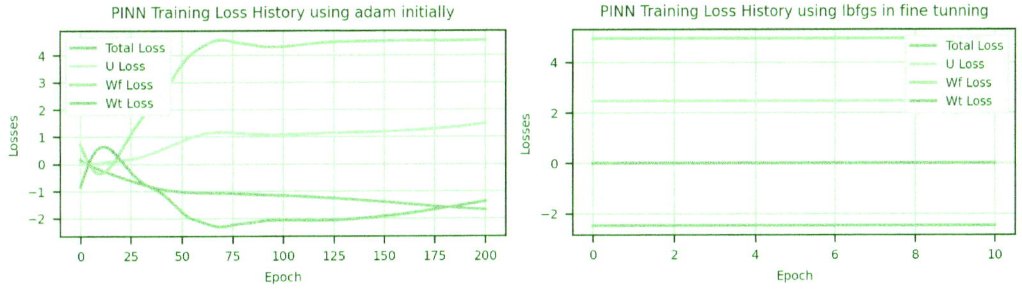

The relative error for this case is on the order of 10^{-3}. The loss convergence is smooth and steady. At the converged state, we again observe that the potential energy U is approximately half of the work done by the force function. The work done by the tractions on the three Neumann boundary edges eventually approaches zero because, at convergence, the solution u on all these edges tends to zero, preventing the NBC functions from performing any work.

It is worth noting that the computation is very fast when using the energy loss, especially with the `lbfgs` optimizer, despite the use of a large number (10,000) of integration points with the trapezoidal rule. In fact, we only need to run `lbfgs` for a single iteration, which may imply that the loss function is very close to a pure quadratic form with respect to the learning parameters.

```
printSummary(pinns, LossTerms, Residuals, totalTimes,error_L2s,errorVIs)
```

```
#  PINN model        Loss_terms residual    total_time error_L2s  errorVIs
==========================================================================
1  2D                DBC      _clctn        1.833e+01  1.008e-03  1.004e+01
2  2D                DBC      _FDM          3.420e+00  7.129e-02  6.082e+00
3  2D                DBC      _intGQ        1.877e+01  1.855e-03  6.090e+00
4  2D                DBC      _intTR        1.108e+01  7.923e-04  9.155e+00
5  2D_4DBC           PDE_only _energyTR     5.013e+00  6.298e-04  3.715e+00
6  2D_4DBC           PDE_only _energyGQ     1.231e+00  7.792e-04  4.079e+00
7  2D_4DBC           PDE_only _clctn        4.566e+00  5.196e-05  4.493e+00
8  2D_4DBC           PDE_only _intTR        5.539e+00  7.298e-06  4.524e+00
9  2D_LeftDBC        NBC      _clctn        1.454e+01  5.140e-04  3.385e+00
10 2D_LeftDBC        NBC      _intGQ        1.961e+01  1.112e-04  4.199e+00
11 2D_LeftDBC        NBC      _intTR        1.383e+01  2.797e-04  4.679e+00
12 2D_LeftDBC        PDE_only _E_NBCTR      4.802e+00  2.669e-03  3.672e+00
```

It is found that the accuracy level of the PINN trained using the energy method is lower than that of the earlier PDE_NBC models. However, the training is notably faster.

The following snippet trains the same model using adam optimizer alone; this run will not be recorded:

```python
# --- Create new model  ---              (not recorded)
pinn = PINN2D_LeftDBC(Nn, NL=NL, a=a, b=b, c=c, d=d)

# Train the model with energy loss using ---adam alone---
losses, totalTime = train_adam_lbfgs(pinn, Trainer, Residual, f_th,
       x_Ω_th, w_s_th, x_DBC_th, u_DBC_th, x_NBC_th, t_NBC_th, w_NBC_th,
                       I_NBC, epochs1=2001, epochs2=11, use_lbfgs=False)
```

```
==================================================
2001 epoches training with Adam:
Epoch     0, Loss: 5.289e-02, U: 1.855e-02, Wf: 6.090e-01, Wt: -6.434e-01
Epoch   700, Loss: -2.442e+00, U: 2.437e+00, Wf: 4.840e+00, Wt: 3.898e-02
Epoch  1400, Loss: -2.462e+00, U: 2.458e+00, Wf: 4.887e+00, Wt: 3.346e-02
Epoch  2000, Loss: -2.466e+00, U: 2.464e+00, Wf: 4.920e+00, Wt: 1.002e-02
Time taken by adam: 18.178243 seconds

Training complete.
```

```python
plot3loss(losses, PDE="U", C1='Wf', C2='Wt', C3='energy', log=False)
```

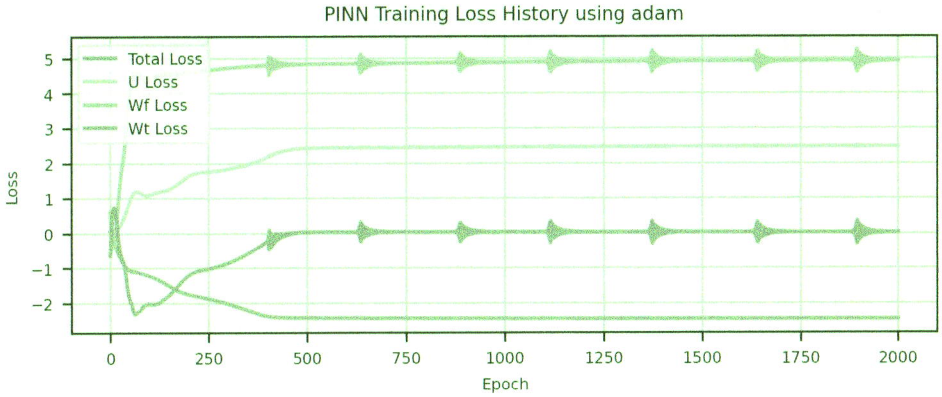


```
# --- Check the solution accuracy ---
error_, VI_, u_diff_, _ = plot_sln_Surface(pinn, exact_u, f_th=f_th,
                    N_plot=50, a=a, b=b, c=c, d=d, dpi=300, plot_it=False)
```

```
L2 mean residual of trained pinn: 6.2458e-01
Maximum Error: 1.3374e-02
Mean Absolute Error: 3.7457e-03
Error Variation Index (ratio of Max/Mean): 3.5705e+00
L2 Relative Error: 1.0052e-02
```

It is found that using `adam` alone can train the model reasonably well within 2000 epochs. The total energy reaches approximately the same level of −2.467 as obtained with the two-stage training. The training is very fast, taking about 10 seconds.

However, oscillations are still observed as the loss approaches its minimum, and the solution accuracy is slightly reduced. Therefore, using a few epochs of `lbfgs` for fine-tuning remains preferable. Nevertheless, a PINN trained with the energy method can often achieve acceptable results using `adam` alone—provided more epochs are used—due to the well-behaved nature of the energy loss function.

The following snippet finds the solution for the same model and problem, but employs Gaussian quadrature for domain integration.

```
set_seed_th(88)                                          # for reproducibility
#                 ---Energy method, Gauss_points in the Domain ---
DI_rule = "GQ"
w_s_th, x_Ω_th = sampleD_int(Gauss_points_weights, N_x=10,N_y=10,
                                                     a=a, b=b, c=c, d=d)
I_NBC = "GQ"
#  --- Gauss_points on NBC boundary ---
x_NBC_th, t_NBC_th, w_NBC_th = sampleNBCs_rect_GQ(t_np, N_NBC=10,
                                                     a=a, b=b, c=c, d=d)
# --- Create new model  ---
pinn = PINN2D_LeftDBC(Nn, NL=NL, a=a, b=b, c=c, d=d)

# Train the model with energy loss
Trainer  = trainerE
```

```
Residual = residual_E_NBC
losses, totalTime = train_adam_lbfgs(pinn, Trainer, Residual, f_th,
       x_Ω_th, w_s_th, x_DBC_th, u_DBC_th, x_NBC_th, t_NBC_th, w_NBC_th,
                         I_NBC, epochs1=201, epochs2=11, use_lbfgs=True)
```

```
==================================================
201 epoches training with Adam:
Epoch     0, Loss: 1.554e-01, U: 3.408e-02, Wf: 7.122e-01, Wt: -8.335e-01
Epoch   100, Loss: -1.167e+00, U: 1.070e+00, Wf: 4.323e+00, Wt: -2.086e+00
Epoch   200, Loss: -1.691e+00, U: 1.480e+00, Wf: 4.553e+00, Wt: -1.381e+00
Time taken by adam: 0.507751 seconds

==================================================
11 epoches fine-tuning with LBFGS:
Epoch     0, Loss: -2.467e+00, U: 2.467e+00, Wf: 4.932e+00, Wt: 1.872e-03
Epoch    10, Loss: -2.467e+00, U: 2.467e+00, Wf: 4.932e+00, Wt: 1.872e-03

Training complete.
Time taken by lbfgs: 0.553772 seconds
Total time: 1.061523 seconds
Ratio (time_lbfgs/time_adam) of time taken per epoch:19.928908
```

In this case, the computation is even faster due to the use of significantly fewer quadrature points, with the converged solution obtained in about one second. The solution closely matches that from the previous case using the trapezoidal rule, as demonstrated below.

```
# --- Compute the solution accurace ---
error_L2, errorVI, u_diff, _= plot_sln_Surface(pinn, exact_u, f_th=f_th,
                  N_plot=50, a=a, b=b, c=c, d=d, dpi=300, plot_it=False)

recordSheet( )     # Update the record
```

```
L2 mean residual of trained pinn: 1.8699e-01
Maximum Error: 4.9998e-03
Mean Absolute Error: 1.4370e-03
Error Variation Index (ratio of Max/Mean): 3.4794e+00
L2 Relative Error: 3.6786e-03
```

The energy is driven by the optimizers very close to the equilibrium state. Let us now compare it against the other models trained so far:

```
printSummary(pinns, LossTerms, Residuals, totalTimes,error_L2s,errorVIs)
```

```
#  PINN model          Loss_terms residual    total_time error_L2s  errorVIs
============================================================================
1  2D                  DBC        _clctn      1.833e+01  1.008e-03  1.004e+01
2  2D                  DBC        _FDM        3.420e+00  7.129e-02  6.082e+00
3  2D                  DBC        _intGQ      1.877e+01  1.855e-03  6.090e+00
4  2D                  DBC        _intTR      1.108e+01  7.923e-04  9.155e+00
5  2D_4DBC             PDE_only   _energyTR   5.013e+00  6.298e-04  3.715e+00
6  2D_4DBC             PDE_only   _energyGQ   1.231e+00  7.792e-04  4.079e+00
7  2D_4DBC             PDE_only   _clctn      4.566e+00  5.196e-05  4.493e+00
```

```
8  2D_4DBC        PDE_only   _intTR     5.539e+00  7.298e-06  4.524e+00
9  2D_LeftDBC     NBC        _clctn     1.454e+01  5.140e-04  3.385e+00
10 2D_LeftDBC     NBC        _intGQ     1.961e+01  1.112e-04  4.199e+00
11 2D_LeftDBC     NBC        _intTR     1.383e+01  2.797e-04  4.679e+00
12 2D_LeftDBC     PDE_only   _E_NBCTR   4.802e+00  2.669e-03  3.672e+00
13 2D_LeftDBC     PDE_only   _E_NBCGQ   1.062e+00  3.679e-03  3.479e+00
```

It is found that the solution accuracy obtained using Gaussian quadrature is lower than that using the trapezoidal rule in this test. However, Gaussian quadrature speeds up the computation due to the fewer integration points required.

In addition to the points discussed in previous sections, we further note the following from this case study:

1. The `PINN2D_LeftDBC` models trained with the energy method are about 10 times less accurate, consistent with earlier observations.

2. The `PINN2D_LeftDBC` models trained with the energy method converge roughly 10 times faster. This may be attributed to the energy-based loss being a single-term loss with favorable (quadratic) behavior, which facilitates easier minimization.

2.12 Alternative Activations

In all examples studied so far, we have used the `tanh()` activation function. For PINN models, however, many other types of activation functions may be employed, including the non-monotonic `sin()` function. The following snippet demonstrates this approach.

```
# Use the non-monotonic `sin()` function:
set_seed_th(88)                                   # for reproducibility

#a, b = 0.0, 1.0; c, d = 0.0, 1.0                 # same as the original setup
#Nn = 50
NL  = 1                          # change to use only one hidden layer  ␣
    ↪

def my_actvtn(x):
    return torch.sin(x)

def f_th(x, y):
    '''Source or force term f in torch in the Poisson's equation'''
    return 2*torch.pi**2 *torch.sin(torch.pi*x) * torch.sin(torch.pi*y)

#               Generate dataset using the trapezoidal rule:
DI_rule = 'TR'
N_x = int(sqrt(N_sample))+1
N_y = N_x
w_s_th, x_Ω_th=sampleD_int(trapz_points_weights, N_x=N_x, N_y=N_y,
                                                 a=a, b=b, c=c, d=d)
print(f"# of weights: {len(w_s_th)}, # of sample points: {len(x_Ω_th)}")

# Create new PINN model with DBC enforced, and integral residual:
```

```
pinn = PINN2D_4DBC(Nn, NL, actit=my_actvtn, a=a, b=b, c=c, d=d)

Trainer  = trainer_int_PDE
Residual = residual_int
losses, totalTime = train_adam_lbfgs(pinn, Trainer, Residual, f_th,
        x_Ω_th, w_s_th, x_DBC_th, u_DBC_th, x_NBC_th, t_NBC_th, w_NBC_th,
                        I_NBC, epochs1=201, epochs2=11, use_lbfgs=True)
```

```
# of weights: 121, # of sample points: 121
==================================================
201 epoches training with Adam:
Epoch     1, Loss: 1.020e+02, PDE: 1.020e+02
Epoch   101, Loss: 6.784e+01, PDE: 6.784e+01
Epoch   201, Loss: 3.279e+01, PDE: 3.279e+01
Time taken by adam: 0.838039 seconds

==================================================
11 epoches fine-tuning with LBFGS:
Epoch     1, Loss: 5.363e-06, PDE: 5.363e-06
Epoch    11, Loss: 5.363e-06, PDE: 5.363e-06

Training complete.
Time taken by lbfgs: 0.586388 seconds
Total time: 1.424427 seconds
Ratio (time_lbfgs/time_adam) of time taken per epoch:12.785693
```

```
error_L2, errorVI, u_diff, _ = plot_sln_rect(pinn, exact_u, f_th=f_th,
            N_plot=200, vmin=0., vmax=1., a=a,b=b,c=c,d=d, plot_it=True)
```

```
L2 mean residual of trained pinn: 4.2906e-06
Maximum Error: 2.5853e-05
Mean Absolute Error: 6.1628e-06
Error Variation Index (ratio of Max/Mean): 4.1951e+00
L2 Relative Error: 1.6409e-05
```

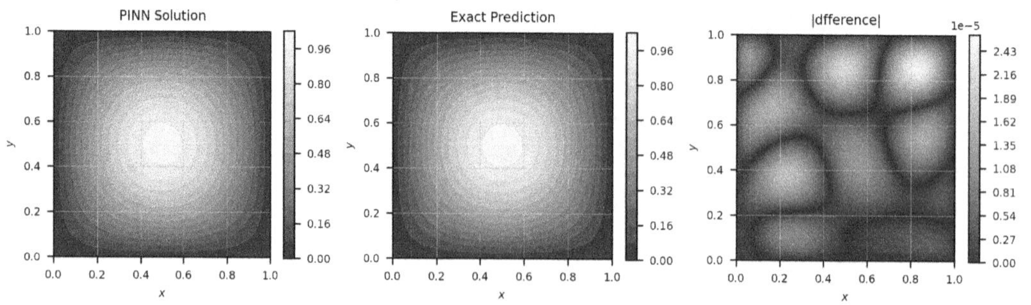

As found, the solution accuracy is also very high, with the relative error on the order of 10^{-6}, despite reducing the number of layers in the neural network to 1. The reason for choosing only 1 layer is due to the use of the sine activation function, which is not monotonic [12]. It is often observed in the author's study that when such a non-monotonic function is used in an ML model, using more layers does not necessarily lead to more accurate results, but it certainly requires more computation time.

Readers are encouraged to experiment with other activation functions and different NN configurations.

So far, our study has focused on the Poisson problem, which admits a relatively simple exact solution. We now turn to a more realistic and mechanically relevant problem—still governed by the Poisson equation—but featuring a much more complicated solution structure.

2.13 Case Study: PINN for Rectangular Bars Subject to Torsion

2.13.1 Problem Setup

Consider uniform straight bars with a rectangular cross-section subjected to torsion. The cross-section has width w and height h, as illustrated in Fig.2.3. This classic problem is well-documented in standard mechanics references [5, 13].

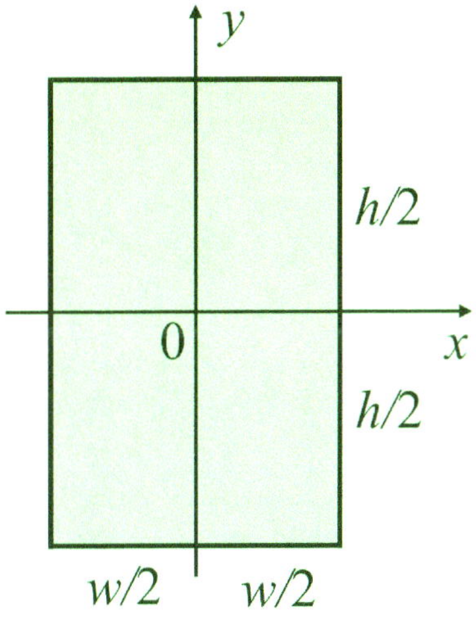

Figure 2.3: A typical rectangular cross-section for uniform straight bars subjected to torsion.

The shear stresses developed on the rectangular cross-section are described via the stress function ϕ, which depends on the twist angle. The governing equation for the stress function ϕ is given by:

$$\begin{aligned} \nabla^2\phi + 2G\,\vartheta_t = 0 \quad & \text{in the domain of the cross-section (PDE)} \\ \phi = 0 \quad & \text{on the boundary of the cross-section} \end{aligned} \tag{2.57}$$

where G is the shear modulus and ϑ_t is the twist angle per unit length. Both parameters are specified by the given torsional problem and serve as the source term for the Poisson equation governing ϕ. To simplify the formulation, we set

$$2G\vartheta_t = 1 \tag{2.58}$$

to solve for ϕ first, and subsequently multiply the result by $2G\vartheta_t$ to obtain the true stress function.

The exact solution for ϕ, along with the corresponding Python code, can be found in Ref. [13]. It is expressed in the series form:

$$\phi = G\vartheta_t \left[\left(\frac{w}{2}\right)^2 - x^2 \right] - \frac{32G\vartheta_t \left(\frac{w}{2}\right)^2}{\pi^3} \sum_{n=1,3,5,\ldots} \frac{(-1)^{(n-1)/2}}{n^3 \cosh \frac{n\pi\left(\frac{h}{2}\right)}{w}} \cos \frac{n\pi x}{w} \cosh \frac{n\pi y}{w} \tag{2.59}$$

The following snippet defines the source function and the exact solution.

```
def f_np(x, y):
    '''Force or source function in numpy'''
    return 1 + 0*x*y                                   # source term set to 1

def f_th(x, y):
    '''Force or source function in torch'''
    return 1 + 0 * x * y                       # scalar source term set to 1

# Exact solution (for reference)
def exact_ϕ(x, y, a=0.0, b=1.0, c=0.0, d=1.0, Gt=0.5, n_end = 21):
    '''Compute the stress function phi for torsion bar with rectangular
    cross-section. It is a series solution to 2D Poisson's PDE
    with a constant source term, and fixed zero DBC.
    b:         half-height of the cross-section
    h:         half-width  of the cross-section
    Gt:        constant source term
    n_end:     terms used in the series of the solution
    x and y: points at which the solution to be found.'''

    wh = (b-a)/2; hh = (d-c)/2
    # wh: half width; hh: half height of the cross-section of the bar

    xp = x - (b+a)/2; yp = y - (d+c)/2              # coordinate translation
    sum1 = 0.0
    for n in range(1, n_end, 2):
        sum1 += (-1)**((n-1)/2)/(n**3*np.cosh(n*np.pi*hh/(2*wh)))*\
                np.cos(n*np.pi*xp/(2*wh))*np.cosh(n*np.pi*yp/(2*wh))

    return Gt*(wh**2-xp**2) - (32*Gt*wh**2/np.pi**3)*sum1
```

2.13.2 PINN Model for the Stress Function

We can utilize any of the previously discussed PINN architectures to train a model for the stress function ϕ. The following code snippet demonstrates how to prepare the training data for this PINN model:

```
# Geometric data for the cross-section of the torsional bar~:
a=-0.5; b=0.5; c=-1.; d=1.
```

```
# Buid a PINN model and train it:
pinns = []                          # reset for logging the key variables:
timeAdam  = []; timeLbfgs = []
Trainers  = []; LossTerms = []; Residuals = []
totalTimes= []
error_L2s = []; errorVIs = []                              # log the error

p = 2; Nn=50; NL=2      # NL hidden layers with nn number of neurons each

# Create new PINN model with DBC enforced in class:
pinn = PINN2D_4DBC(Nn=Nn, NL=NL, a=a, b=b, c=c, d=d)
P_star, P_total = print_model_parameters(pinn, print_table=False)

N_sample_ = 1*P_star
N_x = int(sqrt(N_sample_/2))+1                                  #   h = 2w
N_y = 2*N_x
N_sample = N_x*N_y
print(f"Nn={Nn}, NL={NL}. The actual N_sample in domain = {N_sample}")
print(f"Is N_sample >= P* ? {N_sample >= P_star}")
```

```
P* (Total pseudo-dimension of the affine space) = 105
Total trainable parameters: 2751
Nn=50, NL=2. The actual N_sample in domain = 128
Is N_sample >= P* ? True
```

2.13.3 Use of Trapezoidal Rule

First, we construct a PINN model that employs the trapezoidal rule to compute the residuals.

```
#    --- Sample trapz_points (TR) in the rectangular Domain ---
set_seed_th(88)                                   # for reproducibility
DI_rule = "TR"
w_s_th, x_Ω_th = sampleD_int(trapz_points_weights, N_x=N_x, N_y=N_y,
                                                    a=a, b=b, c=c, d=d)
print(f"Sampled {x_Ω_th.shape[0]} collocation points in the domain.")
LossTerm = 'PDE_only'

# Train the model with integrated squared residual:
Trainer  = trainer_int_PDE
Residual = residual_int
losses, totalTime = train_adam_lbfgs(pinn, Trainer, Residual, f_th,
       x_Ω_th, w_s_th, x_DBC_th, u_DBC_th, x_NBC_th, t_NBC_th, w_NBC_th,
                         I_NBC, epochs1=201, epochs2=11, use_lbfgs=True)
```

```
Sampled 128 collocation points in the domain.
==================================================
201 epoches training with Adam:
Epoch     1, Loss: 1.617e+00, PDE: 1.617e+00
Epoch   101, Loss: 2.093e-01, PDE: 2.093e-01
Epoch   201, Loss: 9.204e-02, PDE: 9.204e-02
Time taken by adam: 1.134126 seconds
```

```
=======================================================
11 epoches fine-tuning with LBFGS:
Epoch      1, Loss: 2.025e-02, PDE: 2.025e-02
Epoch     11, Loss: 1.906e-02, PDE: 1.906e-02

Training complete.
Time taken by lbfgs: 12.881563 seconds
Total time: 14.015689 seconds
Ratio (time_lbfgs/time_adam) of time taken per epoch:207.544174
```

```
error_L2, errorVI, u_diff, residual_pred = plot_sln_rect(pinn, exact_ϕ,
  f_th=f_th, N_plot=100, vmin=0,vmax=.11, a=a,b=b,c=c,d=d, plot_it=True)

plot3loss2(losses, ('adam initially','lbfgs in fine tunning'),
                                                    PDE="PDE", log=False)
```

```
L2 mean residual of trained pinn: 1.7511e-03
Maximum Error: 1.9121e-04
Mean Absolute Error: 1.9936e-05
Error Variation Index (ratio of Max/Mean): 9.5916e+00
L2 Relative Error: 5.1176e-04
```

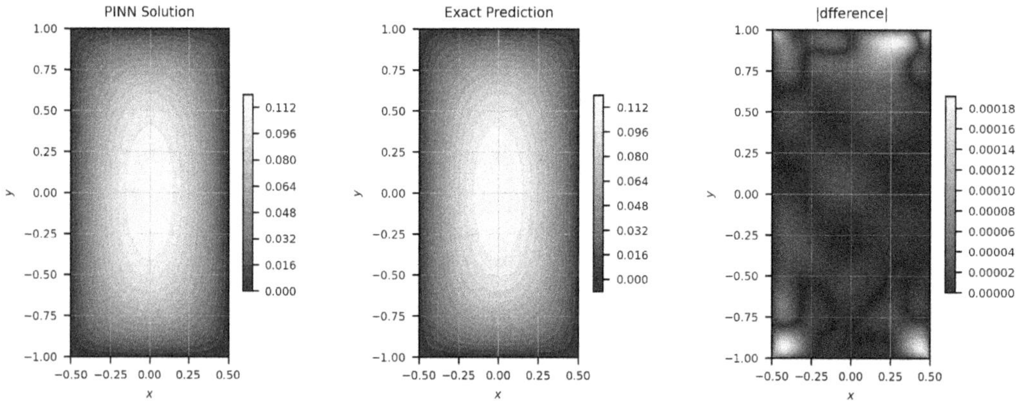

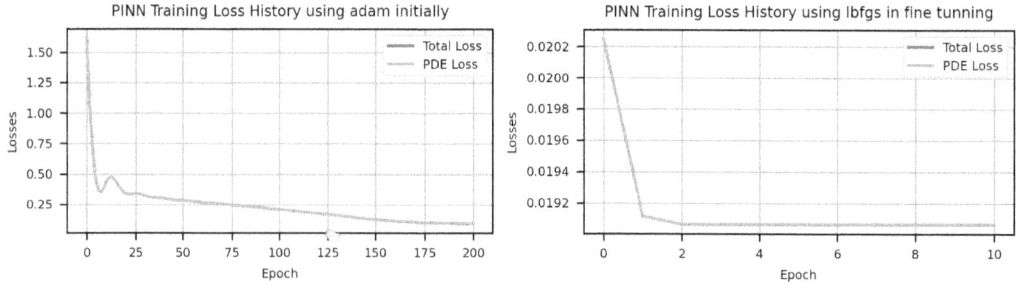


```
# Create the record (exercute this only once!):
recordSheet()
```

We obtained a quite accurate solution using 105 sampling points with the trapezoidal rule, achieving a relative error on the order of 10^{-4}. Let us record the performance of this model for the torsional bar problem.

```
printSummary(pinns, LossTerms, Residuals, totalTimes,error_L2s,errorVIs)
```

```
#  PINN model        Loss_terms residual    total_time error_L2s  errorVIs
==========================================================================
1  2D_4DBC           PDE_only   _intTR      1.402e+01  5.118e-04  9.592e+00
```

2.13.4 Use of Gaussian Quadrature

Next, we apply Gaussian quadrature while keeping all other settings unchanged.

```
set_seed_th(88)                                    # for reproducibility
                   # Generate dataset, ---Gauss points---:
DI_rule = "GQ"
w_s_th, x_Ω_th=sampleD_int(Gauss_points_weights, N_x=6, N_y=11,
                                                    a=a, b=b, c=c, d=d)
print(f"Sampled {x_Ω_th.shape[0]} collocation points in the domain.")
# Create new PINN model with DBC enforced, and train it:
pinn = PINN2D_4DBC(Nn=Nn, NL=NL, a=a, b=b, c=c, d=d)

# Train the model with integrated squared residual:
Trainer  = trainer_int_PDE
Residual = residual_int
losses, totalTime = train_adam_lbfgs(pinn, Trainer, Residual, f_th,
       x_Ω_th, w_s_th, x_DBC_th, u_DBC_th, x_NBC_th, t_NBC_th, w_NBC_th,
                        I_NBC, epochs1=201, epochs2=11, use_lbfgs=True)
```

```
Sampled 66 collocation points in the domain.
==================================================
201 epoches training with Adam:
Epoch     1, Loss: 1.963e+00, PDE: 1.963e+00
Epoch   101, Loss: 2.239e-01, PDE: 2.239e-01
Epoch   201, Loss: 9.313e-02, PDE: 9.313e-02
Time taken by adam: 1.040662 seconds

==================================================
11 epoches fine-tuning with LBFGS:
Epoch     1, Loss: 1.044e-04, PDE: 1.044e-04
Epoch    11, Loss: 6.581e-06, PDE: 6.581e-06

Training complete.
Time taken by lbfgs: 6.377863 seconds
Total time: 7.418526 seconds
Ratio (time_lbfgs/time_adam) of time taken per epoch:111.987276
```

```
error_L2, errorVI, u_diff, _ = plot_sln_rect(pinn, exact_ϕ, f_th=f_th,
           N_plot=100, vmin=0, vmax=.11, a=a,b=b,c=c,d=d, plot_it=False)

# plot3loss2(losses, ('adam initially','lbfgs in fine tunning'),
```

```
#                                   PDE="PDE", log=False)  # optional

recordSheet()             # Update the record (exercute this only once!)
```

```
L2 mean residual of trained pinn: 9.7423e-04
Maximum Error: 1.3315e-04
Mean Absolute Error: 3.4394e-05
Error Variation Index (ratio of Max/Mean): 3.8713e+00
L2 Relative Error: 7.2121e-04
```

```
printSummary(pinns, LossTerms, Residuals, totalTimes,error_L2s,errorVIs)
```

```
#  PINN model        Loss_terms residual    total_time error_L2s  errorVIs
==========================================================================
1  2D_4DBC           PDE_only   _intTR      1.402e+01  5.118e-04  9.592e+00
2  2D_4DBC           PDE_only   _intGQ      7.419e+00  7.212e-04  3.871e+00
```

For this problem, it is observed that using Gaussian quadrature to integrate the residual results in lower solution accuracy, but offers faster computation and a smaller error VI value compared to the trapezoidal rule.

2.13.5 Use of Energy Method

The following snippet trains a PINN using the energy method, where U_T is integrated via the trapezoidal rule.

```
set_seed_th(88)                                  # for reproducibility

#                Generate dataset, using trapezoidal rule:
DI_rule = "TR"
N_x = 50                                # for accurate energy integration
w_s_th, x_Ω_th=sampleD_int(trapz_points_weights, N_x=N_x, N_y=2*N_x,
                                                   a=a, b=b, c=c, d=d)
# Create new PINN model with DBC enforced, and train it:
pinn = PINN2D_4DBC(Nn=Nn, NL=NL, a=a, b=b, c=c, d=d)

# Train the model with energy loss
Trainer  = trainer_int_PDE
Residual = residual_energy
losses, totalTime = train_adam_lbfgs(pinn, Trainer, Residual, f_th,
      x_Ω_th, w_s_th, x_DBC_th, u_DBC_th, x_NBC_th, t_NBC_th, w_NBC_th,
                          I_NBC, epochs1=201, epochs2=11, use_lbfgs=True)
```

```
==================================================
201 epoches training with Adam:
Epoch     1, Loss: -5.780e-03, PDE: -5.780e-03
Epoch   101, Loss: -5.519e-02, PDE: -5.519e-02
Epoch   201, Loss: -5.586e-02, PDE: -5.586e-02
Time taken by adam: 1.323248 seconds

==================================================
11 epoches fine-tuning with LBFGS:
```

```
Epoch     1, Loss: -5.698e-02, PDE: -5.698e-02
Epoch    11, Loss: -5.698e-02, PDE: -5.698e-02

Training complete.
Time taken by lbfgs: 1.855485 seconds
Total time: 3.178733 seconds
Ratio (time_lbfgs/time_adam) of time taken per epoch:25.622375
```

As seen again, the computation using the energy method is fast.

```
error_L2, errorVI, u_diff, _ = plot_sln_rect(pinn, exact_ϕ, f_th=f_th,
           N_plot=100, vmin=0., vmax=1., a=a,b=b,c=c,d=d, plot_it=False)

# plot3loss2(losses, ('adam initially','lbfgs in fine tunning'),
#                                        PDE="PDE", log=False) # optional
recordSheet()                                             # Update the record
```

```
L2 mean residual of trained pinn: 2.2744e-02
Maximum Error: 2.1609e-03
Mean Absolute Error: 4.3126e-04
Error Variation Index (ratio of Max/Mean): 5.0107e+00
L2 Relative Error: 9.1010e-03
```

```
printSummary(pinns, LossTerms, Residuals, totalTimes,error_L2s,errorVIs)
```

```
#  PINN model        Loss_terms residual    total_time error_L2s  errorVIs
=========================================================================
1  2D_4DBC           PDE_only   _intTR      1.402e+01  5.118e-04  9.592e+00
2  2D_4DBC           PDE_only   _intGQ      7.419e+00  7.212e-04  3.871e+00
3  2D_4DBC           PDE_only   _energyTR   3.179e+00  9.101e-03  5.011e+00
```

The PINN trained using the energy method produces a much less accurate result, but the training is faster.

2.13.6 Use of Collocation Method

Next, we will try the collocation method studied earlier.

```
set_seed_th(88)                                        # for reproducibility

DI_rule = ''          # use collocation
w_s_th, x_Ω_th = sampleD_clctn(N_clctn=N_sample, a=a, b=b, c=c, d=d)
print(f"Sampled {x_Ω_th.shape[0]} collocation points in the domain.")

# Create new PINN model with DBC enforced, and train it:
pinn = PINN2D_4DBC(Nn=Nn, NL=NL, a=a, b=b, c=c, d=d)

Trainer  = trainer_int_PDE
Residual = residual_clctn
losses, totalTime = train_adam_lbfgs(pinn, Trainer, Residual, f_th,
      x_Ω_th, w_s_th,  x_DBC_th, u_DBC_th, x_NBC_th, t_NBC_th, w_NBC_th,
                        I_NBC, epochs1=201, epochs2=11, use_lbfgs=True)
```

```
Sampled 128 collocation points in the domain.
==================================================
201 epoches training with Adam:
Epoch     1, Loss: 1.042e+00, PDE: 1.042e+00
Epoch   101, Loss: 1.325e-01, PDE: 1.325e-01
Epoch   201, Loss: 5.172e-02, PDE: 5.172e-02
Time taken by adam: 1.179440 seconds

==================================================
11 epoches fine-tuning with LBFGS:
Epoch     1, Loss: 1.258e-04, PDE: 1.258e-04
Epoch    11, Loss: 9.314e-06, PDE: 9.314e-06

Training complete.
Time taken by lbfgs: 13.500598 seconds
Total time: 14.680038 seconds
Ratio (time_lbfgs/time_adam) of time taken per epoch:209.160874
```

```python
# Plot the contours of both PINN solution and the exact one:
error_L2, errorVI, u_diff, _ = plot_sln_rect(pinn, exact_ϕ, f_th=f_th,
          N_plot=100, vmin=0., vmax=.11, a=a,b=b,c=c,d=d, plot_it=False)

recordSheet()                                         # Update the record
```

```
L2 mean residual of trained pinn: 2.6024e-03
Maximum Error: 2.7452e-04
Mean Absolute Error: 2.9392e-05
Error Variation Index (ratio of Max/Mean): 9.3401e+00
L2 Relative Error: 6.6345e-04
```

```python
printSummary(pinns, LossTerms, Residuals, totalTimes,error_L2s,errorVIs)
```

#	PINN model	Loss_terms	residual	total_time	error_L2s	errorVIs
1	2D_4DBC	PDE_only	_intTR	1.402e+01	5.118e-04	9.592e+00
2	2D_4DBC	PDE_only	_intGQ	7.419e+00	7.212e-04	3.871e+00
3	2D_4DBC	PDE_only	_energyTR	3.179e+00	9.101e-03	5.011e+00
4	2D_4DBC	PDE_only	_clctn	1.468e+01	6.634e-04	9.340e+00

The use of collocation took longer to train compared to the energy method, but it produced a much more accurate solution. The accuracy level is on par with Model-1 and Model-2, as all three models are collocation-based.

2.13.7 Shear Stress Field Plots

After obtaining the PINN model for the stress function, we can easily compute its derivatives with respect to the inputs. The shear stresses are directly related to the first derivatives of the stress function ϕ (see Section 9.5 of Ref. [13]):

$$\tau_{zx} = \frac{\partial \phi}{\partial y}; \quad \tau_{zy} = -\frac{\partial \phi}{\partial x} \tag{2.60}$$

Using these two stress components, the so-called **shear flow** is formed as

$$= \begin{bmatrix} \tau_{zx} & \tau_{zy} \end{bmatrix}^{\top} \tag{2.61}$$

It is a vector field [17] defined in the 2D plane. The following snippet computes these stresses and plots the corresponding vector fields:

```python
def compute_derivatives(pinn, inputs):
    """
    Compute u, du/dx, du/dy, d²u/dx², d²u/dy² using a trained PINN pinn.
    Args: pinn, a PyTorch neural network taking (m, 2) inputs.
          inputs, torch.Tensor of shape (m, 2) with requires_grad=True.
    Returns: u, du_dx, du_dy, d2u_dx2, d2u_dy2: tensors of shape (m, 1)
    """
    x = inputs.clone().detach().requires_grad_(True)     # grad tracking

    u = pinn(x)                                          # output u(x, y)

    # First derivatives
    grads = torch.autograd.grad(u, x, grad_outputs=torch.ones_like(u),
                                create_graph=True, retain_graph=True)[0]
    du_dx = grads[:, 0:1]
    du_dy = grads[:, 1:2]

    # Second derivatives
    d2u_dx2=torch.autograd.grad(du_dx, x,
                         grad_outputs=torch.ones_like(du_dx),
                         create_graph=True, retain_graph=True)[0][:, 0:1]

    d2u_dy2=torch.autograd.grad(du_dy, x,
                         grad_outputs=torch.ones_like(du_dy),
                         create_graph=True, retain_graph=True)[0][:, 1:2]

    return u, du_dx, du_dy, d2u_dx2, d2u_dy2
```

The following function computes the two shear stress components and then plots their distributions, along with the shear flow derived from these stress components.

```python
def plot2D (pinn, N_plot=100, a=0, b=1, c=0, d=1, shrink=0.7, dpi=500):
    """
    Plot shear stress components _zx, _zy and the total shear flow
    computed from the stress function ϕ of a PINN model of over a
    rectangular domain [a,b]×[c,d].
    Args: pinn, trained PINN model in PyTorch
          N_plot, number of points along each axis (int)
          a, b, c, d: the rectangular domain bounds
    """
    x = np.linspace(a, b, N_plot)                        # Create meshgrid
    y = np.linspace(c, d, N_plot)
    X, Y = np.meshgrid(x, y)
```

```
XY = np.hstack((X.flatten()[:, None], Y.flatten()[:, None]))

# Convert to torch tensor
XY_th = torch.tensor(XY, dtype=torch.float32, requires_grad=True)

# Compute derivatives using the PINN model
ϕ, dϕ_dx, dϕ_dy, _, _ = compute_derivatives(pinn, XY_th)

# Extract numpy arrays
 _zx = dϕ_dy.detach().numpy().reshape(N_plot, N_plot)
 _zy = -dϕ_dx.detach().numpy().reshape(N_plot, N_plot)

# Plot shear stress components
fig, axs = plt.subplots(1, 3, figsize=(6, 2.8), dpi=dpi)
magnitude = np.sqrt( _zx**2 +  _zy**2)
 0 = np.zeros_like( _zx)

c0 = axs[0].quiver(X, Y,  _zx,  0, magnitude, scale=1, width=0.01,
                                                           cmap='jet')
axs[0].set_title(r'Shear stress component $ _{zx}$')
plt.colorbar(c0, ax=axs[0], shrink=shrink)
axs[0].axis('equal')

c1 = axs[1].quiver(X, Y,  0,  _zy, magnitude, scale=1, width=0.01,
                                                           cmap='jet')
axs[1].set_title(r'Shear stress component $ _{zy}$')
plt.colorbar(c1, ax=axs[1], shrink=shrink)
axs[1].axis('equal')

c2 = axs[2].quiver(X, Y,  _zx,  _zy, magnitude, scale=1, width=0.01,
                                                           cmap='jet')
axs[2].set_title('Shear stress flow')
plt.colorbar(c2, ax=axs[2], shrink=shrink)
axs[2].axis('equal')

plt.tight_layout()
plt.savefig('imagesPN/shearOnRectangle.png', dpi=500)
plt.show()
```

```
plot2D (pinn, N_plot = 25, a=a, b=b, c=c, d=d, shrink=0.7)
```

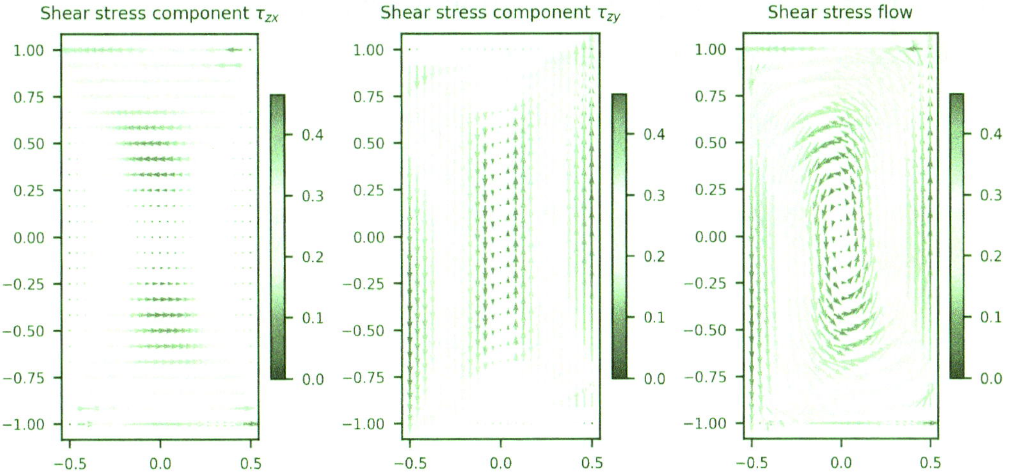

The PINN model has produced shear stress patterns that agree well with those obtained using the analytic method discussed in Ref. [13].

As seen, once the PINN model is trained, computing other solutions related to the function and its various derivatives becomes quite straightforward.

We have explored several PINN models that use integration to evaluate the loss function. Both the trapezoidal rule and Gaussian quadrature were applied without domain discretization, relying on the regularity of the problem domain. When the problem domain becomes irregular, integration using Gaussian quadrature or the standard trapezoidal rule becomes challenging. A more practical approach is to discretize the problem domain into triangles. The following section discusses this approach in detail.

2.14 Triangulation of Domain

Triangulation of a domain is a well-studied topic, with many useful existing techniques available to generate triangular meshes for various problems with irregular domains [11,20]. Such a mesh consists of triangular cells, each with three nodes, hence also called a T3 mesh. The procedure is as follows:

1. Discretize the problem domain into sufficiently small triangular cells of reasonable quality, creating a triangular mesh with nodes distributed inside the domain. These nodes serve as sampling points for a PINN model.
2. For integration, compute the integration weights corresponding to each node using the areas of the triangles connected to that node. This is done by assuming the integrand is linear over each triangular cell, allowing sampling at the centroid of the cell. The weight for a node is then the sum of the areas of its connected triangles divided by three, because one third of each triangle's area contributes to one of its nodes. This approach is known as barycentric averaging or lumped mass .

The following Python code demonstrates how to generate a triangular mesh using the `meshpy` module and then compute the weights for each node. In this section, the domain considered is rectangular so that all techniques can be examined in detail, comparing with exact solutions and PINN models trained earlier. These techniques will be extended to irregular domains in the subsequent chapter.

2.14.1 Integration Weights at Sampling Points

The following set of utility functions generate a T3 (triangular) mesh over a rectangular domain using the `meshpy` module. These functions also compute integration weights at each sampling point based on the areas of the connected triangles, suitable for PINN residual integration.

```
#!pip install meshpy  # this is needed, if it is not already done.
from matplotlib.colors import Normalize, ListedColormap
from meshpy import triangle

def rectangle_meshT3(a=0.0, b=1.0, c=0.0, d=1.0, maxA=0.01):
    """Generate a triangular (T3) mesh for a rectangular domain
       (a,b)×(c,d).
    """
    # Define the 4 corner points for the rectangle:
    points = [(a, c), (b, c), (b, d), (a, d)]
    facets = [(0, 1), (1, 2), (2, 3), (3, 0)]  # Counter-clockwise edges

    info = triangle.MeshInfo()
    info.set_points(points)
    info.set_facets(facets)

    mesh = triangle.build(info, max_volume=maxA)
    return np.array(mesh.points), np.array(mesh.elements)
```

The area of a triangular (T3) cell formed by two vectors $\mathbf{u}$ and $\mathbf{v}$ in 2D is computed as half the magnitude of their cross product. Given:

$$\mathbf{u} = \begin{bmatrix} u_x \\ u_y \end{bmatrix}, \quad \mathbf{v} = \begin{bmatrix} v_x \\ v_y \end{bmatrix}$$

the area A of the triangle is

$$A = \frac{1}{2}\,|\mathbf{u} \times \mathbf{v}| = \frac{1}{2}|u_x v_y - u_y v_x|$$

This formula is fundamental in computational geometry and can be found in Chapter 2 of Ref. [16].

$$\begin{aligned} A &= \frac{1}{2}|\mathbf{u} \times \mathbf{v}| = \frac{1}{2}|u_x v_y - u_y v_x| \\ &= \frac{1}{2}|\underbrace{(x_2 - x_1)}_{u_x}\underbrace{(y_3 - y_1)}_{v_y} - \underbrace{(x_3 - x_1)}_{u_y}\underbrace{(y_2 - y_1)}_{v_x}| \end{aligned} \tag{2.62}$$

The snippet is as follows.

```
def triangle_area(p1, p2, p3):
    '''Compute the area of a triangle with 3 nodes p1, p2, p3 arranged
    counterclockwise. '''
```

```
    return 0.5 * np.abs((p2[0] - p1[0]) * (p3[1] - p1[1]) -
                        (p3[0] - p1[0]) * (p2[1] - p1[1]))
```

2.14.2 Generalized Trapezoidal Rule for T3 Mesh

For numerical integration over a domain discretized into triangular (T3) cells, the nodal integration weights are computed by **barycentric averaging**. Mathematically, for node i, the weight w_i is

$$w_i = \sum_{\text{triangles } T \ni i} \frac{A_T}{3} \tag{2.63}$$

where A_T is the area of triangle T adjacent to node i.

This approach ensures that the integration weights sum to the total domain area, providing a consistent quadrature scheme over irregular triangular meshes.

```
def node_integration_weights(points, elements):
    """
    Compute barycenter-based quadrature weights for each node
    """
    n_nodes = len(points)
    weights = np.zeros(n_nodes)

    for tri in elements:
        p1, p2, p3 = points[tri]
        area = triangle_area(p1, p2, p3)

        # Lumped mass: equally distribute triangle area to its 3 nodes
        for i in tri:
            weights[i] += area / 3.0

    return weights
```

The nodes and weights can then be used to integrate a given function using the nodal integration technique. The formulation is provided in Eq.(2.28). In NumPy, this is simply an element-wise product of three arrays, and thus can be computed very efficiently.

The snippet is as follows. The results are also plotted using the following snippet.

```
# Disply the mesh with elements and nodes colored with weights:
# Generate mesh:
points, elements = rectangle_meshT3(a=0.0, b=1.0, c=0., d=1., maxA=.01)
weights = node_integration_weights(points, elements)

print(points.shape, weights.shape)
print("Integration weights for nodes (first 3 to verify):")
for i, w in enumerate(weights[:3]):
    print(f"Node {i}: weight = {w:.6f}")

plt.figure(figsize=(5, 5), dpi=300)                    # plot weights as size
```

```
plt.tripcolor(points[:, 0], points[:, 1], elements,
              facecolors=np.ones(len(elements)),
              edgecolors='k', linewidth=0.2,
              cmap=ListedColormap(['white']),
              norm=Normalize(vmin=0, vmax=1))

plt.scatter(points[:, 0], points[:, 1], c=weights.reshape(-1,1),
                                                    cmap='jet',s=5)
plt.colorbar(label="Integration Weight", shrink=0.5)
plt.title("Nodal Integration Weights")
plt.gca().set_aspect('equal')
plt.show()
```

```
(88, 2) (88,)
Integration weights for nodes (first 3 to verify):
Node 0: weight = 0.004969
Node 1: weight = 0.004564
Node 2: weight = 0.005709
```

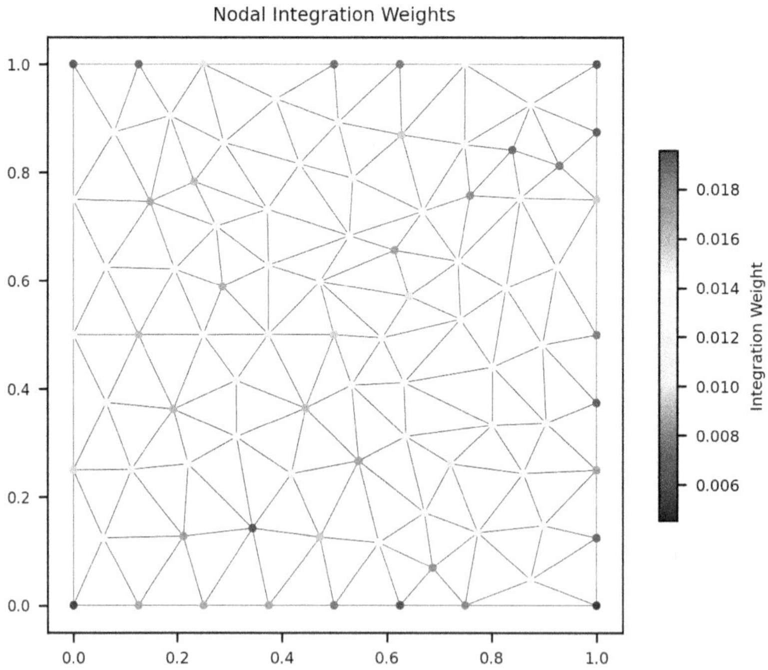

In the foregoing plot, the color of each node represents its integration weight. It can be observed that the total area of all triangles connected to a given node is proportional to the node's weight. In general, nodes located on the edges are connected to fewer triangles, and thus have the smaller weights. The four corner nodes have the smallest weights. The nodal integration based on the T3 mesh is, in fact, a generalized trapezoidal rule applied over an irregular grid.

2.14.3 Gaussian Quadrature for T3 Mesh

Using a T3 mesh, one can also perform Gaussian quadrature, similar to standard FEM techniques [21]. Depending on the desired order of accuracy, the number of Gaussian points within each T3 cell will vary. A reference table is provided in [20], where the Gaussian point coordinates are given in the natural coordinate system of a right-angled triangle with unit base length. These coordinates must be mapped to the physical coordinates of the actual T3 cell. The code snippet for performing this mapping and computing the quadrature is as follows.

```
import matplotlib.tri as mtri

def map_to_physical_T3(bary_coords, p1, p2, p3):
    """Map reference triangle barycentric coordinates
        to physical  triangle. """
    return np.dot(bary_coords, np.array([p1, p2, p3]))

def get_T3_gauss_rule(degree):
    """

    Return barycentric coords and weights for triangle quadrature
    of given degree: 1, 2, 3, or 4.
    """
    if degree == 1:
        # 1-point (degree 1)
        bary_coords = np.array([[1/3, 1/3, 1/3]])
        weights = np.array([1.0])
    elif degree == 2:
        # 3-point (degree 2)
        bary_coords = np.array([ [2/3, 1/6, 1/6],
                                 [1/6, 2/3, 1/6],
                                 [1/6, 1/6, 2/3] ])
        weights = np.array([1/3, 1/3, 1/3])
    elif degree == 3:
        # 4-point (degree 3) - symmetric
        bary_coords = np.array([ [1/3, 1/3, 1/3], [0.6, 0.2, 0.2],
                                 [0.2, 0.6, 0.2], [0.2, 0.2, 0.6] ])
        weights = np.array([-27/48, 25/48, 25/48, 25/48])
    elif degree == 4:
        # 6-point Dunavant rule (degree 4) - 12-digit precision
        bary_coords = np.array([
            [0.445948490915965, 0.445948490915965, 0.108103018168070],
            [0.445948490915965, 0.108103018168070, 0.445948490915965],
            [0.108103018168070, 0.445948490915965, 0.445948490915965],
            [0.091576213509771, 0.091576213509771, 0.816847572980458],
            [0.091576213509771, 0.816847572980458, 0.091576213509771],
            [0.816847572980458, 0.091576213509771, 0.091576213509771]])
        weights = np.array([    0.111690794839005, 0.111690794839005,
                                0.111690794839005, 0.054975871827661,
                                0.054975871827661, 0.054975871827661 ])

    else:
        raise ValueError(f"Degree supported is 1, 2, 3, or 4.")
```

```
    return bary_coords, weights

def gauss_quadrature_T3(points, elements, degree=2):
    """
    Compute Gaussian quadrature points and weights on triangles (T3).
    """
    x_quad = []; w_quad = []

    bary_coords, ref_weights = get_T3_gauss_rule(degree)

    for tri in elements:
        p1, p2, p3 = points[tri]
        area = triangle_area(p1, p2, p3)

        quad_pts = map_to_physical_T3(bary_coords, p1, p2, p3)
        weights = ref_weights * area

        x_quad.append(quad_pts)
        w_quad.append(weights)

    return np.vstack(x_quad), np.hstack(w_quad)
```

The following snippet samples the Gaussian points and produces the corresponding weights.

```
def sampleD_T3_Gauss1_4(points, elements, degree=4):
    """
    Sample points and weights for Gaussian quadrature using T3 mesh
    with points and elements given.
    degree: desired accuracy (1, 2, 3 or 4)
    """
    x_quad, w_quad = gauss_quadrature_T3(points, elements,degree=degree)

    x_Ω_th = torch.tensor(x_quad, dtype=torch.float32)
    w_s_th = torch.tensor(w_quad, dtype=torch.float32).reshape(-1, 1)

    return w_s_th, x_Ω_th, x_quad, w_quad
```

The formula for computing the integration of a general function $g(x, y)$ over its domain Ω is simply as follows.

$$I_g = \int_\Omega g(x, y)\mathrm{d}x\mathrm{d}y = \sum_{i=1}^{n_{int}} w_i\, g(x_i, y_i) \tag{2.64}$$

where n_{int} is the total number of integration points used in the entire T3 mesh. When the Gaussian quadrature rule is applied (see below), this equals the number of integration points per T3 cell multiplied by the total number of T3 cells in the mesh. When the generalized trapezoidal rule is used, n_{int} corresponds to the number of nodes in the T3 mesh. The symbol w_i denotes the weight associated with the ith integration point, and $g(x_i, y_i)$ is the value of the function at that point.

Note also that when Gaussian quadrature employs only one point located at the centroid of a T3 cell, it reduces to the generalized trapezoidal rule.

When the integrand is the L^2 loss, the integration formula is given in Eq.(2.28).

2.14.4 Plot of Gaussian Points in T3 Mesh

The following snippet generates a T3 mesh over a rectangular domain, computes the coordinates and corresponding weights of the Gaussian points, and plots these points within the mesh.

```
def plot_T3_mesh_gauss_points(points, elements, x_quad, dpi=300):
    """
    Plot triangular mesh and Gaussian quadrature points.
    Args: points:   Array of mesh vertices (N_points, 2)
          elements: Array of triangle indices (N_elements, 3)
          x_quad:   Array of Gaussian quadrature points (N_quad, 2)
    """
    points = np.array(points)
    elements = np.array(elements)
    x_quad = np.array(x_quad)

    triangulation=mtri.Triangulation(points[:,0], points[:,1], elements)

    plt.figure(figsize=(4.5, 4.5), dpi=dpi)
    plt.triplot(triangulation, color="black", linewidth=0.6)
    plt.scatter(x_quad[:, 0], x_quad[:, 1], color="blue", s=2,
                                              label="Gaussian points")
    plt.gca().set_aspect("equal")
    plt.title("T3 Mesh with Gaussian Quadrature Points")
    plt.xlabel("x"); plt.ylabel("y")
    plt.legend(); plt.grid(False)
    plt.show()
```

```
# Generate mesh and Gaussian quadrature points
points, elements = rectangle_meshT3(a=0., b=1., c=0., d=1., maxA=0.03)
x_quad, w_quad = gauss_quadrature_T3(points, elements, degree=2)
print(f"shape of the Gaussian points: {x_quad.shape}; "
      f"shape of the weights: {w_quad.shape}")

plot_T3_mesh_gauss_points(points, elements, x_quad, dpi=300)
```

```
shape of the Gaussian points: (180, 2); shape of the weights: (180,)
```

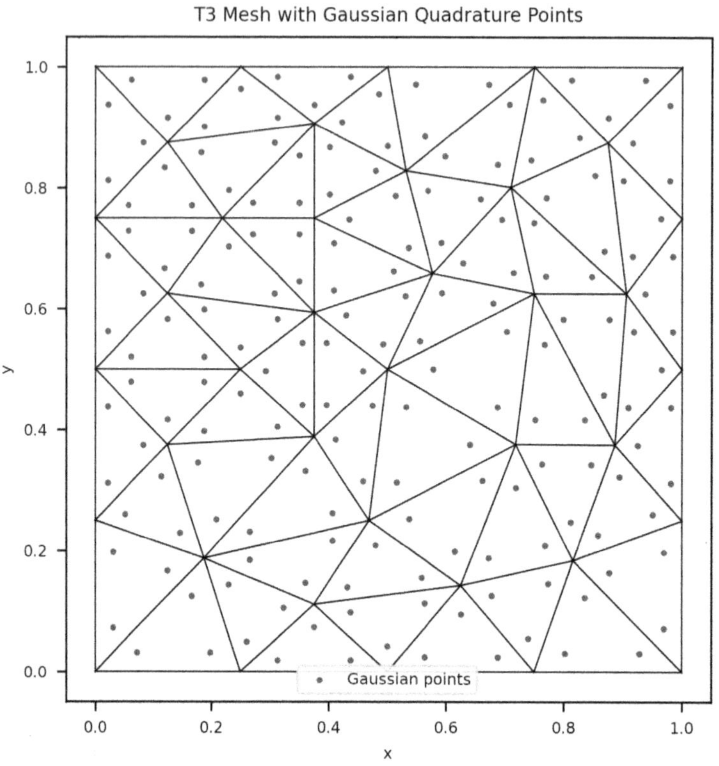

2.14.5 Numerical Integration via T3 Mesh

Gaussian points located in larger triangles carry larger weights. For verification purposes, let us compute the numerical value of the integral of the exact solution given in Eq.(2.15), and compare it with the exact integral value $\frac{4}{\pi^2} \approx 0.405285$, as obtained in Eq.(2.36).

The following SymPy code confirms this result.

```
a, b, c, d = 0.0, 1.0, 0.0, 1.0
points, elements = rectangle_meshT3(a=a, b=b, c=c, d=d, maxA=0.01)
_, _, x_quad, w_quad = sampleD_T3_Gauss1_4(points, elements, degree=2)

f_vals = exact_u(x_quad[:, 0], x_quad[:, 1])          # f at Gauss points

I_approx = np.sum(f_vals * w_quad)                # numerical integral values
I_exact = 4 / np.pi**2                                         # derived above

print(f"Numerical integration result: {I_approx:.6f}")
print(f"Exact integration result:     {I_exact:.6f}")

error_L2 = lg.norm(I_exact - I_approx) / lg.norm(I_exact)
print(f"L2 Relative Error:            {error_L2:.4e}")
```

```
Numerical integration result: 0.405286
Exact integration result:     0.405285
```

```
L2 Relative Error:              2.8640e-06
```

This confirms that Gaussian quadrature using a T3 mesh performs well. Higher accuracy can be achieved by using more Gaussian points per T3 cell and/or refining the mesh.

2.15 PINN Models Trained Using T3 Mesh

The following snippet performs the following tasks:

1. Uniformly samples the four edges of a rectangular domain $[a, b] \times [c, d]$ to generate arrays for boundary conditions (BCs).
2. Meshes the given rectangular domain $[a, b] \times [c, d]$ into a set of triangular elements. The nodes of the triangular mesh become the sampling points for integration, and the integration weight for each node is computed using Eq.(2.63).
3. Returns the dataset for a PINN model.

2.15.1 Sampling on Boundaries of Rectangular Domains

The following snippet computes sampling points on the four boundary segments of a given rectangular domain and assigns zero values for DBCs on these segments.

```python
def sampleB_T3(N_DBC=50, a=0.0, b=1.0, c=0.0, d=1.0):
    '''
    Sample data points on the four edges of a rectangular domain
    for training PINN with integral loss.
    Args: N_DBC: number of Dirichlet BC points per edge
          a, b: x-domain boundaries
          c, d: y-domain boundaries
    Returns: x_DBC_th: BC coordinates      (shape [N_NBC, 2])
             u_DBC_th: BC values           (shape [N_NBC, 1])
    '''
    x_DBCs = []                          # Boundary Condition coordinates
    for s in np.linspace(0, 1, N_DBC):
        xi = a + s * (b - a)
        yi = c + s * (d - c)
        x_DBCs.extend([
            [xi, c],       # bottom edge
            [xi, d],       # top edge
            [a, yi],       # left edge
            [b, yi]])      # right edge

    x_DBC = np.array(x_DBCs)
    u_DBC = np.zeros((x_DBC.shape[0], 1))        # DBC: u=0 on all 4 edges

    # Convert BCs to torch tensors
    x_DBC_th = torch.tensor(x_DBC, dtype=torch.float32)
    u_DBC_th = torch.tensor(u_DBC, dtype=torch.float32)

    return x_DBC_th, u_DBC_th
```

2.15.2 Sampling in Rectangular Domains Using T3 Mesh

The following snippet generates a T3 mesh for a given rectangular domain, producing nodes and elements. The nodes serve as sampling points for a PINN model, and the elements are used to compute integration weights when needed.

```
def sampleD_T3(a=0.0, b=1.0, c=0.0, d=1.0, maxA=0.01):
    '''
    Sample data points in a Domain with T3 mesh for training PINN with
    integral loss via T3 elements over a rectangular domain.
    Args: a, b: x-domain boundaries
          c, d: y-domain boundaries
          maxA: max triangle area for internal mesh resolution
    Returns: w_s_th: integration weights  (shape [N_interior, 1])
             x_Ω_th: interior coordinates (shape [N_interior, 2])
    '''
    # Triangular mesh and integration weights inside the domain
    points, elements = rectangle_meshT3(a=a, b=b, c=c, d=d, maxA=maxA)
    print('No. of points =', len(points), 'No. of elements =',
                                                  len(elements))
    w_s = node_integration_weights(points, elements)

    x_Ω_th = torch.tensor(points, dtype=torch.float32)
    w_s_th = torch.tensor(w_s,    dtype=torch.float32)

    return w_s_th, x_Ω_th
```

2.15.3 Example: PINN for Torsional Bar Meshed with T3

The following code snippets train PINN models using the dataset created via `sampleD_T3`, for the same torsional bar with a rectangular cross-section discussed in the previous section using the regular grids. We can now perform a comparative study of these different PINN models.

2.15.3.1 Use of Generalized Trapezoidal Rule

When the nodes in the T3 mesh are used as sampling points, and the barycentric-averaged area weights are applied, this constitutes the generalized trapezoidal rule for 2D integration. The following snippet uses this approach to construct and train a PINN model.

```
#                    --- use Generalized Trapezoidal Rule for T3 ---
set_seed_th(88)

# Geometric data for the rectangular cross-section of a bar:
a=-0.5; b=0.5;  c=-1.; d=1.         # the same rectangular cross-section

# Buid a PINN model and train it:
p = 2; Nn=50; NL=2        # NL hidden layers with nn number of neurons each
# Create new PINN model with DBC enforced:
pinn = PINN2D_4DBC(Nn=Nn, NL=NL, a=a, b=b, c=c, d=d)
P_star, P_total = print_model_parameters(pinn, print_table=False)
```

```
# Generate the training dateset, via triangulation of the domain:
DI_rule = 'T3'
w_s_th, x_Ω_th = sampleD_T3(a=a, b=b, c=c, d=d, maxA=0.01)
print(f"Sampled {x_Ω_th.shape[0]} collocation points in the domain.")
print(f"Is N_sample >= P* ? {x_Ω_th.shape[0] >= P_star}")
```

```
P* (Total pseudo-dimension of the affine space) = 105
Total trainable parameters: 2751
No. of points = 173 No. of elements = 301
Sampled 173 collocation points in the domain.
Is N_sample >= P* ? True
```

With the maxA parameter controlling the mesh density, the generated T3 mesh contains 173 nodes, which serve as the sampling points. Hence, the condition $m \geq P^*$ holds. Let's train the pinn:

```
LossTerm = 'PDE_only'
x_DBC_th = None; u_DBC_th = None                                # no BCs
x_NBC_th = None; t_NBC_th = None; w_NBC_th = None; I_NBC = None

# Train the model with integrated squared residual:
Trainer  = trainer_int_PDE
Residual = residual_int
losses, totalTime = train_adam_lbfgs(pinn, Trainer, Residual, f_th,
       x_Ω_th, w_s_th, x_DBC_th, u_DBC_th, x_NBC_th, t_NBC_th, w_NBC_th,
                        I_NBC, epochs1=201, epochs2=11, use_lbfgs=True)
```

```
==================================================
201 epoches training with Adam:
Epoch     1, Loss: 3.448e+02, PDE: 3.448e+02
Epoch   101, Loss: 5.206e+01, PDE: 5.206e+01
Epoch   201, Loss: 2.560e+01, PDE: 2.560e+01
Time taken by adam: 1.169521 seconds

==================================================
11 epoches fine-tuning with LBFGS:
Epoch     1, Loss: 8.062e+00, PDE: 8.062e+00
Epoch    11, Loss: 8.001e+00, PDE: 8.001e+00

Training complete.
Time taken by lbfgs: 11.822733 seconds
Total time: 12.992254 seconds
Ratio (time_lbfgs/time_adam) of time taken per epoch:184.719743
```

```
error_L2, errorVI, u_diff, _ = plot_sln_rect(pinn, exact_ϕ, f_th=f_th,
   N_plot=100, vmin=0, vmax=.11, a=a,b=b,c=c,d=d,shrink=1,plot_it=False)

# plot3loss2(losses, ('adam initially','lbfgs in fine tunning'),
#                                                PDE="PDE", log=False)
recordSheet()        # Update the record
```

```
L2 mean residual of trained pinn: 1.6501e-03
Maximum Error: 1.0789e-04
Mean Absolute Error: 7.1390e-06
Error Variation Index (ratio of Max/Mean): 1.5113e+01
L2 Relative Error: 1.8219e-04
```

```
printSummary(pinns, LossTerms, Residuals, totalTimes,error_L2s,errorVIs)
```

```
#  PINN model        Loss_terms residual    total_time error_L2s  errorVIs
=============================================================================
1  2D_4DBC           PDE_only   _intTR      1.402e+01  5.118e-04  9.592e+00
2  2D_4DBC           PDE_only   _intGQ      7.419e+00  7.212e-04  3.871e+00
3  2D_4DBC           PDE_only   _energyTR   3.179e+00  9.101e-03  5.011e+00
4  2D_4DBC           PDE_only   _clctn      1.468e+01  6.634e-04  9.340e+00
5  2D_4DBC           PDE_only   _intT3      1.299e+01  1.822e-04  1.511e+01
```

It is observed that using a T3 mesh for integration with the generalized trapezoidal rule yields the most accurate results compared to all the previously trained models. The computation time is on par with Model-1 using the regular grid and Model-4 using collocation.

2.15.3.2 Use of T3 Element-wise Gaussian Quadrature

Let us now apply Gaussian quadrature to each individual element in the T3 mesh:

```
#                          --- use Gaussian quadrature for T3 ---
set_seed_th(88)

# Generate the training dateset, via triangulation of the domain:
points, elements = rectangle_meshT3(a=a, b=b, c=c, d=d, maxA=0.01)
DI_rule = 'T3GQ'
w_s_th, x_Ω_th,_,_= sampleD_T3_Gauss1_4(points, elements, degree=2)
print(f"Sampled {x_Ω_th.shape[0]} collocation points in the domain.")

# Create new PINN model with DBC enforced:
pinn = PINN2D_4DBC(Nn=Nn, NL=NL, a=a, b=b, c=c, d=d)
print("No. of trainable parameters:", count_parameters(pinn))

# Train the model with integrated squared residual:
Trainer  = trainer_int_PDE
Residual = residual_int
losses, totalTime = train_adam_lbfgs(pinn, Trainer, Residual, f_th,
       x_Ω_th, w_s_th, x_DBC_th, u_DBC_th, x_NBC_th, t_NBC_th, w_NBC_th,
                          I_NBC, epochs1=201, epochs2=11, use_lbfgs=True)
```

```
Sampled 903 collocation points in the domain.
No. of trainable parameters: 2751
==================================================
201 epoches training with Adam:
Epoch     1, Loss: 1.963e+00, PDE: 1.963e+00
Epoch   101, Loss: 2.239e-01, PDE: 2.239e-01
Epoch   201, Loss: 9.312e-02, PDE: 9.312e-02
Time taken by adam: 2.551976 seconds
```

```
=================================================
11 epoches fine-tuning with LBFGS:
Epoch     1, Loss: 4.219e-04, PDE: 4.219e-04
Epoch    11, Loss: 1.638e-05, PDE: 1.638e-05

Training complete.
Time taken by lbfgs: 22.190334 seconds
Total time: 24.742310 seconds
Ratio (time_lbfgs/time_adam) of time taken per epoch:158.887810
```

```
#for item in recordItems: item.clear()     # to clear the entire record
#for item in recordItems: del item[-1]           # to delete one record
```

```
error_L2, errorVI, u_diff, _ = plot_sln_rect(pinn, exact_φ, f_th=f_th,
   N_plot=100, vmin=0, vmax=.11, a=a,b=b,c=c,d=d,shrink=.8,plot_it=True)

plot3loss2(losses, ('adam initially','lbfgs in fine tunning'),
                                                    PDE="PDE", log=False)
recordSheet()       # Update the record
```

```
L2 mean residual of trained pinn: 6.1021e-04
Maximum Error: 5.8069e-05
Mean Absolute Error: 6.3674e-06
Error Variation Index (ratio of Max/Mean): 9.1198e+00
L2 Relative Error: 1.2895e-04
```

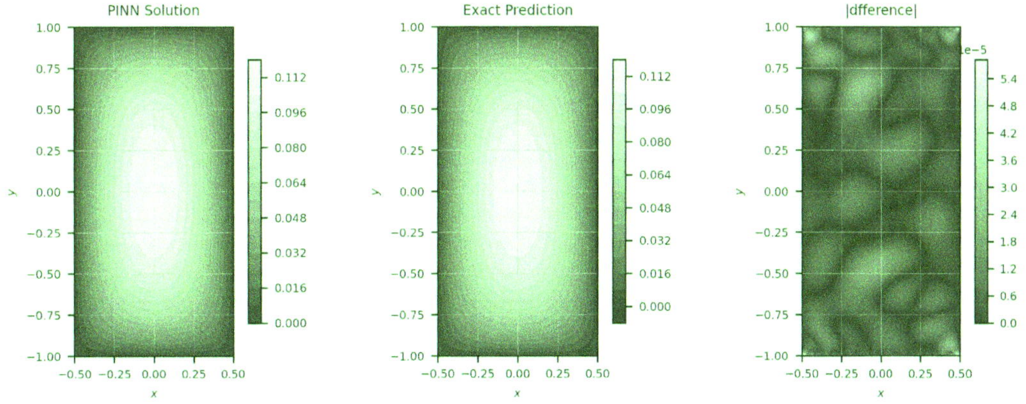

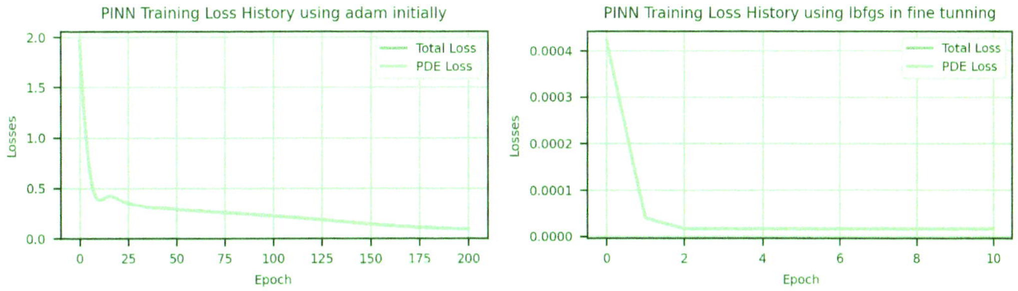

```
printSummary(pinns, LossTerms, Residuals, totalTimes,error_L2s,errorVIs)
```

```
#  PINN model       Loss_terms residual    total_time error_L2s  errorVIs
=========================================================================
1  2D_4DBC          PDE_only   _intTR      1.402e+01  5.118e-04  9.592e+00
2  2D_4DBC          PDE_only   _intGQ      7.419e+00  7.212e-04  3.871e+00
3  2D_4DBC          PDE_only   _energyTR   3.179e+00  9.101e-03  5.011e+00
4  2D_4DBC          PDE_only   _clctn      1.468e+01  6.634e-04  9.340e+00
5  2D_4DBC          PDE_only   _intT3      1.299e+01  1.822e-04  1.511e+01
6  2D_4DBC          PDE_only   _intT3GQ    2.474e+01  1.289e-04  9.120e+00
```

The use of Gaussian quadrature has improved the solution accuracy for this problem. It is the most accurate PINN model trained so far for this problem. However, it took more time to compute compared to the use of the generalized trapezoidal rule (which uses just one Gaussian point per T3 cell).

2.15.3.3 Use of Energy Method with T3 Mesh

Let us also try the energy method together with Gaussian quadrature using the same T3 mesh.

```
#                  --- Energy, Gaussian quadrature for T3 ---
set_seed_th(88)

# Generate the training dateset, via triangulation of the domain:
points, elements = rectangle_meshT3(a=a, b=b, c=c, d=d, maxA=0.01)
DI_rule = 'T3GQ'
w_s_th, x_Ω_th,_,_= sampleD_T3_Gauss1_4(points, elements, degree=2)
print(f"# of sample points: {len(x_Ω_th)}")

# Create new PINN model with DBC enforced:
pinn = PINN2D_4DBC(Nn=Nn, NL=NL, a=a, b=b, c=c, d=d)
print("No. of trainable parameters:", count_parameters(pinn))

# Train the model with energy loss
Trainer  = trainer_int_PDE
Residual = residual_energy
losses, totalTime = train_adam_lbfgs(pinn, Trainer, Residual, f_th,
       x_Ω_th, w_s_th, x_DBC_th, u_DBC_th, x_NBC_th, t_NBC_th, w_NBC_th,
                      I_NBC, epochs1=201, epochs2=11, use_lbfgs=True)
```

```
# of sample points: 903
No. of trainable parameters: 2751
==================================================
201 epoches training with Adam:
Epoch     1, Loss: -5.788e-03, PDE: -5.788e-03
Epoch   101, Loss: -5.528e-02, PDE: -5.528e-02
Epoch   201, Loss: -5.595e-02, PDE: -5.595e-02
Time taken by adam: 0.640747 seconds

==================================================
11 epoches fine-tuning with LBFGS:
```

```
Epoch     1, Loss: -5.712e-02, PDE: -5.712e-02
Epoch    11, Loss: -5.714e-02, PDE: -5.714e-02

Training complete.
Time taken by lbfgs: 2.111811 seconds
Total time: 2.752558 seconds
Ratio (time_lbfgs/time_adam) of time taken per epoch:60.224285
```

```
error_L2, errorVI, u_diff, _ = plot_sln_rect(pinn, exact_ϕ, f_th=f_th,
   N_plot=100, vmin=0, vmax=.11, a=a,b=b,c=c,d=d,shrink=.8,plot_it=True)

plot3loss2(losses, ('adam initially','lbfgs in fine tunning'),
                                              PDE="PDE", log=False)
recordSheet()        # Update the record
```

```
L2 mean residual of trained pinn: 1.6619e-02
Maximum Error: 1.6109e-03
Mean Absolute Error: 2.8734e-04
Error Variation Index (ratio of Max/Mean): 5.6061e+00
L2 Relative Error: 5.8166e-03
```

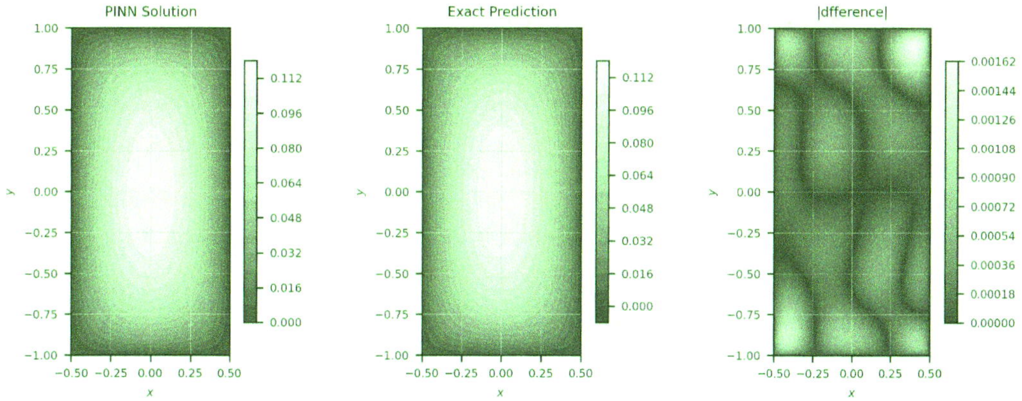

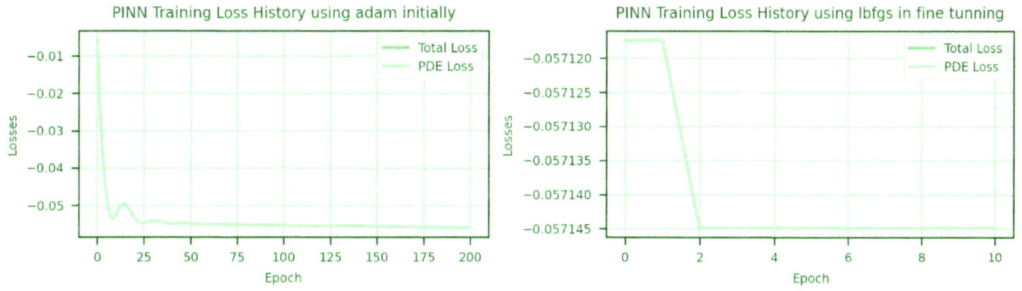


```
printSummary(pinns, LossTerms, Residuals, totalTimes,error_L2s,errorVIs)
```

```
#  PINN model       Loss_terms residual    total_time error_L2s  errorVIs
======================================================================
1  2D_4DBC          PDE_only   _intTR      1.402e+01  5.118e-04  9.592e+00
2  2D_4DBC          PDE_only   _intGQ      7.419e+00  7.212e-04  3.871e+00
```

3	2D_4DBC	PDE_only	_energyTR	3.179e+00	9.101e-03	5.011e+00
4	2D_4DBC	PDE_only	_clctn	1.468e+01	6.634e-04	9.340e+00
5	2D_4DBC	PDE_only	_intT3	1.299e+01	1.822e-04	1.511e+01
6	2D_4DBC	PDE_only	_intT3GQ	2.474e+01	1.289e-04	9.120e+00
7	2D_4DBC	PDE_only	_energyT3GQ	2.753e+00	5.817e-03	5.606e+00

This case study produced the following findings:

1. The PINN models using the energy method yielded less accurate results using PINNs with the same capacity. Their primary advantage lies in fast training.
2. Gaussian quadrature performed well when applied over a T3 mesh, as demonstrated in model-6, where sampling is performed within each T3 cell and the cell density is controlled by the mesh generator.
3. Models 5 and 6 performed the best. Both utilize integrated residuals based on the T3 mesh. Since T3 meshes can be generated with ease, these models can be extended to complex geometries, which will be explored in the following chapter.

We conclude that using triangular meshes offers an effective approach for handling regular domains in PINN models, enabling consistent and accurate PDE solutions.

2.16 On Solution Uniqueness

The examination of solution uniqueness is a subtle and often complex issue across all computational methods, including Physics-Informed Neural Networks (PINNs).

To begin, consider a general partial differential equation (PDE) system consisting of a PDE along with initial and/or boundary conditions. Such a system may not always admit a solution. In our discussion, we restrict attention to well-posed PDE systems that are known to have at least one solution.

2.16.1 PDE Systems with a Unique Solution

Let us first examine PDE systems that are well-posed and admit a unique solution—for example, linear PDEs with proper sufficient boundary and initial conditions. Classical numerical methods, such as the finite element method (FEM), smoothed FEM (S-FEM), or meshfree methods, can typically recover this unique solution through appropriately designed weak formulations.

However, when solving the same problem using a PINN, uniqueness of the obtained solution is not guaranteed, even when the true solution is unique. This is due to the nature of the PINN approach: the solution is represented by a neural network function $\mathcal{N}_{\hat{\mathbf{w}}}$, which is generally a composition of nonlinear activation functions applied to affine transformations of the inputs, with tunable weights $\hat{\mathbf{w}}$ [12]. The resulting approximation is inherently nonlinear, and thus may admit multiple different weight configurations that minimize the loss to a similar extent.

Moreover, training a PINN involves solving a nonlinear optimization problem. The loss function is typically composed of quadratic residual terms involving u, but due to the dependence on $\hat{\mathbf{w}}$, the loss is a highly nonlinear and non-convex function of the training parameters. This means the optimization landscape may contain many local minima. As a result, different random initializations of $\hat{\mathbf{w}}$ can lead to convergence to different local minima, yielding different approximate solutions even though the true solution is unique.

In addition to the loss landscape, the expressiveness of the neural network architecture plays a critical role. Although neural networks are universal function approximators in theory, their ability to represent a given solution depends on the depth, width, and activation functions used. If the network architecture is insufficiently expressive, or if the number and distribution of training points are inadequate, the network may fail to capture the unique solution accurately.

2.16.2 PDE Systems with Multiple Solutions

In many important applications, the underlying PDE system is nonlinear and may admit multiple solutions. In such cases, the non-uniqueness of the PINN solution is expected. A trained PINN may converge to any one of the valid solutions depending on factors such as network architecture, residual definition, loss function formulation, random initialization, training procedure, and optimization algorithm. The use of hybrid approaches, such as combining PINNs with genetic algorithms , has proven effective in identifying multiple solutions [19]. In addition, whether or not a PINN model can be successfully trained depends on many factors, including those mentioned earlier. Some recent studies can be found in Refs. [9, 24, 29, 30].

2.16.3 Summary of Factors Affecting Uniqueness

Several key factors influence the uniqueness (or non-uniqueness) of the solution obtained from a PINN model:

- **Random Initialization:** Different initial guesses for the weights can lead to convergence to different local minima.
- **Residual Definition:** Different definitions of the residual and its evaluation method can change the behavior of the model.
- **Loss Function Design:** The formulation of the residual and loss terms affects the optimization landscape.
- **Training Strategy:** The choice of optimizers, learning rate schedules, and stopping criteria can impact convergence.
- **Network Architecture:** Expressivity must be sufficient to represent the solution space, and the configuration must work hand in hand with the loss function terms and support effective training.
- **Sampling Strategy:** The density and placement of training points influence the approximation quality.

Understanding and managing these factors is essential when using PINNs to solve PDEs, particularly when the goal is to ensure or assess the uniqueness of the obtained solution.

2.17 Remarks

This chapter presents the key concepts, formulations, techniques, and implementations of PINN models with various configurations for solving two-dimensional (2D) Poisson equations under different boundary conditions. Through comprehensive case studies, the performance and characteristics of these PINN models are thoroughly examined. The following key observations are made:

1. The PINN class can be constructed in either a plain form or a trial solution form. Accurate models are more readily achieved when Dirichlet boundary conditions (DBCs) are enforced (in a hard way) via the trial solution form at the class level.

2. The number of collocation samples must satisfy the **samples-neurons theorem**. Specifically, the number of samples should not be less than the pseudo-dimension of the affine space defined by the PINN model's structure.

3. The overall loss function typically consists of a composite form that includes contributions from the PDE residual, DBC loss, and Neumann boundary condition (NBC) loss. The PDE residual component can be evaluated using the following methods:

   - **Pure collocation** : Computes the mean of the L^2 residual over a set of collocation points, which may be randomly sampled or defined over a mesh grid. Model accuracy is highly dependent on the residual magnitude at these points.

   - **Integral residual** or **Integrated squared residual**: Computes the integrated L^2 residual using numerical quadrature techniques such as the trapezoidal rule or Gaussian quadrature. This approach reduces potential biases arising from uneven point distributions and improves integration accuracy.

   - **Energy method**: Uses the total potential energy, expressed as an integral over the domain, as the loss function to be minimized. Unlike squared residuals, this functional may take negative values. The accuracy of this method hinges on the precise evaluation of the energy integral.

   - **Finite Difference Method (FDM)**: Provides a discrete approximation of the derivatives and hence the residuals without the need for `autograd`.

4. Collocation-based methods using L^2 residuals typically yield higher accuracy, but they often require longer training times due to their sensitivity to residual minimization.

5. The energy method tends to converge faster, offering more efficient training, though often at the cost of reduced solution accuracy when measured at points in the domain.

6. Effective training of a PINN model generally requires the combined use of the **Adam** optimizer for initial convergence and the **L-BFGS** optimizer for refining accuracy. L-BFGS is particularly important for achieving high-precision solutions for PINN models.

7. A variety of activation functions can be used in PINN models, including non-monotonic functions, depending on the problem structure and learning objectives.

8. The uniqueness of solutions in PINN models is a subtle and nuanced topic that requires deeper investigation. In general, a PINN model may yield different approximate solutions, even when the true solution is mathematically unique. Finding the global optimum becomes critical and also presents significant challenges.

The next chapter focuses on developing PINN models for Poisson equations defined over irregular domains.

Chapter 3

PINN Models for Irregular Domains

Let us import the necessary external modules or dependencies for later use in this chapter.

```
# Place cursor in this cell, and press Ctrl+Enter to import dependencies.
import os
import sys                              # for accessing the computer system
sys.path.append('../grbin/')              # Add in the path to your system

from commonImports import *         # Import dependencies from '../grbin/'
import grcodes as gr                       # Import the module of the author
importlib.reload(gr)                 # When grcodes is modified, reload it
import time
import math
from matplotlib.tri import Triangulation as mtri # for triangular plots

init_printing(use_unicode=True)          # For latex-like quality printing
np.set_printoptions(precision=4, suppress=True)  # Digits in print-outs
```

In the previous chapter, we introduced PINN models with various structures and demonstrated their effectiveness through a real-life example: modeling the shear stress distribution over a rectangular cross-section of a straight bar under torsion. We showed that a properly constructed and trained PINN model can approximate the stress function with reasonable accuracy. Once the model is trained, computing related quantities—such as derivatives and integrals—is straightforward. The shear stresses and shear flow, derived from the trained stress function, were shown to agree well with analytical results based on classical elasticity theory and the laws of mechanics [13]. Additionally, the trained PINN model was used to compute integral-based quantities, such as strain energy.

This chapter introduces techniques for constructing PINN models defined on irregular domains, enabling us to solve problems with complex geometries. As demonstrated in the previous chapter, triangular meshes (T3 meshes) are effective for developing PINN models. The key idea in this chapter is to discretize irregular domains using T3 meshes. The nodes of these meshes are used for point sampling, and the elements are used to integrate the residuals in the PINN formulation.

Python implementations are provided to build, train, and inference PINN models to provide solution to Poisson equations, focusing on irregular domains. These techniques and codes allow for flexible adaptation to various geometrical settings.

> **Note:** The code cells in this chapter are largely sequentially dependent. For a smooth and coherent study experience, it is recommended to run the cells in order from the beginning of the chapter. Most of the PINN models presented here can be trained in tens of seconds on the author's laptop. Thus, the total runtime for executing all the code cells should be reasonably short.

3.1 Triangulation of Domain

When the problem domain becomes irregular, a practical and effective solution is to discretize the domain into triangular elements. This approach enables both efficient sampling and numerical integration, even for complex geometries. The procedure, similar in spirit to that discussed in Section 2.15, differs in that the domain is now irregular.

The steps are as follows:

1. **Domain Discretization**: Subdivide the irregular domain into sufficiently small triangular elements of reasonably good quality. Each element consists of three nodes, forming what is referred to as a T3 mesh. These nodes serve as the sampling points for the PINN model.
2. **Integration Weights via Barycentric Averaging**: The weight associated with a node in a T3 is computed as the sum of one-third of the areas of all triangles connected to that node, known as *barycentric averaging* or the *lumped mass approach.*

The following Python code snippet demonstrates how to generate a triangular mesh over a general polygonal domain using the `meshpy` library and how to compute the corresponding integration weights for each node.

3.2 Preparation for Coding

The following is a list of Python functions developed in the previous chapter. For convenience and reusability, these functions have been organized into a custom module named `grcodes`. We can import this module to directly utilize its functionalities in this chapter without redefining the same code.

By doing so, we maintain a cleaner workflow and ensure consistency across chapters. The module includes essential tools for building and training PINN models, as well as utilities for sampling, loss evaluation, plotting, and residual computation.

```
from grcodes import train_adam_lbfgs, printSummary, residual_energy,\
             trainer_int_PDE, node_integration_weights, residual_clctn,\
             residual_int, plot3loss2, sampleD_T3_Gauss1_4, sln_error,\
             gauss_quadrature_T3,PINN2D, compute_derivatives, \
             print_model_parameters
```

Readers may review the previous chapter for the functionality and usage of these code snippets. We will not repeat the detailed explanations here.

As in the previous chapter, let us also set the random seed to ensure consistent sampling and initialization behavior in our PINN models. This helps produce reproducible results across different runs:

```
gr.set_seed_th(seed=88)                    # Set seeds for reproducibility
```

The following snippet defines the computing device for training PINN models. It ensures that CUDA is used for accelerated computation if a compatible GPU is available; otherwise, it defaults to using the CPU:

```
# Device management: ensure CUDA is available if you plan to use GPU
device = torch.device('cuda' if torch.cuda.is_available() else 'cpu')
gr.device = device
torch.cuda.empty_cache()                      # Clear previous model state
print(f"Using device: {device}")
```

```
Using device: cpu
```

3.3 PINN Model for Elliptic Domains

3.3.1 Elliptic Cross-Sectional Bar Subject to Torsion

Consider torsion bars with a uniform elliptic cross-section , with a radius r_a along the x-axis and a radius r_b along the y-axis, as shown in Fig.3.1. This is a classical mechanics problem and can be found in textbooks [5, 13].

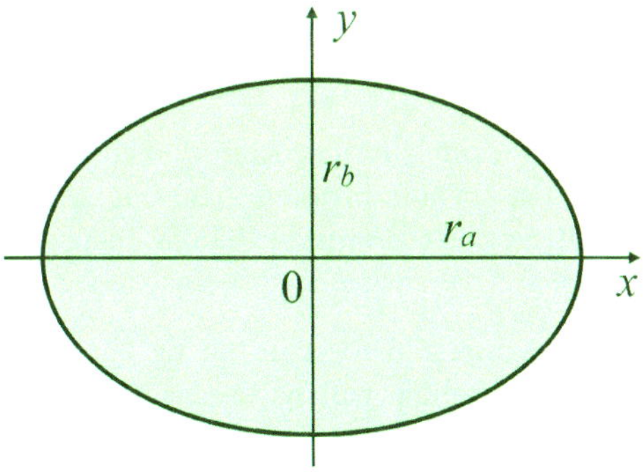

Figure 3.1: A straight bar with a uniform elliptical cross-section subjected to torsion.

The PDE that governs the stress function $u(x, y)$ is a Poisson's equation given in Eq.(2.57). The boundary of the ellipse is defined as

$$F(x, y) = \left(\frac{x}{r_a}\right)^2 + \left(\frac{y}{r_b}\right)^2 - 1 = 0 \tag{3.1}$$

For the torsional bar , the Dirichlet boundary condition (DBC) for the stress function is $u = 0$ at any point on the elliptic boundary. The exact solution can be found in Ref. [13], and is given as follow:

$$u(x,y) = C\left[\left(\frac{x}{r_a}\right)^2 + \left(\frac{x}{r_b}\right)^2 - 1\right] \tag{3.2}$$

where

$$C = \frac{G\vartheta_t r_a^2 r_b^2}{r_a^2 + r_b^2} \tag{3.3}$$

which is a constant for a cross-section with given radii, material shear modulus G, and a specified twist angle per unit length ϑ_t. The following is a SymPy snippet that defines the source function and the exact solution, and confirms the preceding formulas.

```
import sympy as sp
x, y, C = sp.symbols('x, y, C', real=True)       # Define symbolic symbols
ra, rb = sp.symbols('r_a, r_b', positive=True)
G, theta_t = sp.symbols('G, theta_t', positive=True)

C = -(ra**2*rb**2*G*theta_t)/(ra**2 + rb**2)
u = C*((x**2/ra**2+y**2/rb**2)-1)/(2*G*theta_t)
f = -(sp.diff(u, (x,2)) + sp.diff(u, (y,2))).simplify()

display(Math(f" \\text{{source function $f(x,y)$}}=\\small {latex(f)}"))
display(Math(f" \\text{{stress function $u(x,y)$}}=\\small {latex(u)}"))
del x, y, ra, rb, u, f, C, G
```

source function $f(x,y) = 1$

stress function $u(x,y) = -\frac{r_a^2 r_b^2\left(-1 + \frac{y^2}{r_b^2} + \frac{x^2}{r_a^2}\right)}{2\left(r_a^2 + r_b^2\right)}$

Since C is a given constant, we can set it as 1 in the following code for simplicity. When the actual value of C is provided for a specific problem, the true stress function u will simply be Cu. With this setting, we now recover the standard Poisson problem studied in the previous chapter.

The following snippets define the formula for computing the source function and the exact stress function explicitly in Python, which will be used to assess the accuracy of the solution from a trained PINN.

```
# Force or source function in torch for the torsional bar problem:
def f_th(x, y):
    return 1 + 0 * x * y                          # scalar source term (C = 1)

# Exact solution of the stress function (for reference)
def exact_u_ellipse(x, y, ra=1.0, rb=0.5):
    C = -(ra**2*rb**2)/(2*(ra**2 + rb**2))
    return C*((x**2/ra**2+y**2/rb**2)-1)               # the stress function
```

3.3.2 Meshing and Data Generation

As usual, to train a PINN model, we must sample data points inside the elliptical domain and along its boundary. The following Python function generates a T3 mesh for a given ellipse using the `meshpy.triangle` class:

```
#!pip install meshpy  # this is needed, if it is not already done.
from meshpy import triangle

def ellipse_mesh(ra=1.0, rb=0.5, max_area=0.001, N_edge=50):
    """
    Generate triangular mesh inside an ellipse x^2/a^2 + y^2/b^2 <= 1.
    max_area controls the size of the T3 elements/cells.
    """
    def ellipse_boundary():
        '''Create N_edge points on the ellipse boundary arranged
        counterclockwise.'''
        return [(ra*np.cos(2*np.pi*i/N_edge),
                 rb*np.sin(2*np.pi*i/N_edge)) for i in range(N_edge)]

    points = ellipse_boundary()
    facets = [(i, (i + 1) % N_edge) for i in range(N_edge)]

    info = triangle.MeshInfo()       # create a mesh information instance
    info.set_points(points)
    info.set_facets(facets)

    mesh = triangle.build(info, max_volume=max_area)
    return np.array(mesh.points), np.array(mesh.elements)
```

As long as a set of points on the domain boundary is arranged counterclockwise, the previous snippet can be modified to create a T3 mesh for any given geometry. Note that `max_area` is a parameter used in the `meshpy.triangle` class to control the sizes of the T3 cells in the mesh. Our experience shows that this parameter can be quite sensitive and may require careful tuning for a desired density of T3 mesh.

Using the T3 mesh created, we can now sample the data points for training a PINN model:

```
def sampleData_T3_ellipse(ra=1.0, rb=0.5, max_A=0.001, N_DBC=200):
    '''
    Sample data points in domain and on boundary for an ellipse.
    Dirichlet BC: u = 0 on ellipse boundary.'''

    # Boundary conditions (on the ellipse)
    theta = np.linspace(0, 2*np.pi, N_DBC, endpoint=False)
    x_DBCs = np.stack([ra * np.cos(theta), rb * np.sin(theta)], axis=1)
    u_DBC = np.zeros((N_DBC, 1)) # (N_DBC, 1)                  # (N_DBC,2)

    x_DBC_th = torch.tensor(x_DBCs, dtype=torch.float32)      # (N_DBC,2)
    u_DBC_th = torch.tensor(u_DBC , dtype=torch.float32)      # (N_DBC,1)

    # Domain points and integration weights
    points, elements = ellipse_mesh(ra=ra, rb=rb, max_area=max_A)
    points = np.asarray(points).reshape(-1, 2)
    w_s = node_integration_weights(points, elements)
```

```
    w_s = np.asarray(w_s).reshape(-1, 1)                    # (N_PDE, 1)

    x_Ω_th = torch.tensor(points, dtype=torch.float32)       # (N_PDE,2)
    w_s_th = torch.tensor(w_s, dtype=torch.float32)          # (N_PDE,1)

    return w_s_th, x_Ω_th, x_DBC_th, u_DBC_th
```

The following snippet defines a PINN class for elliptic domains with Dirichlet boundary conditions (DBC) enforced via a trial solution inside the class. It accepts a user-defined activation function.

```
class PINN2D_DBC_ellipse(nn.Module):
    '''
    PINN for elliptic domain with Dirichlet BC enforced via trial
    solution. It accepts a custom activation function.
    '''
    def __init__(self, Nn=20, NL=2, ra=1.0, rb=0.5, actit=torch.tanh):
        super(PINN2D_DBC_ellipse, self).__init__()
        self.ra = ra
        self.rb = rb
        self.actit = actit                          # custom actit function

        # Define hidden layers~
        self.hidden = nn.ModuleList([nn.Linear(2, Nn)] +
                        [nn.Linear(Nn, Nn) for _ in range(NL - 1)])

        self.out = nn.Linear(Nn, 1)                         # Output layer~

    def forward(self, x):
        x_, y_ = x[:, 0:1], x[:, 1:2]
        r2 = (x_ ** 2) / (self.ra ** 2) + (y_ ** 2) / (self.rb ** 2)
        DBC = (1 - r2).view(-1, 1)  # DBC bounary, change shape to (m,1)

        z = x
        for layer in self.hidden:
            z = self.actit(layer(z))

        N = self.out(z)
        u_trial = DBC * N
        return u_trial
```

The training of the PINN requires only the PDE loss because the DBC are already enforced in the PINN model construction. The PDE loss is computed using the `residual_int()` function provided in the previous chapter. The trainer, `trainer_int_PDE`, from the previous chapter can also be used without modification.

3.3.3 Example: Bar with Elliptic Cross-Section

We are now ready to build and train a PINN model for a torsional bar with an elliptic cross-section.

3.3.4 PINN Model for Stress Function

The following snippet creates a PINN model for the Poisson's equation that governs the stress function for bars with an elliptic cross-section subjected to torsion.

```
gr.set_seed_th(88)                                      # for reproducibility

# Initialization lists for book-keeping:
pinns = []                                         # for logging key variables
timeAdam  = []; timeLbfgs = []
Trainers  = []; LossTerms = []; Residuals = []
totalTimes= []
error_L2s = []; errorVIs = []                                  # log the error

# Model parameters:
p = 2                                           # for two coordinates, x and y
Nn = 50                                     # Number of neurons in a hidden layer
NL = 2                                               # Numbero (hidden) Layers

# Build the PINN model with DBC enforced:
ra, rb = 0.5, 1.0                               # define the size of the ellipse
pinn = PINN2D_DBC_ellipse(Nn=Nn, NL=NL, ra=ra, rb=rb)
P_star, P_total = print_model_parameters(pinn, print_table=False)

# Sample data over the ellipse
DI_rule = 'T3'                                # Domain Integration uses T3 mesh
max_A  = 0.01                             # constrol the overall sizes of T3 cells
N_DBC  = 1                                      # dymmy, no NBC for this example
w_s_th, x_Ω_th, x_DBC_th, u_DBC_th = sampleData_T3_ellipse(\
                              ra=ra, rb=rb, max_A=max_A, N_DBC=N_DBC)
print(f"Sampled {x_Ω_th.shape[0]} collocation points~ in the domain.")
print(f"Is N_sample >= P* ? {x_Ω_th.shape[0] >= P_star}")
print(f"Sampled {x_DBC_th.shape[0]} boundary points.")
print(f"Weights: {w_s_th.shape[0]}.")
```

```
P* (Total pseudo-dimension~ of the affine space~) = 105
Total trainable parameters: 2751
Sampled 154 collocation points in the domain.
Is N_sample >= P* ? True
Sampled 1 boundary points.
Weights: 154.
```

This `pinn` model has 105 neurons, representing the total pseudo-dimension of the affine space of `pinn` for function approximation. The total number of trainable parameters is 2751. The T3 mesh produces a total of 154 collocation points within the domain, satisfying the condition $m \geq P^*$. This `pinn` is expected to yield stable results if the training is successful. Let's proceed with the training:

```
LossTerm = 'PDE_only'                       # loss term~ in addition to loss_PDE
x_DBC_th, u_DBC_th = None, None                  # no DBC, enforced in the class
x_NBC_th, t_NBC_th, w_NBC_th, I_NBC = None, None, None, ''          # no NBC
```

```
# Train the model with integrated squared residual~~ loss:
Trainer  = trainer_int_PDE
Residual = residual_int                        # from the previous chapter
losses, totalTime = train_adam_lbfgs(pinn, Trainer, Residual, f_th,
       x_Ω_th, w_s_th, x_DBC_th, u_DBC_th, x_NBC_th, t_NBC_th, w_NBC_th,
                          I_NBC, epochs1=201, epochs2=11, use_lbfgs=True)
```

```
==================================================
201 epoches training with Adam:
Epoch     1, Loss: 3.738e+00, PDE: 3.738e+00
Epoch   101, Loss: 4.255e-03, PDE: 4.255e-03
Epoch   201, Loss: 6.490e-04, PDE: 6.490e-04
Time taken by adam: 4.003625 seconds

==================================================
11 epoches fine-tuning~ with LBFGS:
Epoch     1, Loss: 4.454e-06, PDE: 4.454e-06
Epoch    11, Loss: 4.454e-06, PDE: 4.454e-06

Training complete.
Time taken by lbfgs: 1.048113 seconds
Total time: 5.051739 seconds
Ratio (time_lbfgs/time_adam) of time taken per epoch:4.783636
```

The training is very fast for both `adam` and `lbfgs` optimizers. Each took only seconds to reduce the loss by about six orders of magnitude (10^6). The fine-tuning phase using `lbfgs` required just one iteration to reach the minimum, indicating that the loss function is likely very close to quadratic with respect to the learning parameters near its minimum.

The following function can be used to plot the contours of the model's predicted solution alongside the exact solution and the error distribution.

```
def plot2D_ellipse(pinn, exact_u=None, ra=1.0, rb=0.5, N_plot=200,
                                        shrink=0.5, dpi=300, plot_it=True):
    '''
    Plot the PINN prediction, the exact solution (if given), and error.
    For an elliptical domain defined by (x/ra)^2 + (y/rb)^2 <= 1.
    '''
    x = np.linspace(-ra, ra, N_plot)                          # Create grid
    y = np.linspace(-rb, rb, N_plot)
    X, Y = np.meshgrid(x, y)
    XY = np.stack([X.flatten(), Y.flatten()], axis=1)

    # Mask for elliptical domain
    mask = (X**2 / ra**2 + Y**2 / rb**2) <= 1

    # Predict using PINN model
    XY_tensor = torch.tensor(XY, dtype=torch.float32)
    with torch.no_grad():
        u_pred = pinn(XY_tensor).cpu().numpy()
```

```
    u_pred_masked = np.full_like(X, np.nan)
    u_pred_masked[mask] = u_pred.flatten()[mask.flatten()]

    # If exact solution is provided
    if exact_u is not None:
        u_exact = exact_u(X, Y)
        u_exact_masked = np.full_like(X, np.nan)
        u_exact_masked[mask] = u_exact[mask]
        error_L2, errorVI, u_diff = gr.sln_error(u_pred_masked,
                                                 u_exact_masked)
        abs_error = np.abs(u_diff)
    else:
        u_exact_masked = None
        abs_error = None

    if plot_it != True:
        return error_L2, errorVI, u_diff

    width = 7.0; height = width/2.2
    fig, axs = plt.subplots(1, 3, figsize=(width, height), dpi=dpi)
    if u_exact_masked is not None:
        c0 = axs[0].contourf(X, Y, u_exact_masked, levels=20)
        axs[0].set_title('Exact Solution')
        plt.colorbar(c0, ax=axs[0], shrink=shrink)
    else:
        axs[0].set_visible(False)

    c1 = axs[1].contourf(X, Y, u_pred_masked, levels=20)
    axs[1].set_title('PINN Prediction')
    plt.colorbar(c1, ax=axs[1], shrink=shrink)

    if abs_error is not None:
        c2 = axs[2].contourf(X, Y, abs_error, levels=100)
        axs[2].set_title('Absolute Error')
        #plt.colorbar(c2, ax=axs[2], shrink=shrink)
    else:
        axs[2].set_visible(False)

    for ax in axs:
        ax.set_aspect('equal')

    plt.tight_layout()
    plt.show()

    return error_L2, errorVI, u_diff
```

```
# Plot the contours of both PINN solution along with the exact one:
exact_u = lambda x, y: exact_u_ellipse(x, y, ra=ra, rb=rb)
```

```
error_L2, errorVI, u_diff = plot2D_ellipse(pinn, exact_u, ra=ra, rb=rb,
                                N_plot=500, dpi=300, plot_it=True)

plot3loss2(losses, ('adam initially','lbfgs in fine tunning'),
                                           PDE="PDE", log=False)
```

```
Maximum Error: 4.9579e-05
Mean Absolute Error: 1.6253e-05
Error Variation Index (ratio of Max/Mean): 3.0505e+00
L2 Relative Error: 3.5112e-04
```

The relative accuracy is found to be at the 10^{-4} level. The following are the contour plots of the model's predicted solution, the exact solution, and the error distribution. These are followed by the time-history plots of the loss terms from the two-stage optimization process.

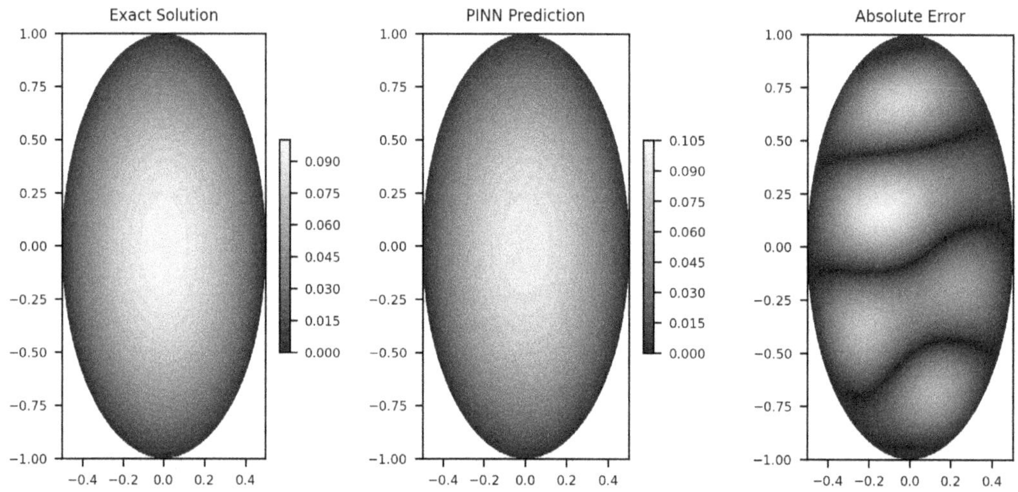

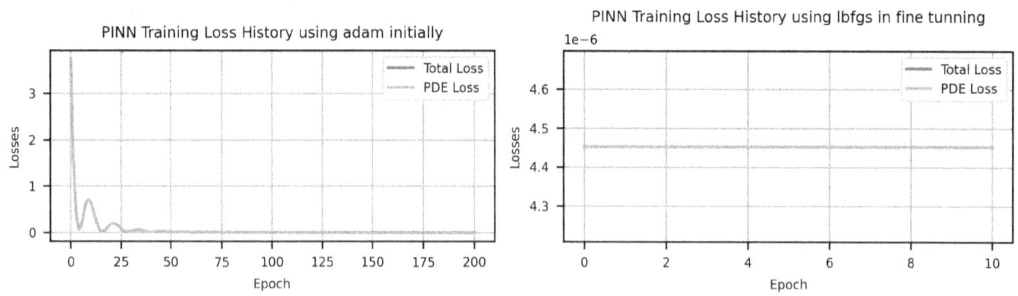

We use the following Python function to log the key performance indicators of the PINN models to be trained for comparison studies. This function is identical to the one used in the previous chapter. Because it needs to assess the names of related variables used in this chapter, it is more convenient to redefine it here.

```
def recordSheet():
    '''
    Log the key performance indicators of PINN models for comparison
    study. '''

    new_error_L2_str = f"{error_L2:.3e}"

    if new_error_L2_str not in error_L2s:
        pinns.append(pinn.__class__.__name__)           # log pinn models
        Residuals.append(Residual.__name__ + DI_rule)     # log residuals
        #Trainers.append(Trainer.__name__)        # log trainers, optional
        LossTerms.append(LossTerm) # additional terms in composite loss
        totalTimes.append(f"{totalTime:.3e}")                 # log time
        error_L2s.append(new_error_L2_str)                # log the error
        errorVIs.append(f"{errorVI:.3e}")        # error variation index
    else:
        print(f"Error_L2 value '{new_error_L2_str}'"\
              f" already exists in error_L2s. Skipping update.")

recordItems= [pinns, LossTerms, Residuals,totalTimes,error_L2s,errorVIs]
```

```
# Create record sheet (exercute this only once!):
recordSheet()
```

```
printSummary(pinns, LossTerms, Residuals, totalTimes,error_L2s,errorVIs)
```

```
#  PINN model        Loss_terms residual     total_time errors_L2  errorVIs
=========================================================================
1  2D_DBC_ellipse    PDE_only   _intT3       5.052e+00  3.511e-04  3.050e+00
```

The use of integration for residual computation via the trapezoidal rule using the T3 mesh has produced an accurate solution for this case. Readers may also try Gaussian quadrature for comparison.

The following snippet trains a PINN model using the energy method , with the energy integration performed using the same T3 mesh.

```
gr.set_seed_th(seed=88)                    # Set seeds for reproducibility

# Build a new PINN model with DBC enforced:
pinn = PINN2D_DBC_ellipse(Nn=Nn, NL=NL, ra=ra, rb=rb)
P_star, P_total = print_model_parameters(pinn, print_table=False)

# create T3 mesh and sample data:
max_A = 0.002
w_s_th, x_Ω_th, x_DBC_th, u_DBC_th = sampleData_T3_ellipse(\
                                ra=ra, rb=rb, max_A=max_A, N_DBC=N_DBC)
print(f"Sampled {x_Ω_th.shape[0]} nodes in the domain.")
print(f"Is N_sample >= P* ? {x_Ω_th.shape[0] >= P_star}")
print(f"Weights for integration: {w_s_th.shape[0]}.")
```

```
# Train the model with energy loss integrated using T3 mesh:
DI_rule  = 'T3'
Trainer  = trainer_int_PDE
Residual = residual_energy
losses, totalTime = train_adam_lbfgs(pinn, Trainer, Residual, f_th,
       x_Ω_th, w_s_th, x_DBC_th, u_DBC_th, x_NBC_th, t_NBC_th, w_NBC_th,
                        I_NBC, epochs1=201, epochs2=11, use_lbfgs=True)
```

```
P* (Total pseudo-dimension of the affine space) = 105
Total trainable parameters: 2751
Sampled 652 nodes in the domain.
Is N_sample >= P* ? True
Weights for integration: 652.
==================================================
201 epoches training with Adam:
Epoch     1, Loss: -6.870e-04, PDE: -6.870e-04
Epoch   101, Loss: -3.891e-02, PDE: -3.891e-02
Epoch   201, Loss: -3.892e-02, PDE: -3.892e-02
Time taken by adam: 0.636217 seconds

==================================================
11 epoches fine-tuning with LBFGS:
Epoch     1, Loss: -3.892e-02, PDE: -3.892e-02
Epoch    11, Loss: -3.892e-02, PDE: -3.892e-02

Training complete.
Time taken by lbfgs: 0.247783 seconds
Total time: 0.883999 seconds
Ratio (time_lbfgs/time_adam) of time taken per epoch:7.116543
```

Since the energy method is used, a sufficiently dense mesh is required to ensure accurate integration of the energies. Therefore, we use more T3 cells. It is found that when using the energy method, the PINN training takes about one second, which is very fast. The accuracy can be assessed using the following snippet.

```
# Plot the contours of both PINN solution and the exact one:
exact_u = lambda x, y: exact_u_ellipse(x, y, ra=ra, rb=rb)
error_L2, errorVI, u_diff = plot2D_ellipse(pinn, exact_u, ra=ra, rb=rb,
                                  N_plot=500, dpi=300, plot_it=True)

plot3loss2(losses, ('adam initially','lbfgs in fine tunning'),
                                          PDE="PDE", log=False)
```

```
Maximum Error: 1.1658e-03
Mean Absolute Error: 4.5384e-04
Error Variation Index (ratio of Max/Mean): 2.5688e+00
L2 Relative Error: 9.4444e-03
```

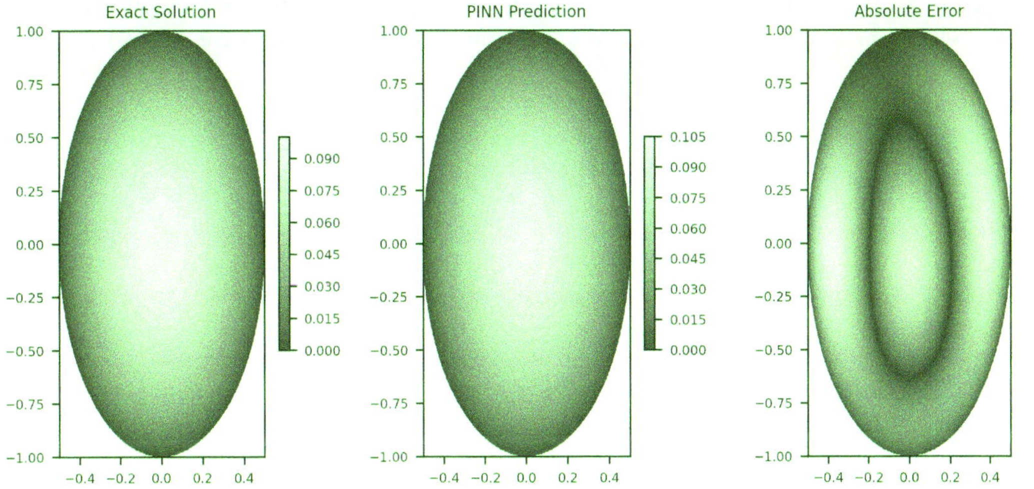

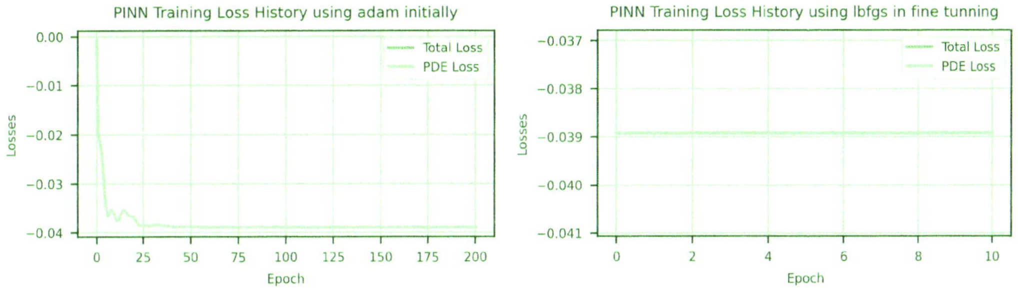


```
recordSheet()                                    # Create record sheet
```

The following snippet is only used to empty the record or remove a record item, if it is created or updated wrongly for some reason.

```
# recordItems=[pinns, LossTerms, Residuals, totalTimes, error_L2s,errorVIs]
# for item in recordItems: item.clear()      # to clear the entire record
# for item in recordItems: item.pop(-1)             # to delete one record
```

```
printSummary(pinns, LossTerms, Residuals, totalTimes,error_L2s,errorVIs)
```

```
#  PINN model        Loss_terms residual    total_time errors_L2  errorVIs
==========================================================================
1  2D_DBC_ellipse    PDE_only   _intT3      5.052e+00  3.511e-04  3.050e+00
2  2D_DBC_ellipse    PDE_only   _energyT3   8.840e-01  9.444e-03  2.569e+00
```

As found in the previous chapter, the PINN model trained using the energy method produces less accurate solutions but can be trained faster. The accuracy may be improved by using a denser mesh for energy integration or by employing Gaussian quadrature of higher order. The following snippet demonstrates the use of quadratic Gaussian quadrature.

```
# Build the PINN model with DBC enforced:
pinn = PINN2D_DBC_ellipse(Nn=Nn, NL=NL, ra=ra, rb=rb)
P_star, P_total = print_model_parameters(pinn, print_table=False)

# create T3 mesh and sample data:
max_A = 0.02
points, elements = ellipse_mesh(ra=ra, rb=rb, max_area=max_A)
DI_rule = 'T3GQ'
w_s_th, x_Ω_th,_,_= sampleD_T3_Gauss1_4(points, elements,degree=2)
print(f"Sampled {x_Ω_th.shape} collocation points in the domain.")
print(f"Weights: {w_s_th.shape}.")

# Train the model with energy loss:
Trainer  = trainer_int_PDE
Residual = residual_energy
losses, totalTime = train_adam_lbfgs(pinn, Trainer, Residual, f_th,
       x_Ω_th, w_s_th, x_DBC_th, u_DBC_th, x_NBC_th, t_NBC_th, w_NBC_th,
                         I_NBC, epochs1=201, epochs2=11, use_lbfgs=True)
```

```
P* (Total pseudo-dimension of the affine space) = 105
Total trainable parameters: 2751
Sampled torch.Size([456, 2]) collocation points in the domain.
Weights: torch.Size([456, 1]).
==================================================
201 epoches training with Adam:
Epoch     1, Loss: -2.461e-02, PDE: -2.461e-02
Epoch   101, Loss: -3.944e-02, PDE: -3.944e-02
Epoch   201, Loss: -3.948e-02, PDE: -3.948e-02
Time taken by adam: 0.712102 seconds

==================================================
11 epoches fine-tuning with LBFGS:
Epoch     1, Loss: -3.948e-02, PDE: -3.948e-02
Epoch    11, Loss: -3.948e-02, PDE: -3.948e-02

Training complete.
Time taken by lbfgs: 0.186804 seconds
Total time: 0.898906 seconds
Ratio (time_lbfgs/time_adam) of time taken per epoch:4.793444
```

```
# Compute the error in the solution:
exact_u = lambda x, y: exact_u_ellipse(x, y, ra=ra, rb=rb)
error_L2, errorVI, u_diff = plot2D_ellipse(pinn, exact_u, ra=ra, rb=rb,
                                   N_plot=500, dpi=300, plot_it=False)

# plot3loss2(losses, ('adam initially','lbfgs in fine tunning'),
#                                       PDE="PDE", log=False) # optional
```

```
Maximum Error: 4.5306e-04
Mean Absolute Error: 2.6747e-04
Error Variation Index (ratio of Max/Mean): 1.6939e+00
L2 Relative Error: 5.1480e-03
```

```
recordSheet()                                          # update the record
```

```
printSummary(pinns, LossTerms, Residuals, totalTimes,error_L2s,errorVIs)
```

```
#  PINN model       Loss_terms residual    total_time errors_L2  errorVIs
=========================================================================
1  2D_DBC_ellipse   PDE_only   _intT3      5.052e+00  3.511e-04  3.050e+00
2  2D_DBC_ellipse   PDE_only   _energyT3   8.840e-01  9.444e-03  2.569e+00
3  2D_DBC_ellipse   PDE_only   _energyT3GQ 8.989e-01  5.148e-03  1.694e+00
```

The use of Gaussian quadrature improved the solution accuracy while reducing the error variation. The first model, which uses collocation residuals integrated over the T3 mesh, provides the most accurate solution.

The following snippet trains the model using purely collocation, with the nodes of the T3 mesh serving as the collocation points. In this case no integration is performed.

```
gr.set_seed_th(seed=88)                      # Set seeds for reproducibility

# Build a new PINN model with DBC enforced:
pinn = PINN2D_DBC_ellipse(Nn=Nn, NL=NL, ra=ra, rb=rb)
P_star, P_total = print_model_parameters(pinn, print_table=False)

max_A = 0.01                                 # create T3 mesh and sample data
w_s_th, x_Ω_th, x_DBC_th, u_DBC_th = sampleData_T3_ellipse(\
                               ra=ra, rb=rb, max_A=max_A, N_DBC=N_DBC)
print(f"Sampled {x_Ω_th.shape[0]} nodes in the domain.")
print(f"Is N_sample >= P* ? {x_Ω_th.shape[0] >= P_star}")
print(f"Weights for integration: {w_s_th.shape[0]}.")

# Train the model with pure collocation~ loss:
DI_rule  = 'T3'
Trainer  = trainer_int_PDE
Residual = residual_clctn                  # pure collocaiton, no integration
losses, totalTime = train_adam_lbfgs(pinn, Trainer, Residual, f_th,
      x_Ω_th, w_s_th, x_DBC_th, u_DBC_th, x_NBC_th, t_NBC_th, w_NBC_th,
                           I_NBC, epochs1=201, epochs2=11, use_lbfgs=True)
```

```
P* (Total pseudo-dimension of the affine space) = 105
Total trainable parameters: 2751
Sampled 154 nodes in the domain.
Is N_sample >= P* ? True
Weights for integration: 154.
==================================================
201 epoches training with Adam:
Epoch     1, Loss: 2.630e+00, PDE: 2.630e+00
Epoch   101, Loss: 2.716e-03, PDE: 2.716e-03
```

```
Epoch   201, Loss: 3.807e-04, PDE: 3.807e-04
Time taken by adam: 1.487220 seconds

==================================================
11 epoches fine-tuning with LBFGS:
Epoch     1, Loss: 4.765e-06, PDE: 4.765e-06
Epoch    11, Loss: 4.765e-06, PDE: 4.765e-06

Training complete.
Time taken by lbfgs: 0.557451 seconds
Total time: 2.044671 seconds
Ratio (time_lbfgs/time_adam) of time taken per epoch:6.849123
```

```
# Plot the contours of both PINN solution and the exact one:
exact_u = lambda x, y: exact_u_ellipse(x, y, ra=ra, rb=rb)
error_L2, errorVI, u_diff = plot2D_ellipse(pinn, exact_u, ra=ra, rb=rb,
                                  N_plot=500, dpi=300, plot_it=True)

plot3loss2(losses, ('adam initially','lbfgs in fine tunning'),
                                           PDE="PDE", log=False)
```

```
Maximum Error: 7.8623e-05
Mean Absolute Error: 2.6494e-05
Error Variation Index (ratio of Max/Mean): 2.9675e+00
L2 Relative Error: 5.7526e-04
```

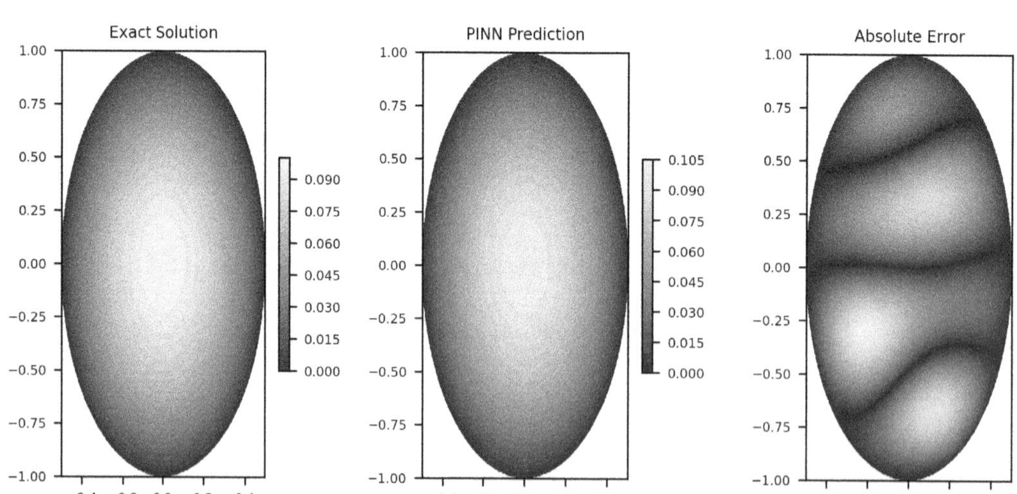

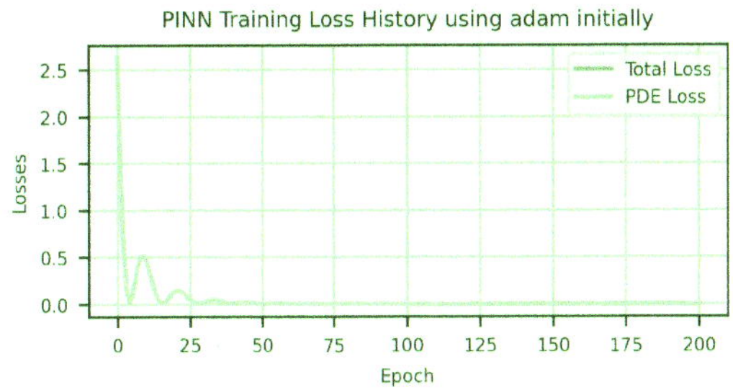

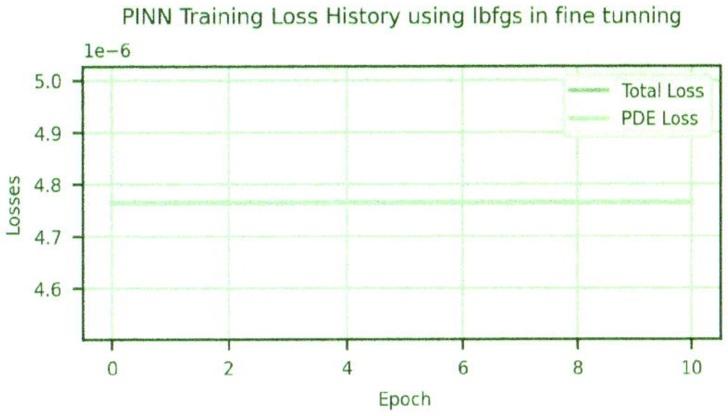


```
recordSheet()                                          # update the record
```

```
printSummary(pinns, LossTerms, Residuals, totalTimes,error_L2s,errorVIs)
```

```
#  PINN model        Loss_terms residual     total_time errors_L2  errorVIs
===========================================================================
1  2D_DBC_ellipse    PDE_only   _intT3       5.052e+00  3.511e-04  3.050e+00
2  2D_DBC_ellipse    PDE_only   _energyT3    8.840e-01  9.444e-03  2.569e+00
3  2D_DBC_ellipse    PDE_only   _energyT3GQ  8.989e-01  5.148e-03  1.694e+00
4  2D_DBC_ellipse    PDE_only   _clctnT3     2.045e+00  5.753e-04  2.968e+00
```

It is again found that using purely collocation with the nodes of the T3 mesh as collocation points produces very accurate results, ranking second only to Model 1, which uses integration of the L^2 residual based on the same T3 mesh.

3.3.5 Shear Flow on Elliptic Cross-Section

Using Eq. (2.60), the shear stresses can be computed from the first derivatives of the stress function. The following snippet computes these stresses and plots the vector fields representing the shear flow on the elliptic cross-section of the bar. It utilizes the `compute_derivatives()` function introduced in the previous chapter:

```
from matplotlib.patches import Ellipse

def plot2D _ellipse(pinn, ra=1.0, rb=0.5, N_plot=200, dpi=300):
    '''Shear stress plots from a PINN model in an elliptical domain. '''

    x = np.linspace(-ra, ra, N_plot)                        # Grid setup
    y = np.linspace(-rb, rb, N_plot)
    X, Y = np.meshgrid(x, y)
    XY = np.hstack((X.flatten()[:, None], Y.flatten()[:, None]))
    XY_th = torch.tensor(XY, dtype=torch.float32, requires_grad=True)

    # Get model derivatives
    ϕ, dϕ_dx, dϕ_dy, _, _ = compute_derivatives(pinn, XY_th)
    zx =  dϕ_dy.detach().numpy().reshape(N_plot, N_plot)
    zy = -dϕ_dx.detach().numpy().reshape(N_plot, N_plot)
```

```
mask = (X**2 / ra**2 + Y**2 / rb**2) <= 1     # Apply elliptical mask
 zx_masked = np.full_like(X, np.nan)
 zy_masked = np.full_like(Y, np.nan)
 zx_masked[mask] = zx.flatten()[mask.flatten()]
 zy_masked[mask] = zy.flatten()[mask.flatten()]
magnitude = np.sqrt( zx_masked**2 + zy_masked**2)
 0 = np.zeros_like( zx_masked)

fig = plt.figure(figsize=(5, 3.5), dpi=dpi)
import matplotlib.gridspec as gridspec
gs = gridspec.GridSpec(1, 3, hspace=-0.1)

titles=[r'$\tau_{zx}$', r'$\tau_{zy}$', r'Shear flow $\vec{\tau}$']
data  =[( zx_masked, 0), ( 0, zy_masked), ( zx_masked, zy_masked)]

for i, (u, v) in enumerate(data):
    ax = fig.add_subplot(gs[i])
    ax.quiver(X, Y, u, v, magnitude, scale=2,width=0.008,cmap='jet')
    ax.add_patch(Ellipse((0, 0), 2*ra, 2*rb, edgecolor='black',
                                   facecolor='none', linewidth=0.8))
    ax.set_title(titles[i], fontsize=5)
    ax.set_aspect('equal')
    ax.set_xlim(-ra*1.01, ra*1.01); ax.set_ylim(-rb*1.01, rb*1.01)
    ax.set_xticks([]); ax.set_yticks([])
    ax.set_frame_on(False)

os.makedirs("imagesPN", exist_ok=True)
plt.savefig("imagesPN/shearOnEllipse.png", dpi=500,
                                              bbox_inches='tight')
plt.show()
```

```
plot2D _ellipse(pinn, ra=0.5, rb=1.0, N_plot=30, dpi=500)
```

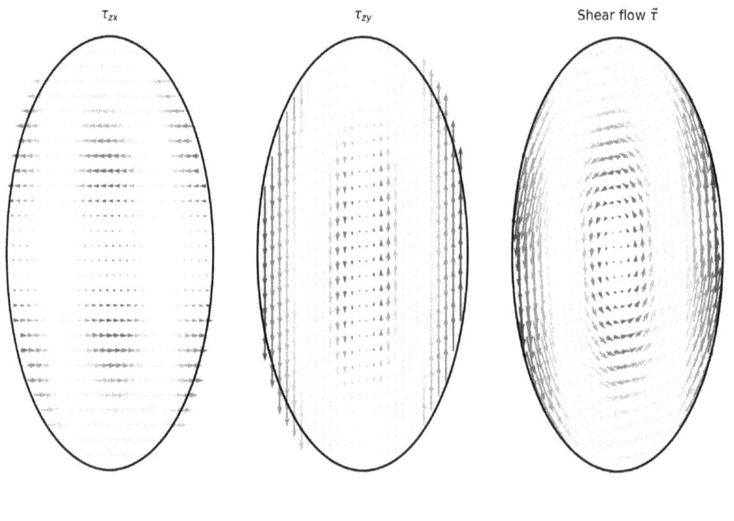

It is observed that once a PINN model is trained, obtaining solutions related to derivatives of the primary function becomes straightforward. In this example, the shear stresses correspond to the first derivatives of the stress function.

We found that the applied torque induces a circulation of shear stresses across the cross-section of the bar. Thus, the torsional mechanics problem has been effectively solved in the form of a trained PINN model, which can be used as a standard function to evaluate both the stress function and the shear stresses.

3.4 PINN for General Polygonal Domains

We can also train a PINN model for general polygonal domains, which can be defined by their vertices. The key requirement is a T3 mesh for the domain; the subsequent operations to train and inference a PINN model are the same as in the previous examples.

3.4.1 Mesh Generation General Polygons

The following snippet generates T3 meshes for general polygons using the `meshpy` module. It also employs the `build()` function for this task:

```
def polygon_mesh(vertices, max_area=0.001):
    '''
    Generate a mshing for a polygon defined for given vertices.
    '''
    mesh_info = MeshInfo()
    mesh_info.set_points(vertices)
    facets=[(i, (i + 1) % len(vertices)) for i in range(len(vertices))]

    mesh_info.set_facets(facets)
    mesh = build(mesh_info, max_volume=max_area)
    points = np.array(mesh.points)
    elements = np.array(mesh.elements)

    return points, elements
```

The `polygon_mesh()` function returns `points` and `elements`, which provide all the necessary information to compute sampling points and corresponding weights for integrating the residuals or energies.

3.4.2 Sample Data for Polygons

To train a PINN model, we need to sample data points both inside the polygon and along all its edges. This is accomplished using the following Python function:

```
from meshpy.triangle import MeshInfo, build

def sampleData_T3_polygon(N_DBC=50, vertices=None, maxA=0.001):
    """
```

```
    Sample data for training PINN using a general polygonal domain.
    u = 0 on boundary. Triangular mesh (T3) is used for the interior.
    Parameters:
        N_DBC:     Number of points per polygon edge for BC.
        vertices: List of polygon corners in counter-clockwise order.
        maxA:      Maximum triangle area for  mesh density control.
    Returns:
        w_s_th:    Integration weights [N_interior, 1]
        x_Ω_th:    Interior coordinates [N_interior, 2]
        x_DBC_th: Dirichlet BC points on boundary [N_DBC * edges, 2]
        u_DBC_th: Dirichlet BC values (all zeros) [N_DBC * edges, 1]
        points:    Numpy array of mesh nodes
        elements: Numpy array of T3 elements (node indices)
    """
    assert vertices is not None and len(vertices) >= 3, \
                                        "Provide >=3 polygon vertices."
    # --- DBC Points (on edges) ---
    x_DBCs = []
    n = len(vertices)
    for i in range(n):
        p_start, p_end = vertices[i], vertices[(i + 1) % n]
        for s in np.linspace(0, 1, N_DBC):
            x = (1 - s) * np.array(p_start) + s * np.array(p_end)
            x_DBCs.append(x)

    x_DBC = np.array(x_DBCs)             # coordinates for DBC on edges
    u_DBC = np.zeros((x_DBC.shape[0], 1))                # u = 0 on edges

    # Convert to PyTorch tensors
    x_DBC_th = torch.tensor(x_DBC, dtype=torch.float32)
    u_DBC_th = torch.tensor(u_DBC, dtype=torch.float32)

    # --- Interior Mesh and Integration Weights ---
    points, elements = polygon_mesh(vertices, max_area=maxA)
    print(f'No. of points = {len(points)} | No. of elements = '
                                            f'{len(elements)}')
    w_s = node_integration_weights(points, elements)
    x_Ω_th = torch.tensor(points, dtype=torch.float32)
    w_s_th = torch.tensor(w_s, dtype=torch.float32).reshape(-1, 1)

    return w_s_th,x_Ω_th,x_DBC_th,u_DBC_th, points,elements
```

Once a T3 mesh is created, data points for the polygonal domain become available for training a PINN model. In a collocation approach, these `points` serve as the sampling points to create the dataset for training. In an integration approach using the trapezoidal rule, these points and elements can be used to compute integration weights for each point to evaluate the residuals. Next, we address the boundary conditions for a polygonal domain.

3.4.3 Boundary Function for Polygons

As emphasized earlier, a stable PINN model is best constructed by enforcing Dirichlet boundary conditions (DBCs) at the PINN class level. Thus, we need to define a function that represents the polygon's boundary. Since a polygon consists of straight line segments, we can use these lines to form the boundary function. Assuming the polygon is **convex**, the formula is as follows.

Let the convex polygon have vertices $(x_0, y_0), (x_1, y_1), \ldots, (x_{n-1}, y_{n-1})$ arranged **counter-clockwise**. For each edge connecting (x_i, y_i) to (x_{i+1}, y_{i+1}), we define the following linear function for the edge:

$$f_{\text{linear}}(x, y) = (x - x_i)(y_{i+1} - y_i) - (y - y_i)(x_{i+1} - x_i) \tag{3.4}$$

This function returns zero on the edge. The boundary function, `f_DBC`, can then be defined by multiplying these linear functions corresponding to all edges of the polygon:

$$f_{\text{DBC}}(x, y) = \prod_{i=0}^{n-1} f_{\text{linear}}^{(i)}(x, y) \tag{3.5}$$

Clearly, this function $f_{\text{DBC}}(x, y)$ vanishes on all polygon edges and is nonzero inside the polygon, provided the polygon is convex. It serves as a valid `f_DBC` function for the PINN model enforcing zero Dirichlet boundary conditions on all edges. When the polygon is a unit square, Eq.(3.5) reduces to Eq.(2.22).

The following Python function generates the `f_DBC` for a convex polygon.

```
def polygon_DBCf_mult(x, vertices):
    '''
    Create the function for the boundary of a convex polygon defined by
    `vertices` arranged counterclockwise. f_DBC(x)=0 defines the boundary.
    '''
    x_ = x[:, 0:1]
    y_ = x[:, 1:2]
    f_DBC = torch.ones_like(x_)

    n = len(vertices)
    for i in range(n):
        x0, y0 = vertices[i]
        x1, y1 = vertices[(i + 1) % n]
        f_linear = (x_ - x0)*(y1 - y0) - (y_ - y0)*(x1 - x0)
        f_DBC *= f_linear

    return f_DBC
```

3.4.4 Class `PINN2D_DBCfun`

Using the `DBCfun`, we can now construct the following `PINN` class in which Dirichlet boundary conditions (DBCs) are enforced on all edges of a polygonal domain:

```
class PINN2D_DBCfun(nn.Module):
    '''
    Define a class of PINN models with externally defined `DBCfun`
    and custom activation passed in for a polygonal domain.
    The polygon is defined by `vertices` arranged counterclockwise.
    '''
    def __init__(self, Nn, NL, vertices, DBCfun, actit=torch.tanh):
        super(PINN2D_DBCfun, self).__init__()
        self.vertices = vertices
        self.DBCfun = DBCfun          # function of the polygon boundary
        self.actit = actit                       # custom actit function

        # Define hidden layers                  #input layer~
        self.hidden = nn.ModuleList([nn.Linear(2, Nn)] +
                    [nn.Linear(Nn, Nn) for _ in range(NL-1)])
        self.out = nn.Linear(Nn, 1)               # output layer of PINN

    def forward(self, x):
        z = x
        for layer in self.hidden:
            z = self.actit(layer(z))

        N = self.out(z)
        DBC = self.DBCfun(x, self.vertices)

        return DBC * N   # trial solution, PINN is now with DBC enforced
```

3.5 Case Study: Equilateral Triangular Bar Subject to Torsion

We are now ready to build and train a PINN model for polygonal domains. Consider a bar with an equilateral triangle cross-section of edge length a, as shown in Fig.3.2. It is subjected to a torque resulting a twist angle per length in the bar.

When the source term in the Poisson's equation is set to 1, the stress function $u(x, y)$ (often denoted as ϕ) can be written as (Chapter 9 of [13]):

$$u = \frac{G\vartheta_t}{2h}\left(x - \sqrt{3}y - \frac{2h}{3}\right)\left(x + \sqrt{3}y - \frac{2h}{3}\right)\left(x + \frac{h}{3}\right) \tag{3.6}$$

where G is the shear modulus of the material and ϑ_t is the twist angle per unit length. Hence, $2G\vartheta_t$ is a constant which plays only a scaling role (see Eq.(2.51)). We can thus set it to 1 without loss of generality, meaning the source term in the Poisson's equation becomes simply 1.

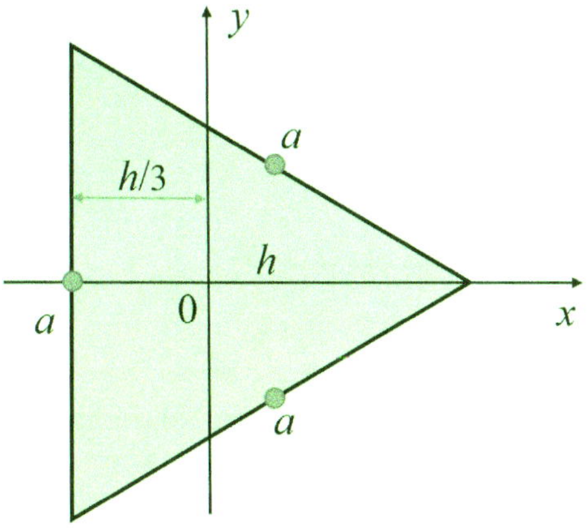

Figure 3.2: Equilateral triangle cross-section for a torsional bar.

Also, from the geometry of the equilateral triangle, we have $h = \frac{\sqrt{3}a}{2}$. Therefore, Eq.(3.6) simplifies to:

$$u = \frac{\sqrt{3}}{6a}\left(x - \sqrt{3}y - \frac{\sqrt{3}a}{3}\right)\left(x + \sqrt{3}y - \frac{\sqrt{3}a}{3}\right)\left(x + \frac{\sqrt{3}a}{6}\right) \tag{3.7}$$

This is the exact solution to validate our PINN model. The following snippet defines an equilateral triangular cross-section, the source term, and the exact solution for the stress function:

```
# Force or source function in torch for a torsional bar problem:
def f_th(x, y):
    return 1 + 0 * x * y                    # constant force term (f = 1)

# Exact Solution for a equilateral triangle cross-section torsional bar:
a = 1.0; h = np.sqrt(3)*a/2.0
vertices = [(-h/3,-a/2), (2*h/3,0.0), (-h/3,a/2)] # equilateral triangle

def exact_u(x, y, a=1.):
    sq3 = np.sqrt(3.0)
    return sq3*a/6 * (x-sq3*y-sq3*a/3)*(x+sq3*y-sq3*a/3)*(x + sq3*a/6)
```

3.5.1 Use Integral Residual by Trapezoidal Rule

Let us first train a PINN model using the generalized trapezoidal rule over the T3 mesh to compute the integral of the residual at the collocation points. The following snippet generates the sampling data:

```
# Initialization:
pinns = []                          # reset for logging the key variables:
timeAdam  = []; timeLbfgs = []
Trainers  = []; Residuals = []
```

```
totalTimes= []
error_L2s = []; errorVIs = []                          # log the error

gr.set_seed_th(88)                                # for reproducibility

# Sampling over an equilateral triangle cross-section:
points, elements = polygon_mesh(vertices)           # These are in NumPy

# Generate dataset: w_s_th, x_Ω_th, used throughout training
DI_rule = 'T3'
w_s_th, x_Ω_th, x_DBC_th, u_DBC_th, points, elements = \
          sampleData_T3_polygon(N_DBC=20, vertices=vertices, maxA=0.003)
print(f"Sampled {x_Ω_th.shape[0]} collocation points in the domain.")
print(f"Is N_sample >= P* ? {x_Ω_th.shape[0] >= P_star}")
print(f"Sampled {x_DBC_th.shape[0]} boundary points.")
print(f"Weights: {w_s_th.shape[0]}.")
```

```
No. of points = 125 | No. of elements = 216
Sampled 125 collocation points in the domain.
Is N_sample >= P* ? True
Sampled 60 boundary points.
Weights: 125.
```

The T3 mesh has 125 nodes, which are used as the collocation points. The following snippet builds and trains the PINN model:

```
w_PDE = 1.                                       # weight for loss_PDE term
p = 2; Nn=50; NL=2

# new ---PINN with DBCfun-- for polygon domain:
pinn = PINN2D_DBCfun(Nn, NL, DBCfun=polygon_DBCf_mult,vertices=vertices)
P_star, P_total = print_model_parameters(pinn, print_table=False)

# Train the pinn model with the residual is computed via integration:
x_DBC_th, u_DBC_th = None, None                                # not used
x_NBC_th, t_NBC_th, w_NBC_th = None, None, None                # not used
I_NBC = ''                               # no Integration is needed for NBC

Trainer  = trainer_int_PDE
Residual = residual_int                          # integrated squared residual
losses, totalTime = train_adam_lbfgs(pinn, Trainer, Residual, f_th,
       x_Ω_th, w_s_th, x_DBC_th, u_DBC_th, x_NBC_th, t_NBC_th, w_NBC_th,
                           I_NBC, epochs1=201, epochs2=11, use_lbfgs=True)
```

```
P* (Total pseudo-dimension of the affine space) = 105
Total trainable parameters: 2751
==================================================
201 epoches training with Adam:
Epoch     1, Loss: 4.608e-01, PDE: 4.608e-01
Epoch   101, Loss: 1.150e-04, PDE: 1.150e-04
Epoch   201, Loss: 9.425e-05, PDE: 9.425e-05
```

```
Time taken by adam: 1.534815 seconds

==================================================
11 epoches fine-tuning with LBFGS:
Epoch     1, Loss: 8.089e-08, PDE: 8.089e-08
Epoch    11, Loss: 8.089e-08, PDE: 8.089e-08

Training complete.
Time taken by lbfgs: 0.364717 seconds
Total time: 1.899532 seconds
Ratio (time_lbfgs/time_adam) of time taken per epoch:4.342136
```

3.5.2 Plots of Stress Function Solution

The training is successful. Let us use the following snippet to plot the solution along with the exact one and the error:

```
def plot_sln2_T3(pinn, exact_u, points, elements, a=1.0, f_name='tmp',
                                                          plot_it=True):
    '''Plot `pinn` model solution over a domain meshed with T3 with
       `elements` and `points` along with the exact solution and error.
    '''
    global error_L2, errorVI

    pinn.eval()                     # Ensure the model is in evaluation mode
    # Compute model predictions at points and convert to NumPy
    points_th = torch.tensor(points, dtype=torch.float32)
    with torch.no_grad():
        u_pred = pinn(points_th).detach().numpy().flatten()

    # Compute exact solution
    x_np = points
    u_exact = exact_u(x_np[:,0], x_np[:,1], a=a)

    error_L2, errorVI, u_diff = sln_error(u_pred, u_exact)

    if plot_it != True:
        return error_L2, errorVI, u_diff

    slns = [u_pred, u_exact, (u_pred - u_exact)]
    titles= ['PINN Solution', 'Exact Prediction', 'Difference']
    # Create triangulation object for plot:
    triang = mtri(points[:, 0], points[:, 1], elements)

    # Plot predicted and exact solution
    width = 8; height = width/3
    fig, axs = plt.subplots(1, 3, figsize=(width, height))
    for i, ax in enumerate(axs):
        ci = ax.tricontourf(triang, slns[i], levels=100, cmap='jet')
        ax.set_title(titles[i])                            # or 'viridis'
```

```
        ax.triplot(triang, color='black', linewidth=0.05)   # mesh lines
        #ax.plot(points[:, 0], points[:, 1], 'ko',markersize=0.1) #nodes
        plt.colorbar(ci, ax=ax, shrink=0.7)

        #ax.set_xlabel('$x$');  ax.set_ylabel('$y$')
        ax.set_aspect('equal'); ax.grid(False)

    plt.tight_layout()
    plt.savefig("imagesPN/"+f_name+".png",dpi=500,bbox_inches='tight')
    plt.show()

    return error_L2, errorVI, u_diff
```

```
# Plot the contours of both PINN solution and the exact one:
error_L2, errorVI, u_diff=plot_sln2_T3(pinn, exact_u, points, elements,
                            a=a, f_name='PoissonSqrT3int', plot_it=True)
plot3loss2(losses, ('adam initially','lbfgs in fine tunning'),
                    PDE='PDE', C1='DBC', C2='NBC', C3='Loss',log=False)
```

```
Maximum Error: 3.3906e-06
Mean Absolute Error: 9.0776e-07
Error Variation Index (ratio of Max/Mean): 3.7352e+00
L2 Relative Error: 9.0077e-05
```

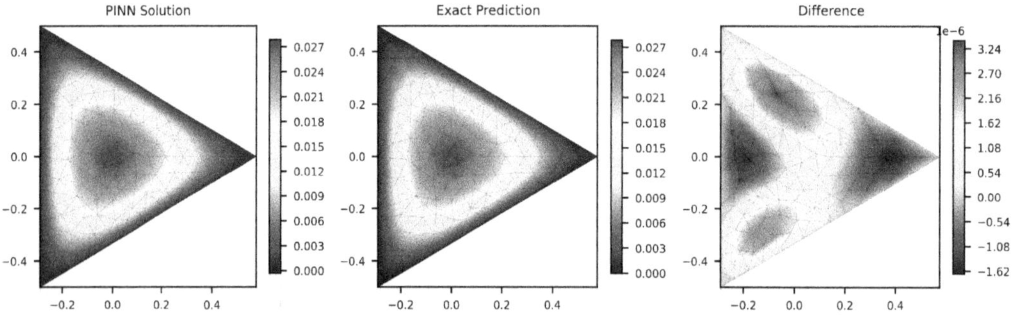

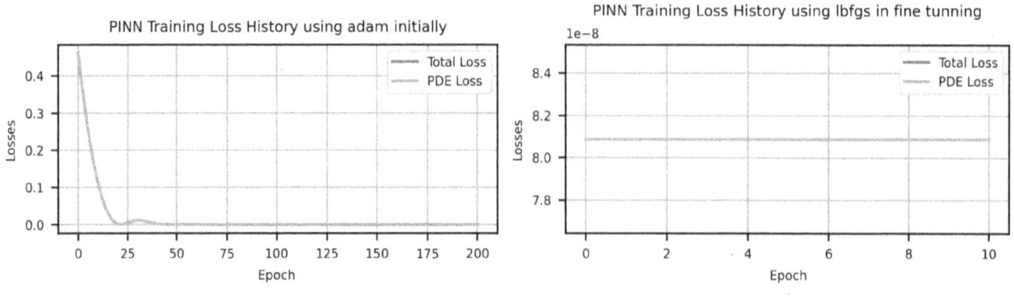

The relative accuracy of the PINN solution is very high—at the order of 10^{-4}—when using 125 points, which are the nodes of the T3 mesh for the equilateral triangle.

```
recordSheet()                                          # Update the record
```

```
printSummary(pinns, LossTerms, Residuals, totalTimes,error_L2s,errorVIs)
```

```
#  PINN model        Loss_terms residual    total_time errors_L2  errorVIs
==========================================================================
1  2D_DBCfun         PDE_only   _intT3      1.900e+00  9.008e-05  3.735e+00
```

3.5.3 Use Integral Residual by Gaussian Quadrature

We now train a PINN model using Gaussian quadrature over the T3 mesh to compute the integral residual at the collocation points. The following snippet generates the sampling data:

```
gr.set_seed_th(88)                                  # for reproducibility

# Generate dataset: w_s_th, x_Ω_th, used for training
DI_rule = 'T3GQ'
x_Ω, w_s  = gauss_quadrature_T3(points, elements, degree=2)
w_s_th = torch.tensor(w_s, dtype=torch.float32).reshape(-1,1)
x_Ω_th = torch.tensor(x_Ω, dtype=torch.float32)
print(w_s_th.shape, x_Ω_th.shape)

pinn = PINN2D_DBCfun(Nn, NL, DBCfun=polygon_DBCf_mult,vertices=vertices)

# The residual_int calculates the integrated squared residual:
Trainer  = trainer_int_PDE
Residual = residual_int
losses, totalTime = train_adam_lbfgs(pinn, Trainer, Residual, f_th,
       x_Ω_th, w_s_th, x_DBC_th, u_DBC_th, x_NBC_th, t_NBC_th, w_NBC_th,
                          I_NBC, epochs1=201, epochs2=11, use_lbfgs=True)
```

```
torch.Size([648, 1]) torch.Size([648, 2])
==================================================
201 epoches training with Adam:
Epoch     1, Loss: 4.608e-01, PDE: 4.608e-01
Epoch   101, Loss: 1.060e-04, PDE: 1.060e-04
Epoch   201, Loss: 8.660e-05, PDE: 8.660e-05
Time taken by adam: 1.698050 seconds

==================================================
11 epoches fine-tuning with LBFGS:
Epoch     1, Loss: 5.782e-08, PDE: 5.782e-08
Epoch    11, Loss: 5.782e-08, PDE: 5.782e-08

Training complete.
Time taken by lbfgs: 0.378931 seconds
Total time: 2.076980 seconds
Ratio (time_lbfgs/time_adam) of time taken per epoch:4.077677
```

The training is very fast and takes only a couple of seconds. The relative error is at the 10^{-5} order, as found using the following snippet:

```
# Compute the solution error:
error_L2, errorVI, u_diff =plot_sln2_T3(pinn, exact_u, points, elements,
                                f_name='PoissonSqrT3int', plot_it=False)
# plot3loss2(losses, ('adam initially','lbfgs in fine tunning'),
#                                         PDE='PDE',log=False) # optional
```

```
Maximum Error: 2.5459e-06
Mean Absolute Error: 7.0000e-07
Error Variation Index (ratio of Max/Mean): 3.6370e+00
L2 Relative Error: 6.9658e-05
```

```
recordSheet()                                       # Update the record
```

```
printSummary(pinns, LossTerms, Residuals, totalTimes,error_L2s,errorVIs)
```

```
#  PINN model        Loss_terms residual     total_time errors_L2  errorVIs
=========================================================================
1  2D_DBCfun         PDE_only   _intT3       1.900e+00  9.008e-05  3.735e+00
2  2D_DBCfun         PDE_only   _intT3GQ     2.077e+00  6.966e-05  3.637e+00
```

The performance of these two models is quite similar. The use of Gaussian quadrature produces a more accurate solution. The computation time and error variation index are at the same level.

3.5.4 Use Energy Residual

Because the energy functional is quadratic, we should use second-order Gaussian quadrature for each of the T3 elements. The snippet to generate the sampling data and train the model is as follows.

```
gr.set_seed_th(88)                                  # for reproducibility

# Generate dataset: x_Ω_th for training
DI_rule = ''
x_Ω, _= gauss_quadrature_T3(points, elements, degree=2)
x_Ω_th = torch.tensor(x_Ω, dtype=torch.float32)
print(x_Ω_th.shape)
print(f"Is N_sample >= P* ? {x_Ω_th.shape[0] >= P_star}")

pinn = PINN2D_DBCfun(Nn, NL, DBCfun=polygon_DBCf_mult,vertices=vertices)

# The residual_energy calculates the energy functional U_T:
x_DBC_th, u_DBC_th = None, None                                 # not used
x_NBC_th, t_NBC_th, w_NBC_th = None, None, None                 # not used
I_NBC = ''                              # no Integration is needed for NBC

Trainer  = trainer_int_PDE
Residual = residual_energy
```

```
losses, totalTime = train_adam_lbfgs(pinn, Trainer, Residual, f_th,
       x_Ω_th, w_s_th, x_DBC_th, u_DBC_th, x_NBC_th, t_NBC_th, w_NBC_th,
                         I_NBC, epochs1=201, epochs2=11, use_lbfgs=True)
```

```
torch.Size([648, 2])
Is N_sample >= P* ? True
==================================================
201 epoches training with Adam:
Epoch     1, Loss: 1.743e-04, PDE: 1.743e-04
Epoch   101, Loss: -2.706e-03, PDE: -2.706e-03
Epoch   201, Loss: -2.706e-03, PDE: -2.706e-03
Time taken by adam: 0.740221 seconds

==================================================
11 epoches fine-tuning with LBFGS:
Epoch     1, Loss: -2.706e-03, PDE: -2.706e-03
Epoch    11, Loss: -2.706e-03, PDE: -2.706e-03

Training complete.
Time taken by lbfgs: 0.034484 seconds
Total time: 0.774705 seconds
Ratio (time_lbfgs/time_adam) of time taken per epoch:0.851255
```

The training is very fast and took only a few seconds.

```
# Solution accuracy:
error_L2, errorVI, u_diff=plot_sln2_T3(pinn, exact_u, points, elements,
                              f_name='PoissonSqrT3int', plot_it=False)
# plot3loss2(losses, ('adam initially','lbfgs in fine tunning'),
#                                       PDE='PDE',log=False) # optional
```

```
Maximum Error: 1.4254e-04
Mean Absolute Error: 3.5858e-05
Error Variation Index (ratio of Max/Mean): 3.9751e+00
L2 Relative Error: 3.6504e-03
```

```
recordSheet()                                    # Update the record
```

```
printSummary(pinns, LossTerms, Residuals, totalTimes,error_L2s,errorVIs)
```

```
#  PINN model        Loss_terms residual    total_time errors_L2  errorVIs
==========================================================================
1  2D_DBCfun         PDE_only   _intT3      1.900e+00  9.008e-05  3.735e+00
2  2D_DBCfun         PDE_only   _intT3GQ    2.077e+00  6.966e-05  3.637e+00
3  2D_DBCfun         PDE_only   _energy     7.747e-01  3.650e-03  3.975e+00
```

The energy method gives a less accurate solution, but the training is faster.

3.5.5 Use Collocation Residual Based on T3 Mesh

Next, let us use a collocation scheme to compute the residual. In this case, we only need the sampling points inside the domain. Since we have a T3 mesh, we can simply use the nodes of this mesh as the collocation points. No integration will be performed. The snippet is as follows.

```
gr.set_seed_th(88)                                        # for reproducibility

# Generate dataset: x_Ω_th for training
DI_rule = ''
x_Ω, _= gauss_quadrature_T3(points, elements, degree=2)  # w_s: not used
x_Ω_th = torch.tensor(x_Ω, dtype=torch.float32)
print(x_Ω_th.shape)

pinn = PINN2D_DBCfun(Nn, NL, DBCfun=polygon_DBCf_mult,vertices=vertices)

x_DBC_th, u_DBC_th = None, None                                    # not used
x_NBC_th, t_NBC_th, w_NBC_th = None, None, None                    # not used
I_NBC = ''                                # no Integration is needed for NBC

Trainer  = trainer_int_PDE
Residual = residual_clctn
losses, totalTime = train_adam_lbfgs(pinn, Trainer, Residual, f_th,
       x_Ω_th, w_s_th, x_DBC_th, u_DBC_th, x_NBC_th, t_NBC_th, w_NBC_th,
                        I_NBC, epochs1=201, epochs2=11, use_lbfgs=True)
```

```
torch.Size([648, 2])
==================================================
201 epoches training with Adam:
Epoch     1, Loss: 1.060e+00, PDE: 1.060e+00
Epoch   101, Loss: 2.468e-04, PDE: 2.468e-04
Epoch   201, Loss: 2.030e-04, PDE: 2.030e-04
Time taken by adam: 1.768312 seconds

==================================================
11 epoches fine-tuning with LBFGS:
Epoch     1, Loss: 1.866e-07, PDE: 1.866e-07
Epoch    11, Loss: 1.866e-07, PDE: 1.866e-07

Training complete.
Time taken by lbfgs: 0.332993 seconds
Total time: 2.101305 seconds
Ratio (time_lbfgs/time_adam) of time taken per epoch:3.440954
```

```
# Compute the solution accuracy:
error_L2, errorVI, u_diff=plot_sln2_T3(pinn, exact_u, points, elements,
                             f_name='PoissonSqrT3int', plot_it=False)

# plot3loss2(losses, ('adam initially','lbfgs in fine tunning'),
#                                                  PDE='PDE',log=False)
```

```
Maximum Error: 3.6319e-06
Mean Absolute Error: 8.8252e-07
Error Variation Index (ratio of Max/Mean): 4.1153e+00
```

```
L2 Relative Error: 8.9014e-05
```

```
recordSheet()                                          # Update the record

printSummary(pinns, LossTerms, Residuals, totalTimes,error_L2s,errorVIs)
```

```
#  PINN model        Loss_terms residual     total_time errors_L2  errorVIs
===========================================================================
1  2D_DBCfun         PDE_only   _intT3       1.900e+00  9.008e-05  3.735e+00
2  2D_DBCfun         PDE_only   _intT3GQ     2.077e+00  6.966e-05  3.637e+00
3  2D_DBCfun         PDE_only   _energy      7.747e-01  3.650e-03  3.975e+00
4  2D_DBCfun         PDE_only   _clctn       2.101e+00  8.901e-05  4.115e+00
```

The accuracy of this model is on par with model-1. Among all four models, model-2, which uses Gaussian quadrature for the squared L^2 residual, produces the most accurate solution. Model-4, which uses pure collocation, yields a slightly less accurate solution but is the simplest model to build. In the following chapters we shall use pure collocation methods.

With the PINN trained for the stress function, we can obtain its derivatives using `compute_derivatives()` from the previous chapter, and then compute the shear stresses distributed over the triangular cross-section.

3.5.6 Computing Shear Stresses Using Trained PINN

We can now compute the shear stress and shear flows on the equilateral triangular cross-section using the trained `pinn`, which involves only differentiation. The following snippet computes all these and plots the results.

```
def plot2D Tri(pinn, points, elements, scale=8, dpi=300, fname_plt='tmp'):
    '''
    Plot shear stresses and shear flow using a PINN model for stress
    function. Two subplots: the vector field and magnitude surface.
    '''
    triang = mtri(points[:, 0], points[:, 1], elements)

    # Convert points to torch tensor
    x_int = torch.tensor(points, dtype=torch.float32)

    # Compute derivatives
    _, dphi_dx, dphi_dy, _, _ = compute_derivatives(pinn, x_int)

    # Compute  zx and  zy
    tau_zx = dphi_dy.detach().numpy().flatten()
    tau_zy = -dphi_dx.detach().numpy().flatten()
    magnitude = np.sqrt(tau_zx**2 + tau_zy**2)                  # 1D array

    width = 7; height = width/2.4
    fig = plt.figure(figsize=(width, height), dpi=dpi)

    ax1 = fig.add_subplot(1, 2, 1)                             # Vector Field
    ax1.triplot(triang, color='b', linestyle='-', linewidth=0.1)
```

```
qv = ax1.quiver(points[:, 0], points[:, 1], tau_zx,tau_zy,magnitude,
                        scale=scale, width=0.006, cmap='Oranges')
ax1.set_xlabel('x')
ax1.set_ylabel('y')
ax1.set_title('Shear flow pattern')
ax1.set_aspect('equal')
fig.colorbar(qv, ax=ax1,shrink=0.7,label=r'$(\tau_{zx},\tau_{zy})$')

ax2 = fig.add_subplot(1, 2, 2, projection='3d') # Surface Plot Mgntd
surf = ax2.plot_trisurf(points[:, 0], points[:, 1], magnitude,
  triangles=elements, cmap='Oranges',linewidth=0.1,antialiased=True)
ax2.set_title('Shear stress magnitude')
ax2.set_xlabel('x')
ax2.set_ylabel('y')
plt.tight_layout()
plt.savefig(fname_plt+'png', dpi=500, bbox_inches='tight')
plt.show()
```

3.5.7 Plots of shear flow

```
# Plot the shear flow on the equilateral trianglar cross-setion:
plot2D Tri(pinn, points, elements, scale=1, dpi=500,
                        fname_plt='imagesPN/ShearFlowOnTriangle')
```

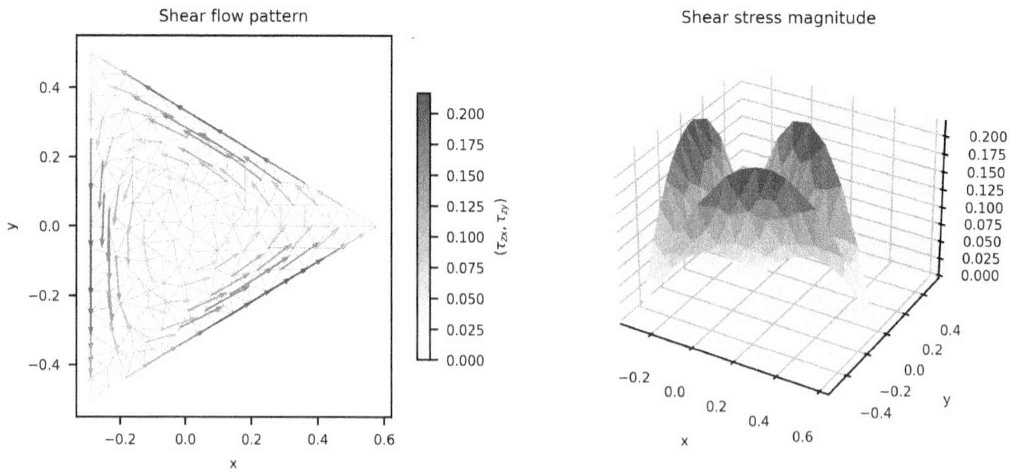

It is seen that the shear stress flows along the edges of the equilateral triangle, with the largest shear flow near the midpoints of each edge. The shear stress is zero at the center and at the vertices. All these observations are consistent with the correct mechanics [13].

3.6 Case Study: Pentagonal Bar Subject to Torsion

For this torsional bar problem governed by Poisson's equation, the force term is a constant 1, and the solution will be the stress function [13]. Although the exact solution is not available, the shear stress pattern over the pentagon can be qualitatively assessed based on solid mechanics principles.

3.6.1 Stress Function Solution Using Integral Residual

Let us first use the integral residual approach to train a PINN model.

```
# Force or source function in torch
def f_th(x, y):
    return 1 + 0*x*y                           # constant force term (C = 1)

p = 2; Nn=50; NL=2
# new PINN with DBCfun for polygon domain:
pinn = PINN2D_DBCfun(Nn, NL, DBCfun=polygon_DBCf_mult,vertices=vertices)
P_star, P_total = print_model_parameters(pinn, print_table=False)

# Define Polygon Vertices of a polygon with arbitrary vertices:
vertices = np.array([(0., 0.), (1., 0.), (1.5, .8), (.5,1.5), (-.5,.8)])
points, elements = polygon_mesh(vertices)             # This is in NumPy

# Generate dataset: w_s_th, x_Ω_th
w_s_th, x_Ω_th, x_DBC_th, u_DBC_th, points, elements = \
         sampleData_T3_polygon(N_DBC=20, vertices=vertices, maxA=0.01)
print(f"Sampled {x_Ω_th.shape[0]} collocation points in the domain.")
print(f"Is N_sample >= P* ? {x_Ω_th.shape[0] >= P_star}")
print(f"Sampled {x_DBC_th.shape[0]} boundary points.")
print(f"Weights: {w_s_th.shape[0]}.")

Trainer  = trainer_int_PDE
Residual = residual_int                        # integrated squared residual
losses, totalTime = train_adam_lbfgs(pinn, Trainer, Residual, f_th,
       x_Ω_th, w_s_th, x_DBC_th, u_DBC_th, x_NBC_th, t_NBC_th, w_NBC_th,
                         I_NBC, epochs1=201, epochs2=11, use_lbfgs=True)
```

```
P* (Total pseudo-dimension of the affine space) = 105
Total trainable parameters: 2751
No. of points = 170 | No. of elements = 303
Sampled 170 collocation points in the domain.
Is N_sample >= P* ? True
Sampled 100 boundary points.
Weights: 170.
==================================================
201 epoches training with Adam:
Epoch     1, Loss: 1.068e+00, PDE: 1.068e+00
Epoch   101, Loss: 5.069e-01, PDE: 5.069e-01
Epoch   201, Loss: 4.565e-01, PDE: 4.565e-01
Time taken by adam: 2.978384 seconds
```

```
==================================================
11 epoches fine-tuning with LBFGS:
Epoch      1, Loss: 7.230e-02, PDE: 7.230e-02
Epoch     11, Loss: 1.153e-02, PDE: 1.153e-02

Training complete.
Time taken by lbfgs: 42.807238 seconds
Total time: 45.785622 seconds
Ratio (time_lbfgs/time_adam) of time taken per epoch:262.627322
```

3.6.2 Plots of Stress Function

The training took a couple of minutes to complete. The following snippet plots the contour of the stress function $u(x, y)$ over the pentagonal domain. Since the exact solution is not available this time, we plot only the predicted solution from the trained PINN model.

```
def plot2_sln1_T3(pinn, points, elements, fname_plt='tmp'):
    '''
    Plot `pinn` result over a domain meshed with T3 using both
    contour (left) and surface (right) plots.
    '''
    pinn.eval()
    points_th = torch.tensor(points, dtype=torch.float32)
    with torch.no_grad():
        u_pred = pinn(points_th).detach().numpy().flatten()

    triang = mtri(points[:, 0], points[:, 1], elements)  # Triangulation

    width = 7; height = width/2.4
    fig = plt.figure(figsize=(width, height))
    ax1 = fig.add_subplot(1, 2, 1)                        # Contour Plot
    tcf = ax1.tricontourf(triang, u_pred, levels=100, cmap='jet')
    ax1.triplot(triang, color='black', linewidth=0.02)
    ax1.set_aspect('equal')
    ax1.set_title("Contour Plot")
    fig.colorbar(tcf, ax=ax1, shrink=0.8, label=r'Solution $u(x, y)$')

    ax2 = fig.add_subplot(1, 2, 2, projection='3d')          # 3D Surface
    surf = ax2.plot_trisurf(triang, u_pred, cmap='jet', linewidth=0.1,
                                                   antialiased=True)
    ax2.set_title("Surface Plot")
    plt.tight_layout()
    plt.savefig(fname_plt+'png', dpi=500, bbox_inches='tight')
    plt.show()
```

```
# Plot the contours of both PINN solution and the exact one:
plot2_sln1_T3(pinn,points,elements,fname_plt='imagesPN/PoissonPenT3int')
plot3loss2(losses, ('adam initially','lbfgs in fine tunning'),
                                                PDE='PDE',log=False)
```

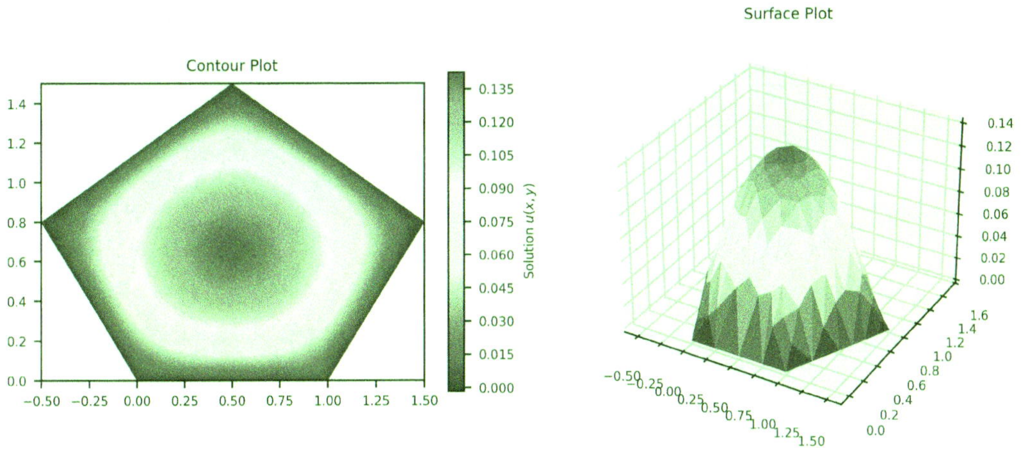

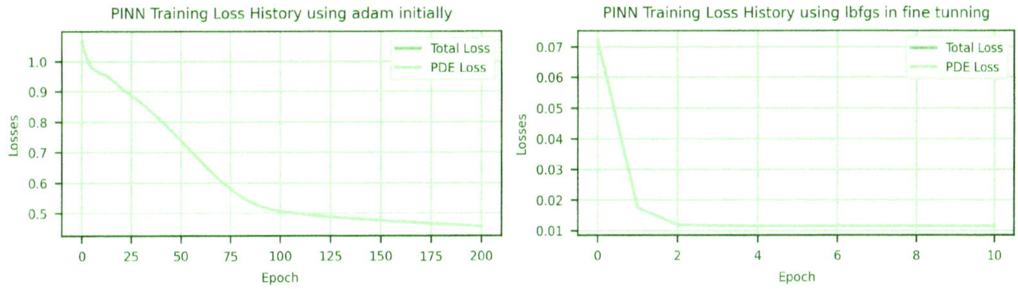

3.6.3 Shear Stress Solution Using Trained PINN

With the PINN trained for the stress function, we can obtain its derivatives using the same `compute_derivatives()` function, and then compute the shear stresses distributed over the pentagonal cross-section.

```
# Plot the shear flow on the Pentagon cross-setion:
plot2DTri(pinn, points, elements, scale=5, dpi=500,
                                        fname_plt='imagesPN/ShearFlowOnPetagon')
```

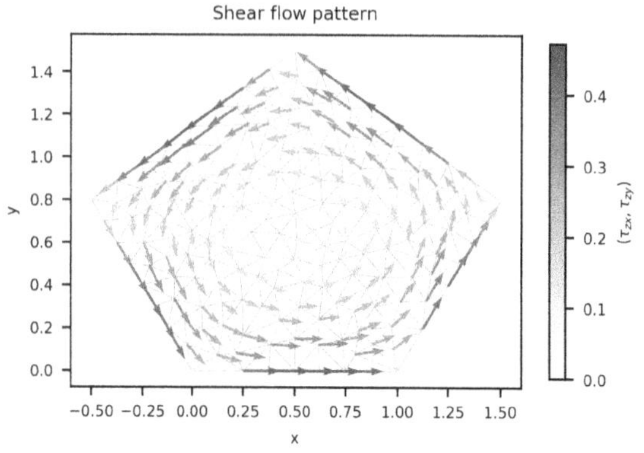

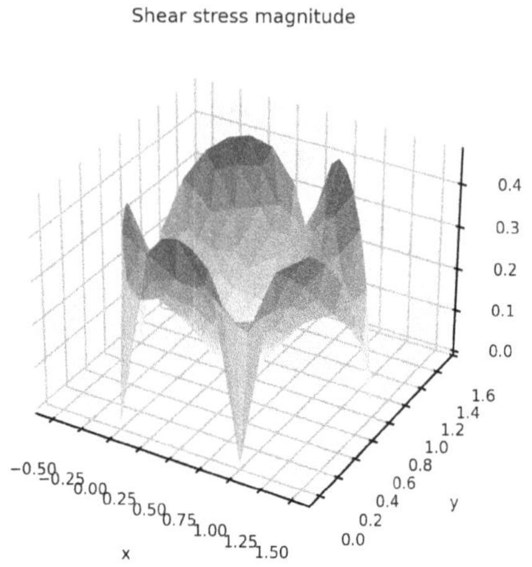

It is seen that the shear stress flows along the edge of the pentagon, the largest shear flow is near the middle points of an edge, and the shear stress is zero at the center. All these present the correct mechanics [13].

3.7 DBC Function Generated Using a Neural Network

For more complicated domain shapes, we may not have an explicit function for the boundary. One effective approach is to use another neural network (NN) model to produce an NN function representing the boundary. We can then use this function to create PINN models.

3.7.1 Training DBC Boundary NN

The following snippet trains an NN function that is a smooth 2D function in (x, y) and is zero along an elliptic boundary. The procedure to train such an NN function is as follows:

1. Use the formula Eq.(3.1) of the ellipse as the target function. Generate a set of labeled values using this target function.
2. Build a plain multilayer perceptron (MLP) neural network that takes (x, y) as input and produces the NN function.

The NN can be built using the plain `PINN2D` class template given in the previous chapter.

```
# Build and Train the model
import torch.optim as optim

# Generate training data
ra = 0.5; rb = 1.0                          # radii of the ellipse

def f_ellipse(x, y):
    '''Target function of ellipse'''
    return 1 - (x**2 / ra**2 + y**2 / rb**2)

N_points = 20000      # number of data points to train the 2D NN function
 = 1.2                                  # for scaling up the domain size
```

```
x_vals = np.random.uniform(-ra *  , ra *  , N_points)
y_vals = np.random.uniform(-rb *  , rb *  , N_points)
z_vals = f_ellipse(x_vals, y_vals)

X_train = np.stack([x_vals, y_vals], axis=1)
y_train = z_vals.reshape(-1, 1)

X_train_tensor = torch.tensor(X_train, dtype=torch.float32)
y_train_tensor = torch.tensor(y_train, dtype=torch.float32)

model = PINN2D(Nn=18, NL=2)        # use the plaen PINN2D to built the NN
optimizer = optim.Adam(model.parameters(), lr=1e-3)
loss_fn = nn.MSELoss()
```

```
epochs = 20001
for epoch in range(epochs):
    model.train()
    optimizer.zero_grad()
    output = model(X_train_tensor)
    loss = loss_fn(output, y_train_tensor)
    loss.backward()
    optimizer.step()
    if epoch % 10000 == 0 or epoch == epochs:
        print(f"Epoch {epoch}: loss = {loss.item():.4e}")

#executed in 1m 24.4s, finished 11:18:17 2025-06-29
```

```
Epoch 0: loss = 3.9988e-01
Epoch 10000: loss = 4.9049e-06
Epoch 20000: loss = 1.4620e-06
```

The model is well trained with the loss function reduced by 5 orders of magnitude. We now save the NN function model for later use, which can then be called as a normal Python function:

```
# Convert to TorchScript and save to file for future use:
example_input = torch.rand(1, 2)
traced_model = torch.jit.trace(model, example_input)
traced_model.save("ellipseH_nn_fun_.pt")
# ellipseH_nn_fun.pt is a correctly trained and saved by the author
```

Let's load the trained model as a regular function and plot both the trained and the exact ellipse.

```
# Load the tained model for the boundary curve, before using it:
loaded_model = torch.jit.load("ellipseH_nn_fun_.pt")
loaded_model.eval()

res = 400                                          # resolution for plots
  = 1.2                                 # for scaling up the domain size
x = np.linspace(-ra *  , ra *  , res)
y = np.linspace(-rb *  , rb *  , res)
```

```
X, Y = np.meshgrid(x, y)
grid = np.stack([X.ravel(), Y.ravel()], axis=1)
grid_tensor = torch.tensor(grid, dtype=torch.float32)

with torch.no_grad():           # call the trained NN function as usual
    Z_pred = loaded_model(grid_tensor).numpy().reshape(X.shape)

plt.figure(figsize=(3, 5))           # Plot trained vs exact zero level
cont = plt.contourf(X, Y, Z_pred, levels=100, cmap='coolwarm')
plt.colorbar(cont, label='$f_{NN}(x, y)$')

# Predicted zero level (f = 0)
plt.contour(X, Y, Z_pred, levels=[0.0], colors='k', linewidths=2, \
                                                    linestyles='--')
# Exact zero level
plt.contour(X, Y, f_ellipse(X, Y), levels=[0.0], colors='lime', \
                                        linewidths=1, linestyles='-')
plt.title("Trained vs Exact Elliptic Curve $f(x, y) = 0$")
plt.xlabel("x");    plt.ylabel("y")
plt.axis('equal'); plt.grid(True)
plt.show()
```

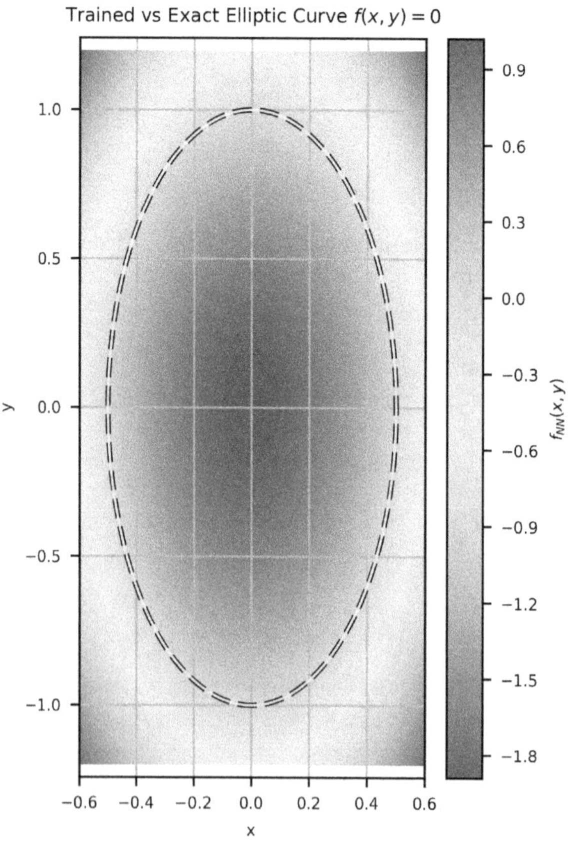

It is evident that the ellipse has been accurately learned by the MLP. We now have a neural network function that defines the boundary of the elliptical domain.

3.7.2 PINN Model Using NN DBC Function

We have trained another NN model to represent the boundary of an ellipse with radii $r_a = 1.0$ and $r_b = 0.5$. The trained NN model was saved to the file named `ellipseH_nn_fun.pt`. Let us use this NN function as a tool to impose the Dirichlet boundary condition (DBC) and train a PINN model for a bar with an elliptic cross-section subjected to torsion.

The following defines a PINN class that uses the NN function to impose the DBC on the edge of the domain.

```
class PINN2D_NN_DBC_ellipse(nn.Module):
    """
    PINN class for elliptic domain with BC enforced via trial solution.
    """
    def __init__(self, Nn, NL, BC_model_fn):
        super().__init__()
        self.hidden = nn.ModuleList(
            [nn.Linear(2, Nn)] + [nn.Linear(Nn,Nn) for _ in range(NL-1)]
        )
        self.out = nn.Linear(Nn, 1)

    def forward(self, x):
        x_, y_ = x[:, 0:1], x[:, 1:2]

        # Equivalent of column-wise flatten and stacking
        grid = torch.cat([x_.reshape(-1, 1), y_.reshape(-1, 1)], dim=1)

        # Compute BC function from the external model/function
        space_DBC = BC_model_fn(grid)   # You must define this elsewhere


        z = x
        for layer in self.hidden:
            z = torch.tanh(layer(z))
        N = self.out(z)

        u_trial = space_DBC * N
        return u_trial
```

The following snippet create the dataset for a PINN model.

```
# Data preparation:
ra, rb = 0.5, 1.0 # Ellipse parameters
# Sample data
w_s_th, x_Ω_th, x_DBC_th, u_DBC_th=sampleData_T3_ellipse(
                                                ra=ra, rb=rb, N_DBC=200)
loaded_NN = torch.jit.load("ellipseH_nn_fun.pt")
loaded_NN.eval()
```

```
with torch.no_grad():
    BC_model_fn = loaded_NN
```

In the preceding snippet, we also define the exact function for the elliptical boundary, which can be used for comparison purposes. The following snippet trains a PINN using the NN function to impose the DBC.

```
#                        --Run PINN2D_NN_DBC_ellipse, NN for DBC--
def f_th(x, y):
    return 1 + 0*x*y                                   # source term, set C = 1

# Sample data
w_s_th, x_Ω_th,_,_,=sampleData_T3_ellipse(ra=ra, rb=rb, N_DBC=200)
print(x_Ω_th.shape, w_s_th.shape)

Nn=50; NL=2
pinn = PINN2D_NN_DBC_ellipse(Nn, NL, BC_model_fn)                # use NN DBC

# Train the pinn:
Trainer  = trainer_int_PDE
Residual = residual_energy
losses, totalTime = train_adam_lbfgs(pinn, Trainer, Residual, f_th,
        x_Ω_th, w_s_th, x_DBC_th, u_DBC_th, x_NBC_th, t_NBC_th, w_NBC_th,
                          I_NBC, epochs1=201, epochs2=11, use_lbfgs=True)
```

```
torch.Size([1255, 2]) torch.Size([1255, 1])
==================================================
201 epoches training with Adam:
Epoch     1, Loss: -1.873e-02, PDE: -1.873e-02
Epoch   101, Loss: -3.919e-02, PDE: -3.919e-02
Epoch   201, Loss: -3.927e-02, PDE: -3.927e-02
Time taken by adam: 1.622839 seconds

==================================================
11 epoches fine-tuning with LBFGS:
Epoch     1, Loss: -3.927e-02, PDE: -3.927e-02
Epoch    11, Loss: -3.928e-02, PDE: -3.928e-02

Training complete.
Time taken by lbfgs: 0.403395 seconds
Total time: 2.026235 seconds
Ratio (time_lbfgs/time_adam) of time taken per epoch:4.542121
```

3.7.3 Plots of Stress Function Solution

The training completes in less than 2 seconds. Let's compute the solution error and plot the result.

```
# Plot results:
exact_u = lambda x, y: exact_u_ellipse(x, y, ra=ra, rb=rb)
```

```
error_L2, errorVI, u_diff = plot2D_ellipse(pinn, exact_u, ra=ra, rb=rb,
                            N_plot=500, shrink=0.2, dpi=300, plot_it=True)

plot3loss2(losses, ('adam initially','lbfgs in fine tunning'),
                   PDE='PDE', C1='DBC', C2='NBC', C3='Loss',log=False)
```

```
Maximum Error: 7.7974e-04
Mean Absolute Error: 2.1673e-04
Error Variation Index (ratio of Max/Mean): 3.5978e+00
L2 Relative Error: 4.8317e-03
```

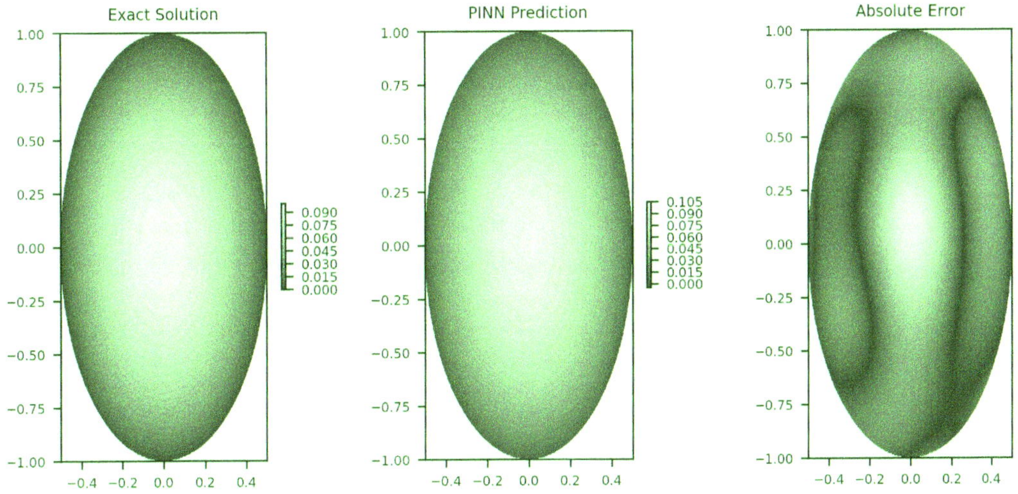

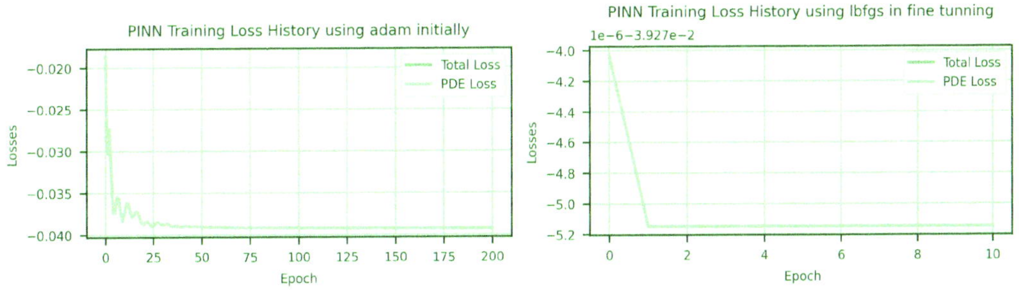

The accuracy of the trained model solution is of order 10^{-3}, which is quite good but less accurate compared to the use of the exact elliptic boundary function studied in Section 3.3.

3.8 Remarks

This chapter presents various PINN models for Poisson's PDEs, which govern a wide range of problems including solid mechanics, steady-state heat transfer , and electrostatics, just to name a few. Our focus is placed on irregular domains, and PINN models for elliptical, triangular, and general polygonal domains are built, trained, and used to infer useful solutions such as shear stresses distributed in bars with various cross-sectional shapes. The following remarks can be drawn:

1. Triangular meshes can be effectively used to train PINN models using the L^2 residual via pure collocation, integrated squared L^2 residual, or energy residual.

2. Accurate PINN models can be built by enforcing DBCs using the trial solution form at the class level.

3. When using the integral residual, the trapezoidal rule and Gaussian quadrature exhibited mixed performance based on the examples studied in the previous chapter and this chapter.

4. A separate neural network can be trained to model the boundary of curved irregular domains. This NN function can then be used to build a PINN model with DBCs enforced via the trial solution form at the class level.

The following chapter builds PINN models for parabolic PDEs . Since the collocation method has been consistently found to be more accurate and the easiest to implement in Chapter 2 and this chapter, it will be used in the following chapters for various PDEs.

Chapter 4

PINN for Transient Heat Transfer Equations

Let us import the necessary external modules or dependencies for later use in this chapter.

```
# Place cursor in this cell, and press Ctrl+Enter to import dependencies.
import os
import sys                                # for accessing the computer system
sys.path.append('../grbin/')                  # Add in the path to your system

from commonImports import *            # Import dependencies from '../grbin/'
import grcodes as gr                        # Import the module of the author
importlib.reload(gr)                     # When grcodes is modified, reload it
import time
import math

init_printing(use_unicode=True)           # For latex-like quality printing
np.set_printoptions(precision=4, suppress=True)  # Digits in print-outs
```

The heat transfer problem is one of the most commonly encountered in science and engineering, and it has been extensively studied using conventional methods, such as the finite element method (Chapter 12 of Ref. [21]) and the smoothed finite element method (Chapter 14 of Ref. [20]).

With the techniques developed in the previous chapters, we now create PINN models for solving heat transfer problems. Given the strong performance and simplicity of the collocation-based models observed earlier, the training data in this chapter is obtained via collocation sampling, guided by the samples–neurons theorem [12].

4.1 Transient Heat Transfer Problems

4.1.1 Governing PDE

Consider a unit square plate with a heat source that may vary over the spatial coordinates and change over time. The temperature field u must also vary in space and time. It is governed by the following **time-dependent heat equation** PDE:

$$\frac{\partial u}{\partial t} = \underbrace{c\left(\frac{\partial^2 u}{\partial x^2} + \frac{\partial^2 u}{\partial y^2}\right)}_{\Delta u} + \underbrace{f(x, y, t)}_{\text{force/source}}, \quad (x, y, t) \in (0, 1)^2 \times (0, t_{\text{end}}] \tag{4.1}$$

in which:

- $f(x, y, t)$ is a given function that determines how the **heat source** or **heat sink** varies in space and time.

- $u(x, y, t)$ is the unknown temperature function resulting from the heat source, and hence depends on both spatial coordinates and time.

- c is a constant that is a material property of the plate, which is known for a given problem. For aluminum, for example, $c = \frac{\text{heat-conductivity}}{\text{density}\times\text{specific-heat-capacity}}$ (Chapter 14 of Ref. [20]), whose nominal value is about 4.2×10^{-4} [m^2/s].

Equation (4.1) is a PDE defined in the **space-time domain** (x, y, t) with $(x, y) \in \Omega$ and $t \in (0, t_{\text{end}})$, where t_{end} is the end time of the transient problem. It can be viewed as a three-dimensional (3D) problem. The derivative of $u(x, y, t)$ with respect to time is first-order, and with respect to spatial coordinates is second-order. This problem is also called **transient heat transfer** problem.

When f is not time-dependent and the boundary conditions are independent of time, u will also not depend on time. In such a case, the problem becomes one of steady-state temperature distribution, and Eq.(4.1) reduces to the Poisson's equation discussed in detail in the previous chapters.

The heat transfer equation like Eq.(4.1) is a type of **parabolic PDE** , which governs thermal diffusion processes. In addition to heat diffusion, there are many other types of diffusion problems in science and engineering, including:

- mass/population diffusion,
- viscous diffusion in fluids,
- magnetic diffusion in conductors,
- neutron diffusion in nuclear reactors,

just to name a few. The techniques for developing PINN models discussed in this chapter are, in principle, applicable to all these types of problems.

Depending on the actual problem, Eq.(4.1) is defined along with Dirichlet boundary conditions (DBCs), initial conditions (ICs), and possibly some Neumann boundary conditions (NBCs).

4.1.2 Dirichlet Boundary Conditions

The unknown temperature function $u(x, y)$ needs to satisfy **Dirichlet boundary conditions** (DBCs), which specify the temperature values of $u(x, y)$ on part or all of the boundary of the spatial domain. It is typically expressed as:

$$u(x, y) = \bar{u}, \quad \text{on } \partial\Omega_{\text{DBC}} \tag{4.2}$$

where $\bar{u}$ is a prescribed temperature value on part of the boundary of Ω, denoted by $\partial\Omega_{\text{DBC}}$. It is often set to zero, though it need not be.

Since the heat equation also involves the Laplace operator, as in Poisson's equation, the formulas for NBCs on $\partial\Omega_{\text{NBC}}$ are the same as those discussed in Chapter 2.

4.1.3 Initial Conditions

Equation (4.1) is time-dependent, and since the unknown function involves a first-order time derivative, we must specify an initial condition (IC) that prescribes the temperature distribution inside the domain at $t = 0$. Assume the initial condition is given as

$$f_{\text{IC}}(x, y) = u_{\text{IC}}(x, y, 0) = \sin(\pi x)\sin(\pi y) \tag{4.3}$$

which is a 2D function of x and y.

The following snippet defines the IC for our heat transfer problem.

```
def f_IC(x, y):
    """ Defines the Initial Condition (IC) for the heat problem. """
    return torch.sin(np.pi*x) * torch.sin(np.pi*y)
```

4.1.4 Exact Solution

When

- $f(x, y, t) = 0$ in Eq.(4.1),
- the DBCs are set as $\bar{u} = 0$ on all edges of the unit square domain, and
- the IC is given in Eq.(4.3),

the problem governed by Eq.(4.1) has the following **exact solution**:

$$u_{\text{exact}}(x, y, t) = \underbrace{\exp(-2\pi^2 c t)}_{\text{amplitude changing in time}} \underbrace{\sin(\pi x)\sin(\pi y)}_{\text{spatial variation}} \tag{4.4}$$

This can be easily confirmed by substituting Eq.(4.4) into Eqs.(4.1), (4.2), and (4.3).

The following snippet defines the exact solution in Python for later use.

```
def exact_u_heat(x, y, t):
    '''Exact solution for the transient heat transfer problem.'''
    return np.exp(-2*np.pi**2 *t) * np.sin(np.pi*x) * np.sin(np.pi*y)
```

4.1.5 Time Variation of the Solution

It is clear from Eq.(4.4) that the amplitude of the solution is controlled by the exponential time-dependent term, since the sine function is bounded by 1. The amplitude of the heat-transfer solution **decays exponentially** with time, which is a characteristic signature of this type of problems.

The constant c should be positive for the solution to remain bounded as time increases. Additionally, $c\,t$ must remain relatively small in practical problems; otherwise, the exponential term $\exp(-2\pi^2 c\,t)$ decays rapidly toward zero.

To simplify the analysis—without loss of generality—we set $c = 1$ in the following studies. This simplification implies that t should be chosen sufficiently small when setting up our PINN models. For predictions in actual problems with a given c value, we can simply replace t in the trained PINN by ct.

To construct proper PINN models for heat-transfer problems, we introduce a **characteristic time** defined as

$$t_{\text{char}} = \frac{1}{2\pi^2}. \tag{4.5}$$

The following snippet computes the value of t_{char} and evaluates the solution amplitude at physical times equal to 1, 5, and 10 times this characteristic value.

```
pi22 = 2*np.pi**2
t_char = 1/pi22                                        # characteristics time
print(f"The characteristics time: {t_char:.6e}")
print(f"u_amplitude at t_char   = {np.exp(-pi22*t_char):.6e}")
print(f"u_amplitude at 5*t_char = {np.exp(-pi22*(5*t_char)):.6e}")
print(f"u_amplitude at 10*t_char= {np.exp(-pi22*(10*t_char)):.6e}")
```

```
The characteristics time: 5.066059e-02
u_amplitude at t_char   = 3.678794e-01
u_amplitude at 5*t_char = 6.737947e-03
u_amplitude at 10*t_char= 4.539993e-05
```

As seen, the solution amplitude becomes practically zero when the time reaches $10\,t_{\text{char}}$. Even at $5\,t_{\text{char}}$, the amplitude is already very small. Therefore, when building a PINN model, we can set $t_{\text{end}} = 5\,t_{\text{char}}$ to avoid sampling unnecessarily many points for values that are close to zero.

Now that we have a well-defined problem with the PDE, DBC, and IC, the setup becomes a unit square plate with an initial temperature distribution given in Eq.(4.3), and all its edges are maintained at zero degrees. The temperature within the plate then decays to zero at an exponential rate. The exact solution to this problem is provided in Eq.(4.4).

This makes it an ideal benchmark problem for evaluating PINN models constructed in this chapter. The accuracy of any resulting PINN solution can be quantitatively assessed.

4.2 Strategy for Building PINN Models

4.2.1 Trial-Solution Form for ICs in a PINN

As shown in the previous chapter, using a trial solution is an effective strategy for training a PINN model for Poisson equations. A similar approach applies to our time-dependent heat-transfer problem, where the initial condition in Eq.(4.3) is enforced exactly with the following **trial solution**:

$$u_{\text{trial}}(x, y, t) = (1-t)\, u_{\text{IC}}(x, y, 0) + t\, \mathcal{N}_{\hat{\mathbf{w}}}(x, y, t) \tag{4.6}$$

Here, $\mathcal{N}_{\hat{\mathbf{w}}}(x, y, t)$ denotes the neural network in the PINN, parameterized by the trainable vector $\hat{\mathbf{w}}$, and producing an output for any input triplet (x, y, t). Clearly, $u_{\text{trial}}(x, y, 0) = f_{\text{IC}}(x, y)$ regardless of what $\mathcal{N}_{\hat{\mathbf{w}}}$ outputs. We therefore take Eq.(4.6) as the PINN's output, automatically satisfying the initial condition.

4.2.2 Trial-Solution Form for Both ICs and BCs

The technique for enforcing Dirichlet boundary conditions (DBCs) is the same as that discussed for Poisson equations. Assuming zero DBCs on all four edges of the square domain $[0, 1] \times [0, 1]$, we multiply the output of $\mathcal{N}_{\hat{\mathbf{w}}}(x, y, t)$ by a polynomial given by

$$f_{\text{BDC}}(x, y) = x(1-x)\, y(1-y) \tag{4.7}$$

Combining the treatments for both ICs and DBCs, we obtain the trial solution for the time-dependent heat-transfer problem:

$$u_{\text{trial}}(x, y, t) = (1-t)\, f_{\text{IC}}(x, y) + t\, \underbrace{x(1-x)\, y(1-y)}_{f_{\text{BDC}}(x,y)}\, \mathcal{N}_{\hat{\mathbf{w}}}(x, y, t) \tag{4.8}$$

The factor $x(1-x)\, y(1-y)$ ensures $u = 0$ along all boundary edges at all times. This polynomial appeared in the previous chapter. Alternatively, one may use the trigonometric function

$$f_{\text{BDC}}(x, y) = \sin(\pi x) \sin(\pi y) \tag{4.9}$$

which also satisfies the DBCs. The neural network $\mathcal{N}_{\hat{\mathbf{w}}}$ must adapt its parameters accordingly when paired with either Eq.(4.7) or Eq.(4.9). When Eq.(4.8) is adopted in constructing a PINN class, both the DBC and IC are satisfied automatically, eliminating the need for separate loss terms for them.

4.3 Procedure and Codes for PINN Models

4.3.1 Preparation for Coding

The following is a list of Python functions developed in the previous chapter. For convenience and reusability, these functions have been organized into a custom module named `grcodes`. We can import this module to directly utilize its functionalities in this chapter without redefining the same code. The following snippet imports them for use in this chapter:

```
from grcodes import train_adam_lbfgs, printSummary, plot3loss2, \
                    sln_error, print_model_parameters
```

Readers may review the previous chapters for the functionality of these snippets. We will not repeat the details here.

Let us also set the seed for consistent random sampling and initialization in our PINN models, as we did in the previous chapter:

```
gr.set_seed_th(seed=88)                    # Set seeds for reproducibility
```

The following snippet defines the device for training PINN models. This ensures that CUDA is used for fast computation if it is available; otherwise, the CPU is used.

```
# Device management: ensure CUDA is available if you plan to use GPU
device = torch.device('cuda' if torch.cuda.is_available() else 'cpu')
gr.device = device                         # inject device to gr module
torch.cuda.empty_cache()                   # Clear previous model state
print(f"Using device: {device}")
```

```
Using device: cpu
```

We are now ready to build and train PINN models.

4.3.2 A PINN Class for Auto-Enforced DBC and IC

The following code snippets define a PINN class, `PINN2D_DBC_IC`, which can be trained to solve 2D time-dependent heat equations. This model automatically satisfies the following conditions through its class-level construction:

- Initial condition (IC)
- Dirichlet boundary condition (DBC)

We begin by defining the necessary functions required for our PINN models.

```
# User defined activation function
def my_actvtn(x):
    return torch.tanh(x) #exp(x) # cos(x), sin(x), ...

# Define the function for the boundary with DBC is prescribed.
def f_DBC(x, y):
    """Defines the boundary function for the unit square for application
       of the Dirichlet Boundary Condition (DBC) on these edges. """
    #return x*(1-x)*y*(1-y)              # alternative
    return torch.sin(np.pi*x) * torch.sin(np.pi*y)
```

The following snippet defines a PINN class for creating PINN models that automatically satisfy the given DBC and IC.

```
class PINN2D_DBC_IC(torch.nn.Module):
    '''Creat a class of PINN models that automatically satisfies
       DBC and IC.
    '''
    def __init__(self, Nn, NL, f_DBC, f_IC, my_actvtn):
        '''Nn:          Number of neurons in a hidden layer~
           NL:          Number of (hidden) Layers
           f_DBC:       function (x, y) of the boundary with zero DBC
           f_IC:        function of (x, y) of u at t=0
           my_actvtn:   function of (x, y, t) user defined activation fun
                        functions are all in torch
        '''
        super().__init__()
        self.f_DBC = f_DBC
        self.f_IC = f_IC
        self.my_actvtn = my_actvtn
        self.hidden = torch.nn.ModuleList([torch.nn.Linear(3, Nn)] +
                      [torch.nn.Linear(Nn, Nn) for _ in range(NL - 1)])
        self.out = torch.nn.Linear(Nn, 1)

    def forward(self, xyt): # xyt: the input data, points in the domain
        z = xyt
        for layer in self.hidden:
            z = self.my_actvtn(layer(z))

        N = self.out(z)
        x = xyt[:, 0:1]
        y = xyt[:, 1:2]
        t = xyt[:, 2:3]
        u_DBC = self.f_DBC(x, y)
        u_IC  = self.f_IC(x, y)
        N_with_IC_DBC = (1 - t)*u_IC + t*N*u_DBC       # enforce IC & DBC

        return N_with_IC_DBC
```

4.3.3 Residual Computation of Heat Equation

Since we are using the PINN2D_DBC_IC model, which automatically enforces both the Dirichlet boundary condition (DBC) and the initial condition (IC), we only need to evaluate the residual of the PDE over the problem domain defined in Eq.(4.1). This process is largely similar to what was done for the Poisson equation, except that we must now include the time-dependent term.

The following code snippet computes the PDE residual for the 2D transient heat transfer problem.

```
def residual_heat(pinn, f_th, xyt, w_s_th=None):
    '''
    Compute the residual in space-time domain for heat transfer problems.
    pinn:    a PINN class object.
```

```
    f_th:    the source/force term in torch (th).
    w_s_th: dummy weight in torch, not used in a collocation sheme.
    '''
    xyt.requires_grad_(True)
    x1 = xyt[:, 0:1]; x2 = xyt[:, 1:2]           # get x and y coordinates
    t  = xyt[:, 2:3]                                    # get t coordinates

    x1_device = x1.to(device)
    x2_device = x2.to(device)
    t_device  = t.to(device)

    u = pinn(xyt)                         # feed forward with the data points
    grads = torch.autograd.grad(u, xyt, grad_outputs=torch.ones_like(u),
                                                     create_graph=True)[0]
    u_x = grads[:, 0:1]   # 1st derivatives
    u_y = grads[:, 1:2]
    u_t = grads[:, 2:3]

    # 2nd derivatives:
    u_xx= torch.autograd.grad(u_x,xyt,grad_outputs=torch.ones_like(u_x),
                                             create_graph=True)[0][:, 0:1]
    u_yy= torch.autograd.grad(u_y,xyt,grad_outputs=torch.ones_like(u_y),
                                             create_graph=True)[0][:, 1:2]

    f_val = f_th(x1_device, x2_device, t_device)      # source/force term
    residuals = (u_t - (u_xx + u_yy) - f_val) ** 2    # The PDE residual

    return residuals.mean(), residuals                     # scalar, tensor
```

4.3.4 Data Sampling over Space-Time Domain

Next, we need to sample data points over the (3D) space-time domain. The following code snippet samples data for a rectangular spatial domain $(a, b) \times (c, d)$ and a time domain $(0, t_{\text{end}}]$, as well as on the spatial boundary. This should be easy to follow, as it is quite similar to the sampling for Poisson's equations.

The only added complexity is the inclusion of the time domain, which results in 3D tensors. We purposely restrict the value of t_{end} due to the exponential decay characteristic of this type of problem. A large t can cause the function to effectively vanish.

```
def sampleData_xyt(N_clctn=100, N_DBC=10, N_IC=10, a=0, b=1, c=0, d=1,
                                                            t_end=0.1):
    '''
    Sampling data in (x,y,t) 3D space-time domain [a,b]×[c,d]×[0,t_end],
    and on the space boundary: bottom y=c, top y=d, and right x=b edges.
    Parameters: a, b: x-domain boundaries
                c, d: y-domain boundaries
    '''
    # Collocation points~ in full space-time domain:
    x_clctn = np.random.rand(N_clctn) * (b - a) + a
```

```
    y_clctn = np.random.rand(N_clctn) * (d - c) + c
    t_clctn = np.random.rand(N_clctn) * t_end
    xyt_clctn = np.stack([x_clctn, y_clctn, t_clctn], axis=1)

    # For Dirichlet boundary conditions (DBCs):
    x = np.random.rand(N_DBC) * (b - a) + a
    y = np.random.rand(N_DBC) * (d - c) + c
    t = np.random.rand(N_DBC) * t_end

    xyt_DBC = np.vstack([
        np.stack([x, np.full(N_DBC, c), t], axis=1), # bottom edge y = c
        np.stack([x, np.full(N_DBC, d), t], axis=1),    # top edge y = d
        np.stack([np.full(N_DBC, b), y, t], axis=1)]) # right edge x = b

    u_DBC = np.zeros((3 * N_DBC, 1))          # DBC: u=0 on all boundaries

    # Convert to torch tensors
    xyt_clctn_th= torch.tensor(xyt_clctn, dtype=torch.float32)
    xyt_DBC_th  = torch.tensor(xyt_DBC,   dtype=torch.float32)
    u_DBC_th    = torch.tensor(u_DBC,     dtype=torch.float32)

    return xyt_clctn_th, xyt_DBC_th, u_DBC_th
```

Note also that in the snippet given previously, we used an **equal-number sampling** approach, meaning that the number of sampling points over the time domain is the same as that for each spatial dimension (along x or y). This allows us to form 3D tensors for (x, y, t), enabling efficient data flow through the neural network.

Alternatively, one can sample the time domain differently from the spatial domains if desired. In such a case, we need to use `meshgrid` or a similar method to create 3D tensors, which will be larger in dimension along axis-0. The definition of the PINN classes remains the same.

4.3.5 PINN Trainer for Heat Equations

The following snippet defines a trainer that supports both first-order optimizers such as `adam` and quasi-second-order optimizers such as `lbfgs`. This code, named `trainerPDE_DBC_NBC`, is largely similar to `trainerPDE_DBC` used in Chapter 2. The difference is that `trainerPDE_DBC_NBC` allows the use of additional loss terms for both Dirichlet boundary conditions (DBCs) and Neumann boundary conditions (NBCs). It logs the training history data for each of these loss terms, including the `loss_PDE`.

```
def trainerPDE_DBC_NBC(pinn, residual, optimizer, f_th, x_Ω_th, w_s_th,
                x_DBC_th, u_DBC_th, x_NBC_th, t_NBC_th, w_NBC_th, I_NBC,
                                              epochs=201, adapt_every=0):
    """Train a PINN using loss_PDE+loss_DBC+loss_NBC.
    Args: pinn:       the PINN model (nn.Module)
          residual:   function to compute residual loss
          optimizer: PyTorch optimizer (e.g. Adam, LBFGS)
          epochs:     number of training iterations for PINN
          adapt_every: dummy, is not used here.
          f_th:      the force/source function~ f(x, y)
```

```
        w_s_th:    weights (dummy for collocation), a tensor on device
        x_Ω_th:    points (tensor, requires_grad=True, on device)
        x_DBC_th: Dirichlet boundary points (tensor,
                   requires_grad=True, on device)
        u_DBC_th: target values on DBC (torch tensor, on device)
        w_DBC:     from outer scope, weights for loss_DBC term.
        w_NBC:     from outer scope, weights for loss_NBC term.
        use_DBC:   from outer scope, if use DBC
        use_NBC:   from outer scope, if use NBC
    """
    loss_history = {"total": [], "PDE": [], "DBC": [], "NBC": []}

    if epochs <= 301: n_prnt = 100
    else:             n_prnt = max(1, round(epochs / 3, -2))

    for epoch in range(epochs):
        # Initialize loss variables for the current epoch
        loss_PDE = None; loss_DBC   = None
        loss_NBC = None; total_loss = None

        def closure():
            nonlocal loss_PDE, loss_NBC, loss_DBC, total_loss
            # Allow modifying variables from outer scope
            optimizer.zero_grad()
            loss_PDE, *_ = residual(pinn, f_th, x_Ω_th, w_s_th)
            total_loss = loss_PDE
            if use_DBC:
                loss_DBC = torch.mean((pinn(x_DBC_th) - u_DBC_th) ** 2)
                total_loss += loss_DBC*w_DBC
            if use_NBC:
                loss_NBC = compute_NBC_loss(pinn, x_NBC_th,
                                            normal_NBC, dudn_val)
                total_loss += loss_NBC*w_NBC

            total_loss.backward()                # Perform backward pass
            return total_loss  # Return the total loss for the optimizer

        if isinstance(optimizer, torch.optim.LBFGS):
            optimizer.step(closure)
            total_loss_val = total_loss.item()
            loss_PDE_val = loss_PDE.item()
            if loss_DBC is not None: loss_DBC_val = loss_DBC.item()
            else: loss_DBC_val = 0
            if loss_NBC is not None: loss_NBC_val = loss_NBC.item()
            else: loss_NBC_val = 0

        else: # Adam or other optimizers
            total_loss_tensor = closure()
            optimizer.step()
            total_loss_val = total_loss_tensor.item()
```

```
            loss_PDE_val = loss_PDE.item()          # set inside closure
            if loss_DBC is not None: loss_DBC_val = loss_DBC.item()
            else: loss_DBC_val = 0
            if loss_NBC is not None: loss_NBC_val = loss_NBC.item()
            else: loss_NBC_val = 0

        # Logging:
        with torch.no_grad():
            loss_history["total"].append(total_loss_val)
            loss_history["PDE"].append(loss_PDE_val)
            loss_history["DBC"].append(loss_DBC_val)
            loss_history["NBC"].append(loss_NBC_val)

        if epoch % n_prnt == 0 or epoch == epochs - 1:
            print(f"Epoch{epoch:5d}, Loss:{total_loss_val:.3e}, "
                  f"PDE:{loss_PDE_val:.3e}, DBC:{loss_DBC_val:.3e}, "
                  f"NBC:{loss_NBC_val:.3e}")

    return loss_history
```

The trainer is called by `train_adam_lbfgs`, which supports two-stage training: using `adam` for initial training, followed by `lbfgs` for fine-tuning. This function was introduced in the previous chapter and is included in the `grcodes` module. We have already imported it at the beginning of this chapter. One can also invoke it directly using `gr.train_adam_lbfgs`.

The procedure for training a PINN model for a heat transfer problem is largely similar to that for Poisson's equation. It is outlined as follows:

1. Define the device for performing the training.
2. Generate the training dataset, which includes the 3D coordinates (x, y, t) inside the space-time domain and on the boundary, along with the values of u on it (if the boundary condition losses are computed separately).
3. Define a PINN object using the `PINN2D_DBC_IC` class.
4. Call `train_adam_lbfgs`, which invokes a trainer to train the model.

4.4 Example: PINN with DBC and IC Enforced

We first create a PINN model with both DBC and IC enforced at the class level. In this case, we only need to minimize `loss_PDE`. Variables related to DBC, NBC, and IC are all set as dummy placeholders. The snippet for training the PINN model for the heat transfer problem is given below:

4.4.1 Training of PINN

The following snippet initiates a PINN model called `pinn` and prepares it for training. Detailed comments are provided, and it is self-explanatory.

```
gr.set_seed_th(88)                       # Ensures reproducibility of results

# Initialization lists for book-keeping: These lists are used to
# log key variables and performance metrics throughout the process.
pinns = []                         # For logging PINN model-related variables
timeAdam = []                        # to log the time used by optimizer adam
timeLbfgs = []                       # to log the time used by optimizer adam
Trainers = []                                       # to log the trainer used
LossTerms = []                                   # to log the loss terms used
Residuals = []                       # to log the residual calculation method
totalTimes = []                      # to log the total time used in training
error_L2s = []                     # To log the L2 error w.r.t exact solution
errorVIs = []                           # To log the error variatin index, VI

# Create a new PINN model: For this specific problem, DBC and IC
# are enforced at the class level.
Nn = 100                             # Number of neurons in each hidden layer
NL = 2                                # Number of (hidden) Layers in the PINN
pinn = PINN2D_DBC_IC(Nn, NL, f_DBC, f_IC, my_actvtn)

# Print model parameters without displaying the full configuration table
P_star, P_total = print_model_parameters(pinn, print_table=False)

# Generate the data points:
t_char = 1 / (2 * np.pi**2)           # Characteristic time for the problem
N_sample = 1 * P_star         # Total sampling points in 3D domain (x,y,t)

# Define points for boundary and initial conditions.
# N_DBC and N_IC are currently set as dummy values in this context.
N_DBC = 2 * int(np.sqrt(N_sample)) + 1          # Points on each DB segment
N_IC = N_sample                                # Points in domain (x,y) for IC

# For generating the dataset over the time interval [0, t_end]:
t_end = 5 * t_char
print(f"End time: {t_end:.4f}")          # Display the calculated end time

# Sample data points for the collocation domain, DBC, and IC
x_Ω_th, x_DBC_th, u_DBC_th = sampleData_xyt(N_clctn = N_sample,
                                            N_DBC   = N_DBC,
                                            N_IC    = N_IC,
                                            t_end   = t_end)

print(f"{x_Ω_th.shape[0]} collocation points in the domain.")

w_s_th = 1.          # Dummy weight, not used in a pure collocation~ scheme

def f_th(x, y, t):
    """
    Source Term Function in PyTorch. It returns a tensor of zeros for
    f(x,y,t) = 0. One may change it to another actual source function.
```

```
    """
    return torch.zeros_like(x)  # tensor of zeros with same shape as 'x'
```

```
P* (Total pseudo-dimension~ of the affine space~) = 206
Total trainable parameters: 10601
End time: 0.2533
206 collocation points in the domain.
```

For dynamic problems, we have doubled the capacity of the PINNs used for Poisson's problems in Chapter 2. This `pinn` is constructed with two hidden layers, each containing 100 neurons. The total number of trainable parameters is $P^* = 206$.

When using a collocation method, the number of sampling points must be equal to or greater than 206. In this example, we use exactly 206 collocation sampling points, which meets the minimum requirement.

The following snippet trains the model. Detailed comments are provided, and it is self-explanatory.

```
# Training execution setup: Initialize variables related to DBC and NBC.
# These are set to dummy values as DBC is not used in this problem setup.

x_DBC_th, t_DBC_th, w_DBC_th = None, None, None
use_DBC = False                                     # Flag to disable DBC
w_DBC = 1.      # Weight for DBC losses, a dummy value as DBC is not used

x_NBC_th, t_NBC_th, w_NBC_th = None, None, None
use_NBC = False                                     # Flag to disable NBC
w_NBC = 1.      # Weight for NBC losses, a dummy value as NBC is not used
I_NBC = ''                     # Indicates no integration for NBC is used

# Define the primary loss term for the training process.
LossTerm = 'PDE_only'     # only the PDE residual contributes to the loss

# Set the trainer and residual functions to be used during training.
Trainer = trainerPDE_DBC_NBC          # handles losses of PDE, DBC, and NBC
Residual = residual_heat      # computes the residual of the heat equation

# Define the domain integration rule.
DI_rule = ''       # Indicates no specific Domain Integration rule is used

# Execute the training process using Adam and L-BFGS optimizers.
losses, totalTime = train_adam_lbfgs(pinn, Trainer, Residual, f_th,
                                     x_Ω_th, w_s_th, x_DBC_th, u_DBC_th,
                                     x_NBC_th, t_NBC_th, w_NBC_th,
                                     I_NBC, epochs1=201, epochs2=11,
                                     use_lbfgs=True)
```

```
==================================================
201 epoches training with Adam:
Epoch    0, Loss:5.688e+01, PDE:5.688e+01, DBC:0.000e+00, NBC:0.000e+00
Epoch  100, Loss:9.484e+00, PDE:9.484e+00, DBC:0.000e+00, NBC:0.000e+00
```

```
Epoch  200, Loss:6.471e+00, PDE:6.471e+00, DBC:0.000e+00, NBC:0.000e+00
Time taken by adam: 2.653845 seconds

==================================================
11 epoches fine-tuning with LBFGS:
Epoch    0, Loss:4.783e-06, PDE:4.783e-06, DBC:0.000e+00, NBC:0.000e+00
Epoch   10, Loss:2.890e-06, PDE:2.890e-06, DBC:0.000e+00, NBC:0.000e+00

Training complete.
Time taken by lbfgs: 8.043192 seconds
Total time: 10.697037 seconds
Ratio (time_lbfgs/time_adam) of time taken per epoch:55.380410
```

The two-stage training was successful and took only about 10 seconds to complete on the author's laptop. The L^2-loss was reduced from 10^1 to 10^{-6}. Because the DBC is enforced within the `PINN2D_DBC_IC` class, `loss_DBC` and `loss_NBC` remain zero throughout the training. The model is trained solely by minimizing `loss_PDE`.

It is observed that the second-order optimizer `lbfgs` runs significantly slower than the first-order optimizer `adam`, for reasons discussed in the previous chapters. In this example, the time taken per epoch by `lbfgs` is approximately 50 times that of `adam`. Therefore, it is common practice to first train the model using `adam` for a sufficient number of epochs to bring the learning parameters closer to optimal values, and then fine-tune using `lbfgs` with only a few iterations. For the present example, only one iteration of `lbfgs` was needed.

This suggests that the L^2-loss landscape for the heat transfer problem becomes nearly quadratic near the minimum, provided the PINN is constructed with DBC and IC enforced internally. This finding is consistent with observations made for the Poisson equation in the previous chapter.

4.4.2 Results Accuracy Evaluation

If the exact (or reference) solution is available, we can assess the error in the PINN solutions. This can be done using the following utility snippet:

```python
# Evaluate at t_obs -- time slice for observation (plot)
t_obs = 0.5 * t_end                    # time to observe the predition results
N_plot = 100                                        # resolution for the plots

# Sampling over rectangular domain (a, b) × (c, d):
a = 0.0; b = 1.0
c = 0.0; d = 1.0

x = np.linspace(a, b, N_plot)             # Create meshgrid over (a,b)x(c,d)
y = np.linspace(c, d, N_plot)

X, Y = np.meshgrid(x, y)
T = t_obs * np.ones_like(X)
XYT = np.stack([X.flatten(), Y.flatten(), T.flatten()], axis=1)
XYT_tensor = torch.tensor(XYT, dtype=torch.float32)

u_pred = pinn(XYT_tensor).detach().numpy().reshape(N_plot, N_plot)
```

```
u_exact = exact_u_heat(X, Y, t_obs)

error_L2, errorVI, u_diff = sln_error(u_pred, u_exact)
```

```
Maximum Error: 2.7888e-05
Mean Absolute Error: 1.1044e-05
Error Variation Index (ratio of Max/Mean): 2.5251e+00
L2 Relative Error: 3.2832e-04
```

We define this following function to built a record sheet for key performance indications of the trained PINN models.

```
def recordSheet():
    '''
    Log the key performance indicators of PINN models for
    comparison study.
    '''
    new_error_L2_str = f"{error_L2:.3e}"

    if new_error_L2_str not in error_L2s:
        pinns.append(pinn.__class__.__name__)          # log pinn models
        Residuals.append(Residual.__name__ + DI_rule)    # log residuals
        #Trainers.append(Trainer.__name__)                # log trainers
        LossTerms.append(LossTerm) # additional terms in composite loss
        totalTimes.append(f"{totalTime:.3e}")                # log time
        error_L2s.append(new_error_L2_str)                 # log the error
        errorVIs.append(f"{errorVI:.3e}")          # error variation index
    else:
        print(f"Error_L2 value '{new_error_L2_str}'"\
              f" already exists in error_L2s. Skipping update.")

recordItems=[pinns, LossTerms, Residuals,totalTimes,error_L2s,errorVIs]
```

```
# Create record sheet (exercute this only once!):
recordSheet()
```

The following snippet clears the entire record sheet or deletes specific records from it, if needed.

```
# recordItems=[pinns, LossTerms,Residuals,totalTimes,error_L2s,errorVIs]
# for item in recordItems: item.clear()      # to clear the entire record
# for item in recordItems: item.pop(-1)            # to delete one record
```

```
printSummary(pinns, LossTerms, Residuals, totalTimes,error_L2s,errorVIs)
```

```
#  PINN model        Loss_terms residual    total_time error_L2s  errorVIs
========================================================================
1  2D_DBC_IC         PDE_only   _heat        1.070e+01  3.283e-04  2.525e+00
```

The solution accuracy is high, with the relative error on the order of 10^{-4} (this may vary slightly due to the randomness involved in training the model).

4.4.3 Plot the Solution Contours

Let us now assess the accuracy of the solution and plot the contours of the PINN solution evaluated at a fixed time slice $t = 0.5\,t_{\text{end}}$, along with the exact solution.

The following snippet computes and plots the contours of the PINN solution and the exact solution, as well as the relative error norm distribution, using the `plot_sln_rect()` function introduced in the previous chapter.

```
def plot_sln_XY(X, Y, u_pred, u_exact, vmin=0, vmax=1, dpi=300):
    '''
    Plots the predicted solution, exact solution, and their absolute
    difference over a 2D rectangular domain.

    Args:
        X (numpy.ndarray): Meshgrid X coordinates of the domain.
        Y (numpy.ndarray): Meshgrid Y coordinates of the domain.
        u_pred (numpy.ndarray):  Predicted solution values.
        u_exact (numpy.ndarray): Exact solution values.
        vmin (float, optional):  Minimum value for the color scale of
                                 u_pred and u_exact. Defaults to 0.
        vmax (float, optional):  Maximum value for the color scale of
                                 u_pred and u_exact. Defaults to 1.
        dpi (int, optional):     Resolution of the figure in dots per
                                                inch. Defaults to 300.
    '''
    # Define figure dimensions and aspect ratio for consistent plotting.
    width = 8
    height = width / 3.6

    # Create a figure and a set of subplots arranged in a 1x3 grid.
    fig, axs = plt.subplots(1, 3, figsize=(width, height), dpi=dpi)

    # --- Plot PINN Prediction ---
    c0 = axs[0].contourf(X, Y, u_pred, vmin=vmin, vmax=vmax, levels=15)
    axs[0].set_title('PINN Prediction')
    plt.colorbar(c0, ax=axs[0]) # Add color bar specific to this subplot.

    # --- Plot Exact Solution ---
    c1 = axs[1].contourf(X, Y, u_exact, vmin=vmin, vmax=vmax, levels=15)
    axs[1].set_title('Exact Solution')
    plt.colorbar(c1, ax=axs[1]) # Add color bar specific to this subplot.

    # --- Plot absolute difference between prediction and exact solution
    c2 = axs[2].contourf(X, Y, np.abs(u_pred - u_exact), levels=100)
    axs[2].set_title('|Difference|')
    plt.colorbar(c2, ax=axs[2]) # Add color bar specific to this subplot.

    # --- Apply consistent formatting to all subplots ---
    for ax in axs:
        ax.set_xlabel('$x$')       # Set x-axis label using LaTeX notation.
```

```
        ax.set_ylabel('$y$')        # Set y-axis label using LaTeX notation.
        ax.set_aspect('equal')          # Ensure the aspect ratio is equal.
        ax.grid(linewidth=0.3)   # Subtle grid for better visual guidance.

    plt.tight_layout()  # Adjust subplot parameters to avoid overlapping.
    plt.show()                                         # Display the plot.
```

```
# Plot the contours of both PINN solution and the exact one:
plot_sln_XY(X, Y, u_pred, u_exact, vmin=0, vmax=0.08)
```

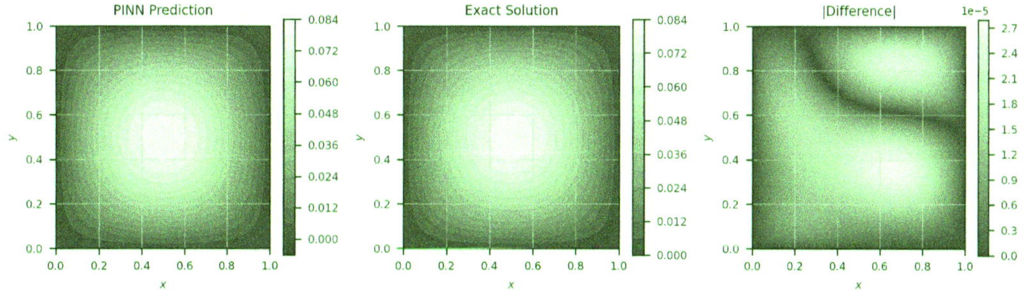

We also note that the distribution of the error is smooth due to the nature of a neural network, which is composed of chained differentiable elementary functions.

4.4.4 Plot of Loss History

The following utility snippet plots the history of the loss terms during training. Since our model is trained using both `adam` and `lbfgs`, we will have two sets of loss function values. We use the function `plot3loss2()` to generate the plots, as introduced in the previous chapter.

If both `adam` and `lbfgs` are used consecutively in training, we may use the following function to plot the loss history, which includes the histories of all loss terms contributing to the composite loss:

```
plot3loss2(losses,('adam initially','lbfgs in fine tunning'), PDE='PDE',
                                        C1='DBC', C2='NBC', CL='Loss',log=False)
```

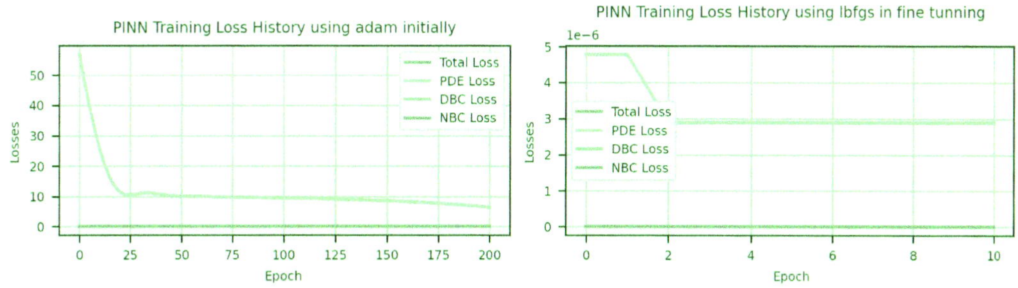

4.5 Example: PINN Model with IC Enforced

This time, we enforce only the initial condition (IC) at the PINN class level. This is done by feeding the PINN class a dummy Dirichlet boundary condition (DBC) function, which is simply the constant 1. In this case, the `loss_DBC` is computed separately, resulting in a composite loss consisting of two terms: `loss_PDE` and `loss_DBC`. The optimizer minimizes the composite loss, which is the sum of L^2 functionals.

All other settings remain unchanged. Therefore, the underlying problem itself is not altered—only the way it is posed differs. As a result, the exact solution remains the same. However, the behavior of the PINN model constructed under this setup may differ.

4.5.1 Training of the Model

The snippet to build and train the new PINN model is as follows:

```
gr.set_seed_th(88)                                    # for reproducibility

def f_DBC(x, y):    # --- DBC not enforced in PINN class ---
    return torch.ones_like(x)*torch.ones_like(y)  # dummy, f_DBC(x, y)=1

def  f_IC(x, y):    # --- only IC is enforced in PINN class ---
    return torch.sin(np.pi*x)*torch.sin(np.pi*y)  # the true IC function

# Create a new PINN model, only IC is enforced internally:
pinn = PINN2D_DBC_IC(Nn, NL, f_DBC, f_IC, my_actvtn)

use_DBC = True                                        # Flag to disable DBC
w_DBC = 1.      # Weight for DBC losses, a dummy value as DBC is not used
# Sample data points for the collocation domain, DBC, and IC
x_Ω_th, x_DBC_th, u_DBC_th = sampleData_xyt(N_clctn=N_sample,
                                N_DBC=N_DBC, N_IC=N_IC, t_end=t_end)

print(f"{x_Ω_th.shape[0]} collocation points in the domain.")
print(f"{x_DBC_th.shape[0]} boundary points.")          # not used here

x_NBC_th, t_NBC_th, w_NBC_th = None, None, None
I_NBC = ''                                # no Integration for NBC is used
use_NBC = False

Trainer  = trainerPDE_DBC_NBC
LossTerm = 'DBC'                         # loss term in addition to loss_PDE
Residual = residual_heat
DI_rule = ''                              # no Domain Integration is used

losses, totalTime = train_adam_lbfgs(pinn, Trainer, Residual, f_th,
       x_Ω_th, w_s_th, x_DBC_th, u_DBC_th, x_NBC_th, t_NBC_th, w_NBC_th,
                          I_NBC, epochs1=201, epochs2=11, use_lbfgs=True)
```

```
206 collocation points in the domain.
87 boundary points.
```

```
==================================================
201 epoches training with Adam:
Epoch    0, Loss:5.584e+01, PDE:5.584e+01, DBC:2.887e-04, NBC:0.000e+00
Epoch  100, Loss:2.349e+01, PDE:2.349e+01, DBC:6.810e-01, NBC:0.000e+00
Epoch  200, Loss:1.347e+01, PDE:1.347e+01, DBC:4.315e-01, NBC:0.000e+00
Time taken by adam: 1.757462 seconds

==================================================
11 epoches fine-tuning with LBFGS:
Epoch    0, Loss:1.967e-02, PDE:1.967e-02, DBC:3.108e-03, NBC:0.000e+00
Epoch   10, Loss:2.910e-05, PDE:2.910e-05, DBC:1.137e-05, NBC:0.000e+00

Training complete.
Time taken by lbfgs: 87.410082 seconds
Total time: 89.167545 seconds
Ratio (time_lbfgs/time_adam) of time taken per epoch:908.822162
```

As observed, the DBC value is no longer exactly zero and is instead driven toward smaller values by the optimizer during training. The training took approximately 10 times longer (this may vary depending on random initialization and system performance).

4.5.2 Assessment of the Solution

The accuracy is also roughly 1000 times lower compared to the case with DBCs enforced at the class level. This can be confirmed using the following snippet:

```
# Evaluate the solution:
u_pred = pinn(XYT_tensor).detach().numpy().reshape(N_plot, N_plot)
u_exact = exact_u_heat(X, Y, t_obs)

error_L2, errorVI, u_diff = sln_error(u_pred, u_exact)
```

```
Maximum Error: 4.4815e-02
Mean Absolute Error: 8.3289e-03
Error Variation Index (ratio of Max/Mean): 5.3807e+00
L2 Relative Error: 3.1182e-01
```

As seen again, enforcing the DBC externally not only makes the training significantly slower, but also leads to much lower solution accuracy. This outcome largely supports the findings for static problems studied in the previous chapter for Poisson's equations.

Let us now plot the history of various loss terms:

```
plot3loss2(losses, ('adam initially','lbfgs in fine tunning'),
                              PDE='PDE', C1='DBC', CL='Loss',log=False)
```

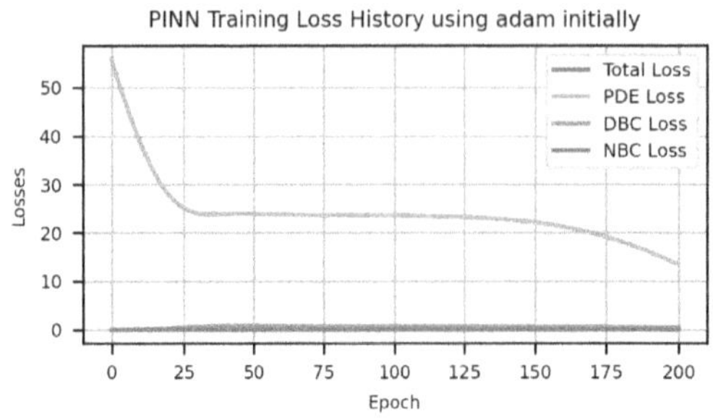

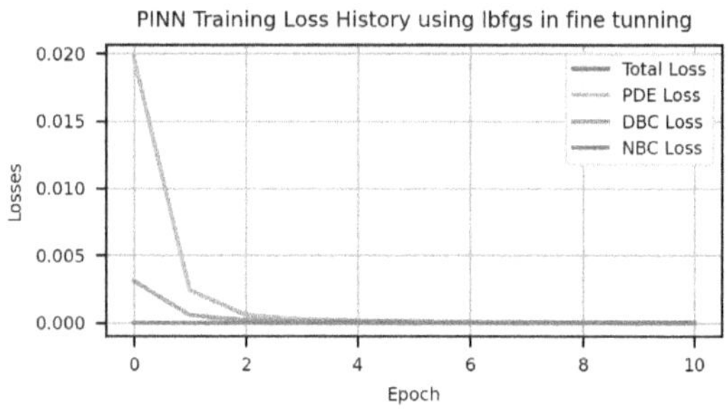

These loss history curves indicate that **lbfgs** requires a longer duration to approach a near-minimum of the composite L^2 loss. If `loss_IC` is also treated separately, the training process may become even more challenging. Readers are encouraged to explore this behavior by making slight modifications to the provided code snippet.

The preceding loss history plots further demonstrate that, while **adam** offers an initial optimization boost, the substantial reduction of the loss is ultimately achieved by **lbfgs**. Readers may attempt training the same PINN model using **adam** alone by disabling the **lbfgs** phase. In such cases, the training process will likely become significantly more difficult, and oscillatory behavior may also emerge—consistent with observations discussed in the previous chapter.

We emphasize that minimizing the L^2 loss terms as much as possible is crucial for training a high-quality PINN model. Additionally, the use of a second-order optimizer plays an important role in achieving this objective.

```
# Create record sheet (exercute this only once!):
recordSheet()
```

```
printSummary(pinns, LossTerms, Residuals, totalTimes,error_L2s,errorVIs)
```

```
#  PINN model       Loss_terms residual    total_time error_L2s  errorVIs
=========================================================================
1  2D_DBC_IC        PDE_only   _heat       1.070e+01  3.283e-04  2.525e+00
2  2D_DBC_IC        DBC        _heat       8.917e+01  3.118e-01  5.381e+00
```

Note that the difference between these two models lies in how the same problem is formulated. Model-1 implements both DBC and IC at the class level using the trial solution trick. In Model-2, the same class template is used; however, the definition of `f_DBC` is set as a dummy constant function equal to 1, and hence multiplying it by $\mathcal{N}_{\hat{\mathbf{w}}}$ has no effect at all. As a result, the DBC is enforced externally via a composite loss function. This leads to a significant reduction in accuracy.

This outcome aligns with the findings discussed in Chapter 2: implementing DBC internally within the class (i.e., using a hard constraint) is the preferred approach.

4.6 Case Study: Heated Plate with Hole and NBCs

4.6.1 Problem Setup

This study trains a PINN model for an idealized heat transfer problem, aiming to compute the temperature distribution in a square plate with an elliptical hole at the center. The setup is as follows. The domain spans $[-1, 1] \times [-1, 1]$, excluding an elliptical hole, and the time interval is defined as $[0, t_{\text{end}}]$. The boundary of the elliptical hole is described using the following function:

$$F(x, y) \equiv \frac{(x - x_0)^2}{r_a^2} + \frac{(y - y_0)^2}{r_b^2} - 1 = 0 \tag{4.10}$$

where (x_0, y_0) is the center of the ellipse, and r_a and r_b are the radii of the ellipse along the x- and y-axes, respectively.

1. Initially, the plate has a temperature of zero degrees everywhere.

2. It is then exposed to sunlight, which heats the plate. The heat source is assumed to vary linearly with time, with a slope of 1.

3. The left edge of the plate is connected to a base, maintaining a constant unit temperature on that edge. This implies that the base supplies or absorbs heat from the plate.

4. The remaining three edges (bottom, top, and right) are insulated, resulting in zero heat flux.

The task of the PINN model is to predict the temperature distribution over the plate at any given time.

4.6.2 Analysis of the Problem

Based on the problem setup, the plate initially has zero temperature everywhere; thus, the initial condition is given by $f_{\text{IC}}(x, y, 0) = 0$.

Since the heat source is uniform in space and varies linearly with time with a slope of 1, we have

$$f(x, y, t) = t \tag{4.11}$$

The left edge has a fixed unit temperature, corresponding to a Dirichlet boundary condition (DBC).

All other edges have zero heat flux, implying Neumann boundary conditions (NBC) on those boundaries. Therefore, we have

$$\frac{\partial u}{\partial n} = 0 \tag{4.12}$$

on all edges except the left one, where n denotes the outward unit normal on the boundary.

4.6.3 Strategy to Study the Problem

We should employ a PINN class that automatically satisfies the DBC on the left edge and the initial condition for the entire plate at $t = 0$. The function for the DBC along $x = 0$ should be

$$f_{\text{DBC}}(x, y, t) = x \tag{4.13}$$

The NBCs should be handled separately by computing the L^2 loss associated with the Neumann conditions. Consequently, the composite loss should be the sum of `loss_PDE` and `loss_NBC`.

The computation of the PDE loss is the same as in previous examples. The NBC loss is computed by taking the dot product of the gradient of the model's output ∇u (which is a vector) and the normal vector $\mathbf{n}$.

$$(\nabla u \cdot \mathbf{n}) = 0 \tag{4.14}$$

This condition must be enforced separately on each of the three Neumann boundary condition (NBC) edges, since each edge has a distinct outward normal vector. According to Eq.(2.18), the NBCs on the bottom, top, and right edges should be expressed as:

- Bottom $(y = c)$ $\frac{\partial u}{\partial y} = 0$
- Top $(y = d)$ $\frac{\partial u}{\partial y} = 0$
- Right $(x = b)$ $\frac{\partial u}{\partial x} = 0$

where both $\frac{\partial u}{\partial x}$ and $\frac{\partial u}{\partial y}$ can be computed from the output of the neural network using `autograd`.

The NBC on the boundary of the ellipse becomes

$$\nabla u \cdot \mathbf{n} = \frac{\partial u}{\partial x} n_x + \frac{\partial u}{\partial y} n_y = 0 \tag{4.15}$$

where n_x and n_y are the components of the outward unit normal vector on the elliptical boundary, which vary along the edge. These components can be computed from the single-piece function defining the ellipse using the following formula:

$$n_x = -\frac{\partial F}{\partial y}\Big|_{x=x_e} = -2\frac{(x_e - x_0)}{r_a^2} \tag{4.16}$$

$$n_y = -\frac{\partial F}{\partial x}\Big|_{y=y_e} = -2\frac{(y_e - y_0)}{r_b^2} \tag{4.17}$$

where (x_e, y_e) denotes a coordinate point on the boundary of the ellipse.

4.6.4 Code for Sampling Points for Plates with Hole

Due to the geometric complexity, the random sampling strategy for residual points must be adjusted. The approach is to randomly sample points within the square plate and then exclude those that lie inside the elliptical hole. The following snippet defines the function used to exclude points inside the ellipse:

```
def sample_xyt_D_ellipseNBC(f_NBC, N_clctn=1000, N_DBC=50, N_NBC=50,
         N_ellipse=200, x0=0, y0=0, ra=0.2, rb=0.1, a=0, b=1, c=0, d=1,\
                                                            t_end=0.1):
    '''Generate training data for a PINN model for transient
      heat transfer problems defined in a rectangular domain with hole.
```

```
    - PDE over Domain [a,b]x[c,d]x[0,t_end]:
    - zero NBCs on y=c, y=d, x=b
    - NBC on the edge of an elliptical hole of radii (ra,rb)
      centered at (x0, y0), ((x-x0)/ra)^2+((y-y0)/rb)^2-1
    - f_NBC: list [u/ x,  u/ y]
    - the left edge is fixed and implemented at the class level.
'''
# --- 1. Interior Collocation Points ---
colo = []
while len(colo) < N_clctn:
    x = np.random.rand() * (b - a) + a
    y = np.random.rand() * (d - c) + c
    # Exclude ellipse:
    if ((x - x0)**2 / ra**2 + (y - y0)**2 / rb**2) >= 1.0:
        t = np.random.rand() * t_end
        colo.append([x, y, t])
xyt_clctn = np.array(colo)

# --- 2. For Dirichlet boundary conditions (DBCs), left edge x = a:
#x = np.random.rand(N_DBC) * (b - a) + a
y = np.random.rand(N_DBC) * (d - c) + c
t = np.random.rand(N_DBC) * t_end

xyt_DBC = np.vstack([np.stack([np.full(N_DBC, a), y, t], axis=1)])
u_DBC = np.zeros((N_DBC, 1))             # DBC: u=0 on left edge x = a

# --- 3. Neumann BCs on edges y=c, y=d, x=b ---
x = np.random.rand(N_NBC) * (b - a) + a
y = np.random.rand(N_NBC) * (d - c) + c
t = np.random.rand(N_NBC) * t_end

xyt_NBC = np.vstack([
    np.stack([x,  np.full(N_NBC, c), t], axis=1),           # bottom
    np.stack([x,  np.full(N_NBC, d), t], axis=1),              # top
    np.stack([np.full(N_NBC, b),  y, t], axis=1)])           # right

normal_NBC = np.vstack([np.tile([0.0, -1.0], (N_NBC, 1)),  # bottom
                        np.tile([0.0,  1.0], (N_NBC, 1)),     # top
                        np.tile([1.0,  0.0], (N_NBC, 1))])  # right

# --- 4. NBC on ellipse edge ---
theta = 2 * np.pi * np.random.rand(N_ellipse)
xe = x0 + ra * np.cos(theta)
ye = y0 + rb * np.sin(theta)
te = np.random.rand(N_ellipse) * t_end
xyt_ellipse = np.stack([xe, ye, te], axis=1)    # points on ellipse

# Outward normal to ellipse, grad. of ((x-x0)/ra)^2+((y-y0)/rb)^2-1
grad_phi = np.stack([-2 * (xe - x0) / ra**2,
                     -2 * (ye - y0) / rb**2], axis=1)    # outward n
```

```
    norm_mag = np.linalg.norm(grad_phi, axis=1, keepdims=True)
    normal_NBC_ellipse = grad_phi/norm_mag    # normalized, unit normals

    # --- Combine all NBC ---
    xyt_NBC_all    = np.vstack([xyt_NBC, xyt_ellipse])
    normal_NBC_all = np.vstack([normal_NBC, normal_NBC_ellipse])
    dudn_val       = f_NBC(xyt_NBC_all)[0] * normal_NBC_all[:,0:1] +\
                     f_NBC(xyt_NBC_all)[1] * normal_NBC_all[:,1:2]

    # --- Convert to torch tensors ---
    xyt_clctn_th  = torch.tensor(xyt_clctn, dtype=torch.float32)

    xyt_DBC_th    = torch.tensor(xyt_DBC, dtype=torch.float32)
    u_DBC_th      = torch.tensor(u_DBC, dtype=torch.float32)

    xyt_NBC_th    = torch.tensor(xyt_NBC_all, dtype=torch.float32)
    normal_NBC_th= torch.tensor(normal_NBC_all, dtype=torch.float32)
    dudn_val_th   = torch.tensor(dudn_val,dtype=torch.float32)

    return xyt_clctn_th, xyt_DBC_th, u_DBC_th, \
                                 xyt_NBC_th, normal_NBC_th, dudn_val_th
```

4.6.5 Code for Computing NBC Loss

This case study requires handling the Neumann boundary conditions (NBCs). The following snippet computes the NBC loss at a given set of points on the NBC boundaries using Eq.(4.15):

```
def compute_NBC_loss(model, xyt_NBC, normal_NBC, dudn_val):
    '''
    Compute the loss for NBC boundary conditions for transient heat
    transfer problems.
    Inputs: - model: neural network u(x, y, t)
            - xyt_NBC: tensor of shape (N, 3)
            - normal_NBC: tensor of shape (N, 2), each row is [n_x, n_y]
            - dudn_val: tensor of shape (N, 1), usually zeros
    '''
    xyt_NBC.requires_grad_(True)
    u = model(xyt_NBC)                                    # u shape: (N, 1)

    grads = torch.autograd.grad(outputs=u, inputs=xyt_NBC,
        grad_outputs=torch.ones_like(u),             # du/dx, du/dy, du/dt
        retain_graph=True, create_graph=True)[0]     # grads shape: (N, 3)

    grad_u_xy = grads[:, :2]                 # Extract du/dx and du/dy, (N, 2)

    # Dot product with normal vectors: ∇u   n = du/dn
    du_dn = torch.sum(grad_u_xy*normal_NBC, dim=1, keepdim=True) # (N,1)

    # Compute MSE loss for NBC condition
```

```
    loss_NBC = torch.mean((du_dn - dudn_val) ** 2)

    return loss_NBC
```

4.6.6 Plot of Sample Points in Plate with Hole

To visualize the problem domain and the sampled points distributed both inside the domain and on its boundaries, we first generate these points and then plot the results.

```
gr.set_seed_th(88)                                    # for reproducibility

a=0; b=1; c=0; d=1                                            # domain
x0=0.5; y0=0.5; ra=0.3; rb=0.2                        # ellipse in center
t_end=5.*t_char

def f_NBC(xyt):                # list of 2 functions, [f/x, f/y], for NBC
    x = xyt[:,0:1]
    y = xyt[:,1:2]
    t = xyt[:,2:3];
    dfdx = 0.*x + 0.*y + 0.*t                          # the flux is zero
    dfdy = 0.*x + 0.*y + 0.*t
    return [dfdx, dfdy]

xyt_clctn, xyt_DBC, u_DBC, xyt_NBC, normal_NBC, dudn_val = \
     sample_xyt_D_ellipseNBC(f_NBC, N_clctn=N_sample, N_NBC=N_DBC,
     N_ellipse=2*N_DBC, x0=x0, y0=y0, ra=ra, rb=rb, a=a, b=b, c=c, d=d,
                                                            t_end=t_end)
print(f"Sampled {xyt_clctn.shape[0]} collocation points in the domain.")
print(f"Sampled {xyt_DBC.shape[0]} boundary points for DBCs.")
print(f"Sampled {xyt_NBC.shape[0]} boundary points for NBCs.")
```

```
Sampled 206 collocation points in the domain.
Sampled 50 boundary points for DBCs.
Sampled 145 boundary points for NBCs.
```

The following is a Python function that plots a given set of sampled points inside a rectangular domain, along with a set of sampled points on its NBC boundaries.

```
def visualize_sampling(a , b , c , d,                   # Rectangle bounds
                       x0, y0, ra, rb,           # Ellipse center and radii
                       xyt_clctn,              # collocation points in domain
                       xyt_NBC):                              # points on NBC

    # Extract spatial components
    x_clctn, y_clctn = xyt_clctn[:, 0], xyt_clctn[:, 1]
    x_NBC,  y_NBC   = xyt_NBC [:, 0], xyt_NBC [:, 1]

    fig, ax = plt.subplots(figsize=(5, 5), dpi=300)
    # Collocation points inside domain
    ax.scatter(x_clctn, y_clctn, s=5, color='blue', alpha=0.4,
```

```
                                                label='Points inside domain')
    # Neumann boundary points
    ax.scatter(x_NBC, y_NBC, s=8, color='red', label='Points with NBC')

    rect = plt.Rectangle((a, c), b - a, d - c, edgecolor='black',
                                                facecolor='none', linewidth=2)
    ax.add_patch(rect)  # Draw rectangle boundary

    theta = np.linspace(0, 2 * np.pi, 200)                  # Draw ellipse
    xe = x0 + ra * np.cos(theta)
    ye = y0 + rb * np.sin(theta)
    ax.plot(xe, ye, 'k--', label='Elliptic Hole')
    ax.set_xlim(a - 0.1, b + 0.1)
    ax.set_ylim(c - 0.1, d + 0.1)
    ax.set_aspect('equal')
    ax.set_title('Sampled Points In Plate Clamped on the Left Edge '+ \
                                                'with Elliptic Hole')
    ax.legend()
    plt.grid(linewidth=0.5); plt.show()
```

```
# Plot the sampling points inside the problem domain and on its boundary:
visualize_sampling(a, b, c, d, x0, y0, ra, rb, xyt_clctn, xyt_NBC)
```

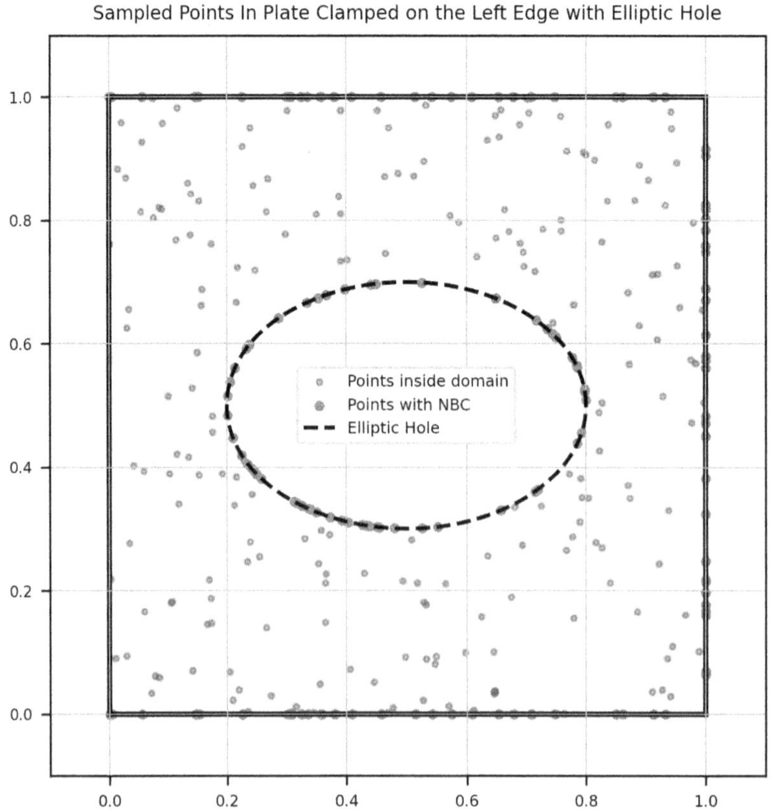

4.6.7 PINN for Plate with Hole and Nonzero NBCs

We now train a PINN model for heat transfer with nonzero Neumann boundary conditions (NBCs) applied on the bottom, top, and right edges, as well as on the boundary of the elliptical hole. The NBC values are computed using the exact solution given in Eq.(4.4) via differentiations. Consequently, the original problem remains unchanged; only the boundary condition setting is modified.

We still have the same exact solution available to evaluate the accuracy of the new PINN model trained under these nonzero NBCs. If the PINN is successfully trained, it should reproduce the exact solution presented in Eq.(4.4).

The code snippet is as follows:

```
# Generate dataset and input functions:
# a=0; b=1; c=0; d=1                                  # the same domain
# x0=0.5; y0=0.5; ra=0.2; rb=0.1                    # ellipse in center

gr.set_seed_th(88)                                # for reproducibility

def f_th(x, y, t):
    '''Source Term Function for the PDE in torch
    one may replace the dummy with actual f_th. '''
    return 0.*x + 0.*y + 0.*t

def  f_IC(x, y):
    return torch.sin(np.pi*x) * torch.sin(np.pi*y) # 0.*x + 0.*y

def f_DBC(x, y): return x        # one may change it, used in PINN class

def f_NBC(xyt):                  # list functions, [f/x, f/y], for NBC
    x = xyt[:,0:1]                         # computed from the exact solution
    y = xyt[:,1:2]
    t = xyt[:,2:3];
    dfdx = np.exp(-2*np.pi**2 *t)*np.pi*np.sin(np.pi*y)*np.cos(np.pi*x)
    dfdy = np.exp(-2*np.pi**2 *t)*np.pi*np.sin(np.pi*x)*np.cos(np.pi*y)
    return [dfdx, dfdy]

# Create a new PINN model, DBC & IC are same function for this problem:
pinn = PINN2D_DBC_IC(Nn, NL, f_DBC, f_IC, my_actvtn)
P_star, P_total = print_model_parameters(pinn, print_table=False)

use_NBC = True          # this problem has NBCs. Its loss is added to total
LossTerm = 'NBC'          # loss term in addition to loss_PDE (always used)

x_Ω_th, x_DBC_th, u_DBC_th, x_NBC_th,normal_NBC, dudn_val=\
     sample_xyt_D_ellipseNBC(f_NBC, N_clctn=N_sample, N_DBC=N_DBC,\
      N_NBC=N_DBC, N_ellipse=2*N_DBC, x0=x0,y0=y0,ra=ra,rb=rb,
                                              a=a,b=b,c=c,d=d,t_end=t_end)

print(f"Sampled {x_Ω_th.shape[0]} collocation points in the domain.")
```

```
print(f"Sampled {x_DBC_th.shape[0]} DBC boundary points.")
print(f"Sampled {x_NBC_th.shape[0]} NBC boundary points.")

w_s_th = 1.            # dummy weight, not used in a collocation scheme

# Training excecution:
Trainer  = trainerPDE_DBC_NBC
Residual = residual_heat
losses, totalTime = train_adam_lbfgs(pinn, Trainer, Residual, f_th,
       x_Ω_th, w_s_th, x_DBC_th, u_DBC_th, x_NBC_th, t_NBC_th, w_NBC_th,
                        I_NBC, epochs1=201, epochs2=11, use_lbfgs=True)
```

```
P* (Total pseudo-dimension of the affine space) = 206
Total trainable parameters: 10601
Sampled 206 collocation points in the domain.
Sampled 29 DBC boundary points.
Sampled 145 NBC boundary points.
==================================================
201 epoches training with Adam:
Epoch    0, Loss:3.758e+01, PDE:3.758e+01, DBC:0.000e+00, NBC:2.401e+00
Epoch  100, Loss:1.421e+01, PDE:1.421e+01, DBC:0.000e+00, NBC:1.504e+00
Epoch  200, Loss:1.009e+01, PDE:1.009e+01, DBC:0.000e+00, NBC:1.238e+00
Time taken by adam: 3.307008 seconds

==================================================
11 epoches fine-tuning with LBFGS:
Epoch    0, Loss:2.656e-02, PDE:2.656e-02, DBC:0.000e+00, NBC:9.294e-03
Epoch   10, Loss:2.116e-05, PDE:2.116e-05, DBC:0.000e+00, NBC:9.944e-06

Training complete.
Time taken by lbfgs: 98.275012 seconds
Total time: 101.582020 seconds
Ratio (time_lbfgs/time_adam) of time taken per epoch:543.014291
```

With the new setting of the same problem, we observe that `loss_NBC` is no longer zero and is progressively minimized by the optimizer as part of the composite loss. Additionally, the training becomes significantly slower, taking approximately 100 seconds.

Let us now evaluate the solution accuracy using the following code:

```
# Evaluate the results at a fixed observation time t_obs
t_obs  = 0.5 * t_end
N_plot = 100                              # resolution for the plots
x = np.linspace(a, b, N_plot)          # Create meshgrid over (a,b)x(c,d)
y = np.linspace(c, d, N_plot)

X, Y = np.meshgrid(x, y)
T = t_obs * np.ones_like(X)
XYT = np.stack([X.flatten(), Y.flatten(), T.flatten()], axis=1)
XYT_tensor = torch.tensor(XYT, dtype=torch.float32)
```

```
u_pred  = pinn(XYT_tensor).detach().numpy().reshape(N_plot, N_plot)
u_exact = exact_u_heat(X, Y, t_obs)

error_L2, errorVI, u_diff = sln_error(u_pred, u_exact)
```

```
Maximum Error: 6.1984e-03
Mean Absolute Error: 1.5209e-03
Error Variation Index (ratio of Max/Mean): 4.0755e+00
L2 Relative Error: 5.2408e-02
```

The accuracy is on the order of 10^{-2}. Let us now plot the temperature distribution predicted by the new PINN model alongside the exact solution for comparison.

```
# Plot the contours of both PINN solution and the exact one:
plot_sln_XY(X, Y, u_pred, u_exact, vmin=0, vmax=0.08)
```

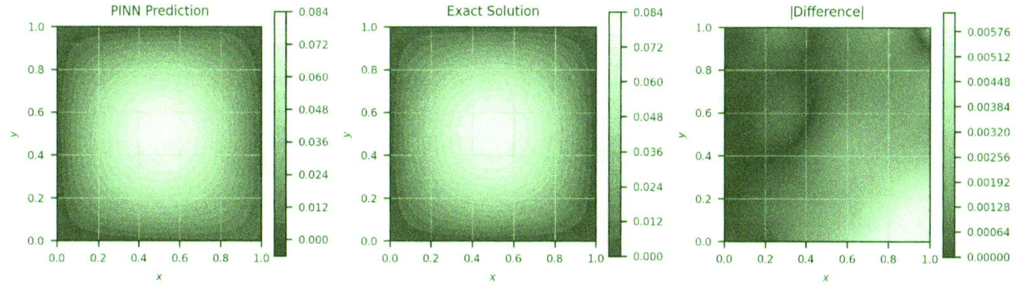

Let's also plot the histories of the individual loss terms using the following snippet:

```
plot3loss2(losses, ('adam initially','lbfgs in fine tunning'),
                    PDE='PDE', C1='DBC', C2='NBC', CL='Loss',log=False)
```

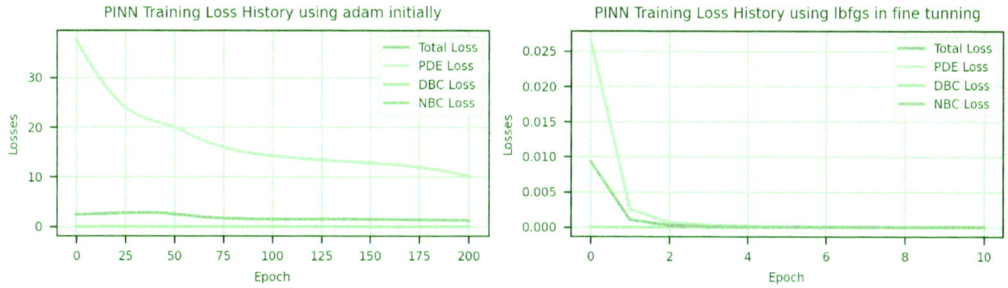

We observe that both the first-order and quasi-second-order optimizers managed to minimize the loss effectively. Let us now add the results to the record sheet introduced earlier:

```
recordSheet()                                    # Update the record sheet
```

```
printSummary(pinns, LossTerms, Residuals, totalTimes,error_L2s,errorVIs)
```

```
#  PINN model        Loss_terms residual    total_time error_L2s  errorVIs
==========================================================================
1  2D_DBC_IC         PDE_only   _heat       1.070e+01  3.283e-04  2.525e+00
```

```
2  2D_DBC_IC        DBC         _heat        8.917e+01  3.118e-01  5.381e+00
3  2D_DBC_IC        NBC         _heat        1.016e+02  5.241e-02  4.075e+00
```

This is a list of three PINN models developed for the same heat transfer problem, each with a different setting. The distinguishing feature of Model-3 is the inclusion of Neumann boundary conditions (NBCs). Since the target values `t_NBC_th` are all derived from the exact solution, the solution remains the same for all three models.

Among them, Model-1 is the most accurate and computationally efficient. Its accuracy is approximately two orders of magnitude higher, training time is also shorter than the others, and the error variation index is smaller.

This study clearly demonstrates that different modeling approaches for the same problem can lead to significantly different levels of accuracy. We emphasize the following key point:

> In building a PINN model, it is generally good practice to reduce the number of terms in the loss function. One effective strategy is to incorporate the trial solution directly in the PINN class definition, when possible.

4.7 Case Study: Time History of Solution of PINNs

Since the problem is time-dependent, this section investigates the time history of the solution produced by the PINN models. In this study, we set $t_{\text{end}} = 10t_{\text{char}}$ and evaluate how well the PINN predicts the solution as it evolves over time.

We consider the same heat transfer problem whose exact solution is given in Eq.(4.4). The problem domain is the unit square, with all four edges subject to zero Dirichlet boundary conditions (DBCs). The class `PINN2D_DBC_IC` is used to enforce both the DBC and the initial condition (IC) directly within the class definition. This setup enables us to focus solely on analyzing the time history of the PINN solution over the time domain $(0, t_{\text{end}}]$.

4.7.1 Train the PINN Model

The following snippet trains the PINN model using the `PINN2D_DBC_IC` class, where both the DBC and IC are enforced at the class level:

```
gr.set_seed_th(88)                                  # for reproducibility

# Generate dataset:
t_end = 10./(2*np.pi**2)
x_Ω_th, x_DBC_th, u_DBC_th = sampleData_xyt(N_clctn=N_sample,
                                N_DBC=N_DBC, N_IC=N_DBC, t_end=t_end)

print(f"{x_Ω_th.shape[0]} collocation points in the domain.")
print(f"{x_DBC_th.shape[0]} boundary points.")

w_s_th = 1.              # dummy weight, not used in a collocation scheme

def f_th(x, y, t):
    '''Source Term Function for the Poisson's equation in torch
    one may replace the dummy with actual f_th. '''
    return torch.zeros_like(x) # 0.*x + 0.*y + 0.*t;          set to zero
```

```
def f_DBC(x, y):
    return torch.sin(np.pi*x) * torch.sin(np.pi*y)   # true DBC boundary
def  f_IC(x, y):
    return torch.sin(np.pi*x) * torch.sin(np.pi*y)            # true IC

# A new model with DBC & IC enforced:
pinn_tanh = PINN2D_DBC_IC(Nn, NL, f_DBC, f_IC, my_actvtn)

# Training excecution:
x_DBC_th, t_DBC_th, w_DBC_th = None, None, None
use_DBC = False

x_NBC_th, t_NBC_th, w_NBC_th = None, None, None
I_NBC = ''                              # no Integration for NBC is used
use_NBC = False
w_NBC   = 1                                      # weight for the losses
LossTerm = 'PDE_only'                 # loss term in addition to loss_PDE

Trainer  = trainerPDE_DBC_NBC
Residual = residual_heat
DI_rule = ''                           # no Domain Integration is used
losses, totalTime = train_adam_lbfgs(pinn_tanh, Trainer, Residual, f_th,
       x_Ω_th, w_s_th, x_DBC_th, u_DBC_th, x_NBC_th, t_NBC_th, w_NBC_th,
                         I_NBC, epochs1=201, epochs2=11, use_lbfgs=True)
```

```
206 collocation points in the domain.
87 boundary points.
==================================================
201 epoches training with Adam:
Epoch    0, Loss:4.234e+01, PDE:4.234e+01, DBC:0.000e+00, NBC:0.000e+00
Epoch  100, Loss:6.989e+00, PDE:6.989e+00, DBC:0.000e+00, NBC:0.000e+00
Epoch  200, Loss:3.404e+00, PDE:3.404e+00, DBC:0.000e+00, NBC:0.000e+00
Time taken by adam: 2.720479 seconds

==================================================
11 epoches fine-tuning with LBFGS:
Epoch    0, Loss:7.566e-06, PDE:7.566e-06, DBC:0.000e+00, NBC:0.000e+00
Epoch   10, Loss:5.362e-06, PDE:5.362e-06, DBC:0.000e+00, NBC:0.000e+00

Training complete.
Time taken by lbfgs: 7.315880 seconds
Total time: 10.036359 seconds
Ratio (time_lbfgs/time_adam) of time taken per epoch:49.138808
```

As expected, both `loss_DBC` and `loss_NBC` are zero, and only the `loss_PDE` is minimized. The training is very fast, and the loss is reduced by approximately seven orders of magnitude.

4.7.2 Results Accuracy Evaluation

The following snippet computes the temperature distribution and the associated error in the domain at a fixed observation time slice:

```
# Evaluate the results:
N_plot = 100                                                # Grid density
# Sampling over rectangular domain (a, b)×(c, d):
a = 0.0; b = 1.0; c = 0.0; d = 1.0

x = np.linspace(a, b, N_plot)          # Create meshgrid over (a,b)x(c,d)
y = np.linspace(c, d, N_plot)

X, Y = np.meshgrid(x, y)
t_obs = 0.5*t_end                                  # the observation time
print(f"t_obs = {t_obs}")
T = t_obs * np.ones_like(X)                           # a section on time

XYT = np.stack([X.flatten(), Y.flatten(), T.flatten()], axis=1)
XYT_th = torch.tensor(XYT, dtype=torch.float32)

u_pred = pinn_tanh(XYT_th).detach().numpy().reshape(N_plot, N_plot)
u_exact = exact_u_heat(X, Y, t_obs)

error_L2, errorVI, u_diff = sln_error(u_pred, u_exact)
```

```
t_obs = 0.25330295910584444
Maximum Error: 1.7211e-04
Mean Absolute Error: 5.4637e-05
Error Variation Index (ratio of Max/Mean): 3.1501e+00
L2 Relative Error: 2.1982e-02
```

4.7.3 Plot Time History of the Solution

We now plot the time history of the solution predicted by the trained PINN model, alongside the exact solution. The following function is defined for this purpose:

```
def plot_time_history(model, u_exact_f, x_obs, y_obs, N_plot=200,
            dpi=300, t_min=0.0, t_max=1.0, device=device, f_name='tmp'):
    """
    Plots the time history of u(x, y, t) at a fixed (x_obs, y_obs)
    using a trained PINN model and compares it with the exact solution.
    Parameters:
    - model          : Trained PINN model (shape [N, 3] of [x, y, t])
    - u_exact_f      : Callable, exact solution u(x, y, t)
    - x_obs, y_obs : Fixed spatial coordinates
    - t_min, t_max : Time range
    - N_plot         : Number of time samples
    - device         : 'cpu' or 'cuda'
    """
    t_vals = np.linspace(t_min, t_max, N_plot)              # Time samples
```

```
    xtyt=np.array([[x_obs, y_obs, t] for t in t_vals], dtype=np.float32)
    xtyt_tensor = torch.tensor(xtyt).to(device)

    model.eval()                                       # Model prediction
    with torch.no_grad():
        u_pred = model(xtyt_tensor).cpu().numpy().flatten()

    # Exact solution
    u_exact_vals = np.array([u_exact_f(x_obs,y_obs,t) for t in t_vals])

    plt.figure(figsize=(5, 3), dpi=dpi)                            # Plot
    plt.plot(t_vals, u_exact_vals, 'g-',
                       label='Exact Solution $u_{exact}$', zorder=1)
    plt.plot(t_vals, u_pred, 'r--', label='PINN Prediction $u_{pred}$',
                                                             zorder=2)
    plt.xlabel('Time $t$')
    plt.ylabel('$u(x, y, t)$')
    plt.title(f'Time History at $(x={x_obs}, y={y_obs})$')
    plt.legend()
    plt.grid(True)
    plt.tight_layout()
    plt.savefig(f_name+'.png', dpi=500)
    plt.show()
```

Let us plot the time history curves at $x = 0.5$ and $y = 0.5$, which corresponds to the center of the unit square.

```
# Plot the time history of solutions of the PINN and the exact:
x_obs, y_obs = 0.5, 0.5
plot_time_history(pinn_tanh, exact_u_heat, x_obs, y_obs, N_plot=200,
                  dpi=300, t_min=0.0, t_max=t_end, device=device,
                                       f_name='imagesPN/u_t_xy0d5tnah')
```

As shown in the plot on the next page, good agreement is observed between the time-history results from the trained PINN model and the exact solution over the same time interval in which the model is trained.

As discussed in Section 4.3, the time history of the solution should decay exponentially. The trained PINN model correctly captures this behavior. At $t = 0.3$ and beyond, the solution is practically zero.

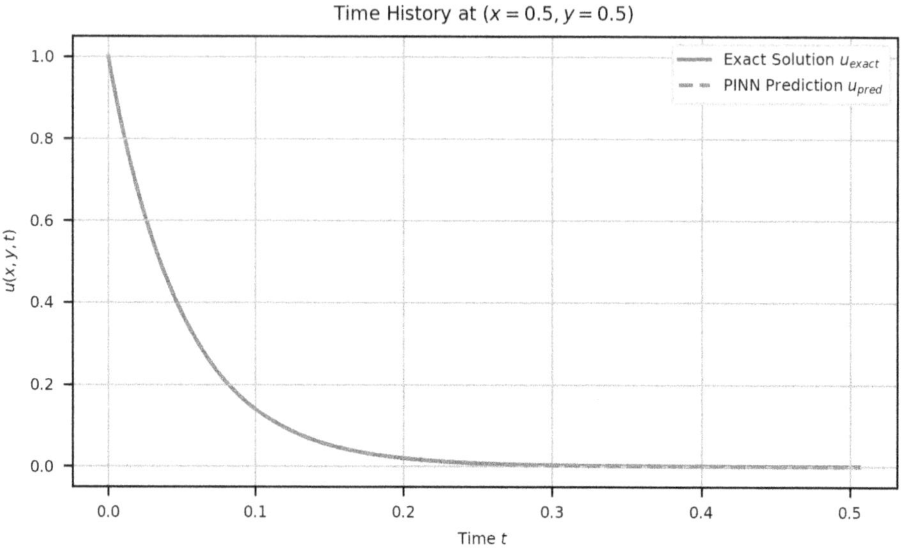

4.7.4 Tests on Alternative Activation

PINN models can employ various activation functions. So far, we have used `tanh()` as the activation function. Let us now test an alternative activation function, such as `exp()`, for this transient heat transfer problem, which exhibits behavior quite different from that of `tanh()`. The corresponding snippet is as follows:

```
gr.set_seed_th(88)                                    # for reproducibility

# User defined activation function
def my_actvtn(x):
    return torch.exp(x) # alternatives: sigmoid(x), sin(x) ...

pinn_exp = PINN2D_DBC_IC(Nn, NL, f_DBC, f_IC, my_actvtn)

# Training excecution:
losses, totalTime = train_adam_lbfgs(pinn_exp, Trainer, Residual, f_th,
      x_Ω_th, w_s_th, x_DBC_th, u_DBC_th, x_NBC_th, t_NBC_th, w_NBC_th,
                         I_NBC, epochs1=201, epochs2=11, use_lbfgs=True)
```

```
==================================================
201 epoches training with Adam:
Epoch    0, Loss:4.165e+01, PDE:4.165e+01, DBC:0.000e+00, NBC:0.000e+00
Epoch  100, Loss:8.160e+00, PDE:8.160e+00, DBC:0.000e+00, NBC:0.000e+00
Epoch  200, Loss:2.324e+00, PDE:2.324e+00, DBC:0.000e+00, NBC:0.000e+00
Time taken by adam: 2.080608 seconds

==================================================
11 epoches fine-tuning with LBFGS:
Epoch    0, Loss:1.229e-04, PDE:1.229e-04, DBC:0.000e+00, NBC:0.000e+00
Epoch   10, Loss:1.229e-04, PDE:1.229e-04, DBC:0.000e+00, NBC:0.000e+00
```

```
Training complete.
Time taken by lbfgs: 4.202962 seconds
Total time: 6.283571 seconds
Ratio (time_lbfgs/time_adam) of time taken per epoch:36.912084
```

Let's compute the solution accuracy:

```
# Evaluate the results:
u_pred = pinn_exp(XYT_th).detach().numpy().reshape(N_plot, N_plot)
u_exact = exact_u_heat(X, Y, t_obs)

error_L2, errorVI, u_diff = sln_error(u_pred, u_exact)
```

```
Maximum Error: 3.1885e-04
Mean Absolute Error: 1.0289e-04
Error Variation Index (ratio of Max/Mean): 3.0990e+00
L2 Relative Error: 4.0462e-02
```

We have successfully trained a new PINN model. The accuracy is somewhat lower than that of the previous model but remains largely comparable. The training is slightly faster, which may be due to the fact that `exp()` is computationally simpler than `tanh()`. We also tested the sigmoid activation function and found that it yielded similarly good results. However, when testing `relu`, the results deviated significantly. This may be attributed to the nonsmooth nature of the ReLU function.

The following code plots the time history curve of the solution predicted by the trained PINN model at $x = 0.5$ and $y = 0.5$, alongside the exact solution.

```
# Plot the time history of solutions of PINN and exact:
x_obs, y_obs = 0.5, 0.5
plot_time_history(pinn_exp, exact_u_heat, x_obs, y_obs, N_plot=200,
                  dpi=300, t_min=0.0, t_max=t_end, device=device,
                                          f_name='imagesPN/u_t_xy0d5exp')
```

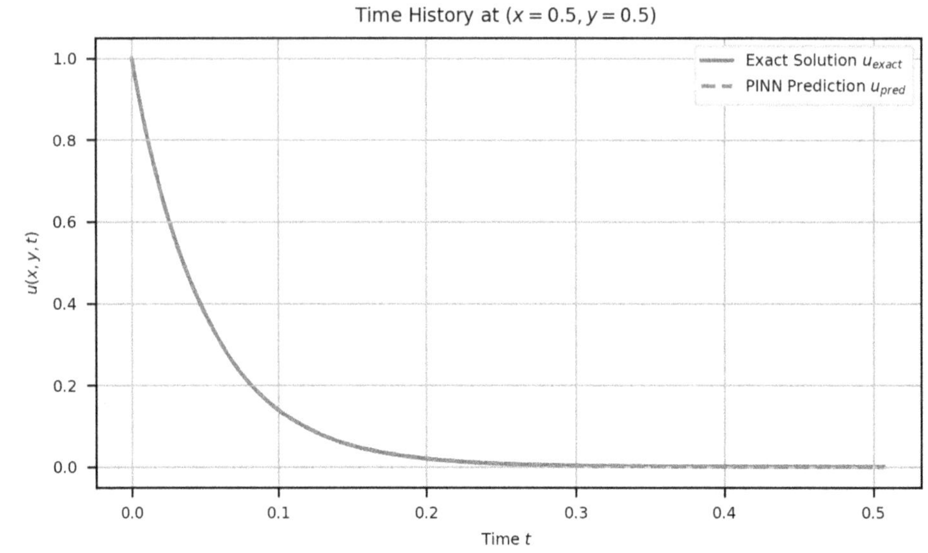

The foregoing figure shows the time history of u at the center of the unit square, obtained using the PINN model trained with the `exp()` activation function. We note the following:

> For training a PINN model, various types of activation functions that are nonlinear and differentiable can be used.

Differentiability is essential because PINNs rely on computing derivatives during the back-propagation process to train the model. Nonlinearity is required because, without it, the neural network reduces to a simple linear regressor [12]. The accuracy and efficiency of the model can depend heavily on how well the activation function matches the nature of the true solution of the problem. Readers are encouraged to try other activation functions using the code provided earlier.

4.8 Case Study: PINN for Plate with Hole and Free Edges

Let us now study another problem: heat transfer in a square plate with an elliptical hole at its center. This time, we assume:

- The left edge of the plate is fixed with the temperature set to zero.
- The bottom, top, and right edges of the plate are free (insulated).
- The edge of the ellipse is also free.

The setting of this problem is quite similar to the previous case of a square plate with a hole; however, the target values for the NBCs, `t_NBC_th`, are zero.

With this setup, the physical problem itself is changed, and no exact solution is available. We shall build and train PINN models to obtain an approximate solution.

4.8.1 Training of the PINN Model

The following snippet defines this problem in Python:

```
# Device management:
device = torch.device('cuda' if torch.cuda.is_available() else 'cpu')

# Initialization lists for book-keeping:
pinns = []                                        # for logging key variables
timeAdam  = []; timeLbfgs = []
Trainers  = []; Residuals = []
totalTimes= []
error_L2s = []; errorVIs = []                                  # log the error

# Generate dataset and input functions:
a=0; b=1.; c=0; d=1.                                               # domain
x0=0.5; y0=0.5; ra=0.3; rb=0.2                           # ellipse in center
t_end = 5*t_char

def is_inside_ellipse(x, y, x0=x0, y0=y0, ra=ra, rb=rb):
    '''Elliptical hole exclusion function'''
    return ((x - x0)**2 / ra**2 + (y - y0)**2 / rb**2) < 1

def f_th(x, y, t):
    '''Source Term Function for the Poisson's equation in torch. One may
    replace the dummy with actual f_th depending the study objective.'''
    return 0.*x + 0.*y + 1.*t

def f_IC(x, y):
    return 0.*x + 0.*y                              # zero initial condition

def f_DBC(x, y): return x   # one may change it to fit the problem setup

def f_NBC(xyt):              # list of 2 functions, [f/x, f/y], for NBC
    x = xyt[:,0:1];  y = xyt[:,1:2]; t = xyt[:,2:3];
    dfdx = 0.*x + 0.*y + 0.*t                                   # zero flux
    dfdy = 0.*x + 0.*y + 0.*t
    return [dfdx, dfdy]

use_NBC = True
```

Next, sample the data inside the domain and on the NBC boundary.

```
# Create a new PINN model for this problem:
gr.set_seed_th(88)                                   # for reproducibility

Nn = 200                                # Number of neurons in a hidden layer
NL = 4                                   # Number of (hidden) Layers in PINN
pinn = PINN2D_DBC_IC(Nn, NL, f_DBC, f_IC, my_actvtn)
P_star, P_total = print_model_parameters(pinn, print_table=False)
```

```
N_sample = 1*P_star                  # sampling points in 3D domain (x,y,t)
N_DBC= 2*int(sqrt(N_sample))+1   # points on each DB segement, dummy here
N_IC = N_sample                  # points in domain (x,y) for IC, dummy here

x_Ω_th, x_DBC_th, u_DBC_th, x_NBC_th, normal_NBC,dudn_val=\
     sample_xyt_D_ellipseNBC(f_NBC, N_clctn=N_sample, N_DBC=N_DBC,\
      N_NBC=N_DBC, N_ellipse=2*N_DBC, x0=x0, y0=y0, ra=ra,rb=rb,
                                         a=a,b=b,c=c,d=d, t_end=t_end)

print(f"Sampled {x_Ω_th.shape[0]} collocation points in the domain.")
print(f"Sampled {x_DBC_th.shape[0]} DBC boundary points.")
print(f"Sampled {x_NBC_th.shape[0]} NBC boundary points.")

w_s_th = 1.             # dummy weight, not used in a collocation scheme
```

```
P* (Total pseudo-dimension of the affine space) = 808
Total trainable parameters: 121601
Sampled 808 collocation points in the domain.
Sampled 57 DBC boundary points.
Sampled 285 NBC boundary points.
```

For this problem, the temperature distribution is expected to be significantly more complex due to the presence of the hole with zero-flux Neumann boundary conditions (NBCs). To capture this complexity, we construct a PINN model with 4 hidden layers, each containing 200 neurons. The resulting P^* number is 808.

According to the Samples–Neurons Theorem [12], we use 808 collocation points inside the domain to ensure the minimum model capacity appropriately.

```
# Training excecution:
w_NBC = 100.                                          # weight for loss_NBC
Trainer  = trainerPDE_DBC_NBC
Residual = residual_heat
losses, totalTime = train_adam_lbfgs(pinn, Trainer, Residual, f_th,
       x_Ω_th, w_s_th, x_DBC_th, u_DBC_th, x_NBC_th, t_NBC_th, w_NBC_th,
                        I_NBC, epochs1=201, epochs2=11, use_lbfgs=True)
```

```
==================================================
201 epoches training with Adam:
Epoch    0, Loss:8.295e-01, PDE:8.295e-01, DBC:0.000e+00, NBC:6.947e-03
Epoch  100, Loss:7.552e-03, PDE:7.552e-03, DBC:0.000e+00, NBC:3.058e-05
Epoch  200, Loss:8.631e-03, PDE:8.631e-03, DBC:0.000e+00, NBC:2.095e-05
Time taken by adam: 7.884431 seconds

==================================================
11 epoches fine-tuning with LBFGS:
Epoch    0, Loss:1.818e-03, PDE:1.818e-03, DBC:0.000e+00, NBC:1.324e-05
Epoch   10, Loss:1.757e-03, PDE:1.757e-03, DBC:0.000e+00, NBC:1.270e-05

Training complete.
Time taken by lbfgs: 52.371116 seconds
Total time: 60.255548 seconds
```

```
Ratio (time_lbfgs/time_adam) of time taken per epoch:121.373765
```

The PINN model took approximately one minutes to train. Since there is no exact solution available for this problem, direct comparison is not possible. However, the total loss and loss_PDE were reduced from the order of 10^0 to the order of 10^{-3}, indicating that the training was likely successful. Both the Dirichlet boundary condition (loss_DBC) and the initial condition were fully satisfied. Moreover, loss_NBC was reduced from the order of 10^{-3} to 10^{-5}, suggesting that the Neumann boundary conditions were also enforced with small residuals. Based on this, we can expect that the PINN model is reasonably well-trained, with an accuracy around the order of 10^{-2}.

4.8.2 On Tuning of the Weights for Loss Terms

It is worth noting that in training the foregoing model, we used $w_{\text{NBC}} = 100$ to magnify the contribution of loss_NBC, ensuring that it had sufficient influence on the total loss for the optimizer to minimize it effectively alongside loss_PDE. This weight was selected through trial and error.

Finding appropriate weights for loss terms is nontrivial and highly problem-dependent. In the author's practice, the weight is chosen so that the loss terms contribute roughly at the same level to the composite loss. This also underscores why it is typically advisable to minimize the number of distinct loss terms, in order to avoid the additional complexity of tuning multiple hyperparameters in the composite loss.

Weights may require tuning as training progresses, particularly if the loss term scales evolve differently. Adaptive weighting strategies (e.g., using dynamic reweighting or uncertainty-based schemes) are active areas of research. Interested readers are referred to Refs. [7,8,28,32]. Such adaptive weighting strategies may be useful for PINN models built with composite loss. This book will not discuss this topic further.

4.8.3 Plots of the Solution

We now examine the predicted solution by plotting the results using the following snippets:

```
def data4plot_XYT(pinn, hole, exact_u_xyt=None,
                  x0=0, y0=0, ra=0.2, rb=.1, N_plot=100, t_obs=.5):
    '''
    Prepare data for plots for a trained `pinn` model over a 2D domain
    with a `hole` at a fixed time, together with `exact_u_XY`.
    N_plot: resolution in one-direction.
    t_obs:  time of observation slice.
    '''
    x = y = np.linspace(0, 1, N_plot)
    X, Y = np.meshgrid(x, y)
    X_flat = X.flatten()
    Y_flat = Y.flatten()

    mask = ~hole(X_flat, Y_flat, x0=x0, y0=y0, ra=ra, rb=rb)
    T = t_obs * np.ones_like(X_flat)
    XYT = np.stack([X_flat, Y_flat, T], axis=1)
    XYT_masked = XYT[mask]
```

```
    XYT_tensor = torch.tensor(XYT_masked, dtype=torch.float32)
    pinn.eval()
    with torch.no_grad():
        u_pred_flat = pinn(XYT_tensor).detach().numpy().flatten()
    u_pred_full = np.full(X.shape, np.nan)
    idx = np.where(mask)
    u_pred_full.flat[idx] = u_pred_flat

    if exact_u_xyt is None:
         return X, Y, u_pred_full
    else:
        u_exact_flat=exact_u_xyt(XYT_masked[:,0],XYT_masked[:,1], t_obs)
        u_exact_full = np.full(X.shape, np.nan)
        u_exact_full.flat[idx] = u_exact_flat

    return X, Y, u_pred_full, u_exact_full
```

```
def plot2Dt_hole2(X, Y, u_pred, t_obs, vmin=0, vmax=1):
    '''
    Plot the contours and 3D surface of solutions for a plate with an
    hole. '''
    width = 6
    height = width / 2.0
    fig = plt.figure(figsize=(width, height))

    # Left: Contour plot
    ax1 = fig.add_subplot(1, 2, 1)
    c0 = ax1.contourf(X, Y, u_pred, cmap='viridis', levels=15)
    ax1.set_title(f"PINN Prediction at t={t_obs:.4e}", fontsize=8)
    ax1.set_aspect('equal')
    ax1.grid(linewidth=0.2)
    plt.colorbar(c0, ax=ax1)

    # Right: 3D surface plot
    ax2 = fig.add_subplot(1, 2, 2, projection='3d')
    surf= ax2.plot_surface(X,Y, u_pred, cmap='viridis',edgecolor='none')
    ax2.set_title(f'PINN Surface at t={t_obs:.4e}')
    ax2.set_xlabel('x')
    ax2.set_ylabel('y')
    ax2.set_zlabel('u')
    # fig.colorbar(surf, ax=ax2, shrink=0.5, aspect=10)       # Optional

    plt.tight_layout()
    plt.show()
```

```
t_obs=.5*t_end                      # take a time observation slice at t_obs
X, Y, u_pred = data4plot_XYT(pinn, is_inside_ellipse, exact_u_xyt=None,
                    x0=x0, y0=y0, ra=ra, rb=rb, N_plot=500, t_obs=t_obs)
```

```
plot2Dt_hole2(X, Y, u_pred, t_obs, vmin=0, vmax=0.005)
```

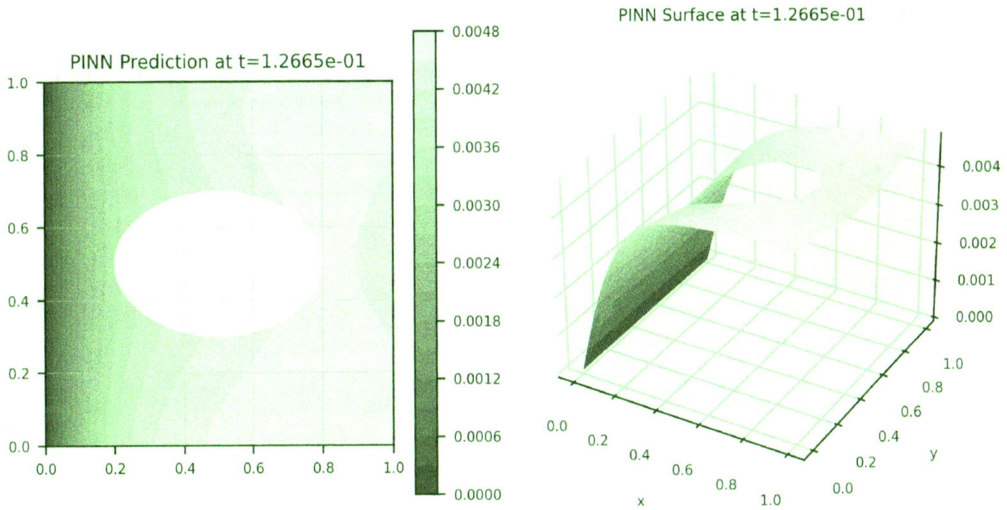

Let us also plot the history of the loss terms for both the **adam** and **lbfgs** optimizers:

```
plot3loss2(losses, ('adam initially','lbfgs in fine tunning'),
                   PDE='PDE', C1='DBC', C2='NBC', CL='Loss',log=False)
```

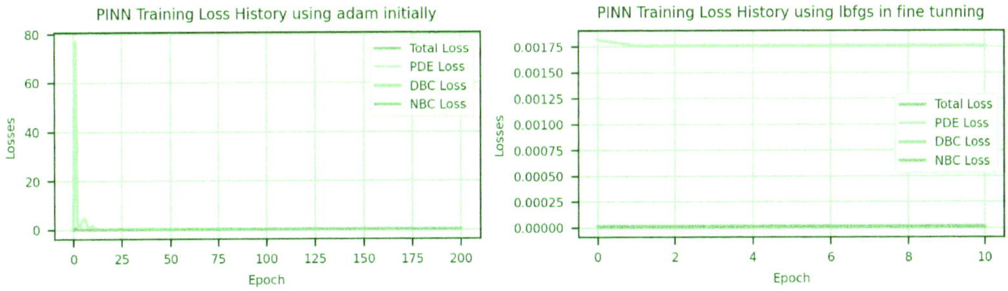

These loss terms converged nicely to a steady state after fine-tuning with **lbfgs**, implying that we found at least a locally optimal solution.

For this problem, the temperature distribution makes physical sense for the plate with a hole under uniform sunshine: the temperature is zero at the left edge, indicating a sink that continuously extracts heat to maintain the fixed temperature (Dirichlet boundary condition). The temperature gradually builds up toward the right edge due to heat insulation (Neumann conditions) on the remaining edges and along the boundary of the elliptical hole, which prevents heat from escaping.

4.9 Case Study: PINN with Fixed Left and Hole Edges

Let us now train a PINN model for a square plate with fixed left and elliptic edges, while keeping all other settings unchanged. This can be easily achieved by adding the function for the ellipse to the trial solution to properly enforce the DBCs. In this setting, the temperature solution is expected to become even more complex.

4.9.1 Training of the PINN Model

The following snippet defines the necessary functions, samples the data, and builds a PINN model.

```
gr.set_seed_th(88)                                   # for reproducibility

# Generate dataset and input functions:
a=0; b=1; c=0; d=1                                            # domain
x0=0.5; y0=0.5; ra=0.3; rb=0.2                        # ellipse in center
t_end=1.*t_char

def is_inside_ellipse(x, y, x0=x0, y0=y0, ra=ra, rb=rb):
    '''Elliptical hole exclusion function'''
    return ((x - x0)**2 / ra**2 + (y - y0)**2 / rb**2) < 1

def f_th(x, y, t):
    '''Source Term Function for the Poisson's equation in torch
    one may replace the dummy with actual f_th. '''
    return 0.*x + 0.*y + 1.*t

def f_IC(x, y):
    return 0.*x + 0.*y

def f_DBC(x, y):  # for trial solution for DBC on left and ellipse edge:
    return x*(((x-x0)**2 / ra**2 + (y-y0)**2 / rb**2)- 1)

def f_NBC(xyt):                    # list functions, [ f/ x,  f/ y], for NBC
    x = xyt[:,0:1];  y = xyt[:,1:2]; t = xyt[:,2:3];
    dfdx = 0.*x + 0.*y + 0.*t
    dfdy = 0.*x + 0.*y + 0.*t
    return [dfdx, dfdy]

use_NBC = True

x_Ω_th, x_DBC_th, u_DBC_th, x_NBC_th, normal_NBC,dudn_val=\
     sample_xyt_D_ellipseNBC(f_NBC, N_clctn=N_sample, N_DBC=N_DBC, \
     N_NBC=N_DBC, N_ellipse=2*N_DBC, x0=x0,y0=y0,ra=ra,rb=rb,
                                             a=a,b=b,c=c,d=d,t_end=t_end)

print(f"Sampled {x_Ω_th.shape[0]} collocation points in the domain.")
print(f"Is N_sample   P* ? {N_sample >= P_star}")
print(f"Sampled {x_DBC_th.shape[0]} DBC boundary points.")
print(f"Sampled {x_NBC_th.shape[0]} NBC boundary points.")

w_s_th = 1.              # dummy weight, not used in a collocation scheme

# Create a new PINN model, DBC & IC are same function for this problem:
```

```
pinn800 = PINN2D_DBC_IC(Nn, NL, f_DBC, f_IC, my_actvtn)
```

```
Sampled 808 collocation points in the domain.
Is N_sample  P* ? True
Sampled 57 DBC boundary points.
Sampled 285 NBC boundary points.
```

The following snippet trains the model, pinn800.

```
# Training excecution:
w_NBC = 1000.                                         # weight for loss_NBC
Trainer  = trainerPDE_DBC_NBC
Residual = residual_heat
losses, totalTime = train_adam_lbfgs(pinn800, Trainer, Residual, f_th,
       x_Ω_th, w_s_th, x_DBC_th, u_DBC_th, x_NBC_th, t_NBC_th, w_NBC_th,
                         I_NBC, epochs1=201, epochs2=11, use_lbfgs=True)
```

```
==================================================
201 epoches training with Adam:
Epoch    0, Loss:8.977e+01, PDE:8.977e+01, DBC:0.000e+00, NBC:8.856e-02
Epoch  100, Loss:8.424e-02, PDE:8.424e-02, DBC:0.000e+00, NBC:6.896e-05
Epoch  200, Loss:4.629e-02, PDE:4.629e-02, DBC:0.000e+00, NBC:3.753e-05
Time taken by adam: 9.314534 seconds

==================================================
11 epoches fine-tuning with LBFGS:
Epoch    0, Loss:7.102e-04, PDE:7.102e-04, DBC:0.000e+00, NBC:1.212e-07
Epoch   10, Loss:7.102e-04, PDE:7.102e-04, DBC:0.000e+00, NBC:1.212e-07

Training complete.
Time taken by lbfgs: 10.718145 seconds
Total time: 20.032679 seconds
Ratio (time_lbfgs/time_adam) of time taken per epoch:21.026253
```

This time, we used `w_NBC = 1000` to successfully train the PINN. For this problem, there is also no exact solution to compare with. Judging from the reduction of the loss terms, we can expect that this PINN model is reasonably well-trained with an accuracy on the order of 10^{-3}. Since all the loss terms have converged to a steady state, the trained model should produce at least a locally optimal solution.

4.9.2 Plots of the Solution

The following snippet plots the results and the history of these loss terms.

```
t_obs= 0.5* t_end                   # take a time observation slice at t_obs
X, Y, u_pred = data4plot_XYT(pinn800, is_inside_ellipse,
  exact_u_xyt=None, x0=x0, y0=y0, ra=ra, rb=rb, N_plot=500, t_obs=t_obs)

plot2Dt_hole2(X, Y, u_pred, t_obs, vmin=-6.e-4, vmax=1.2e-3)

plot3loss2(losses, ('adam initially','lbfgs in fine tunning'),
```

```
                    PDE='PDE', C1='DBC', C2='NBC', CL='Loss',log=False)
```

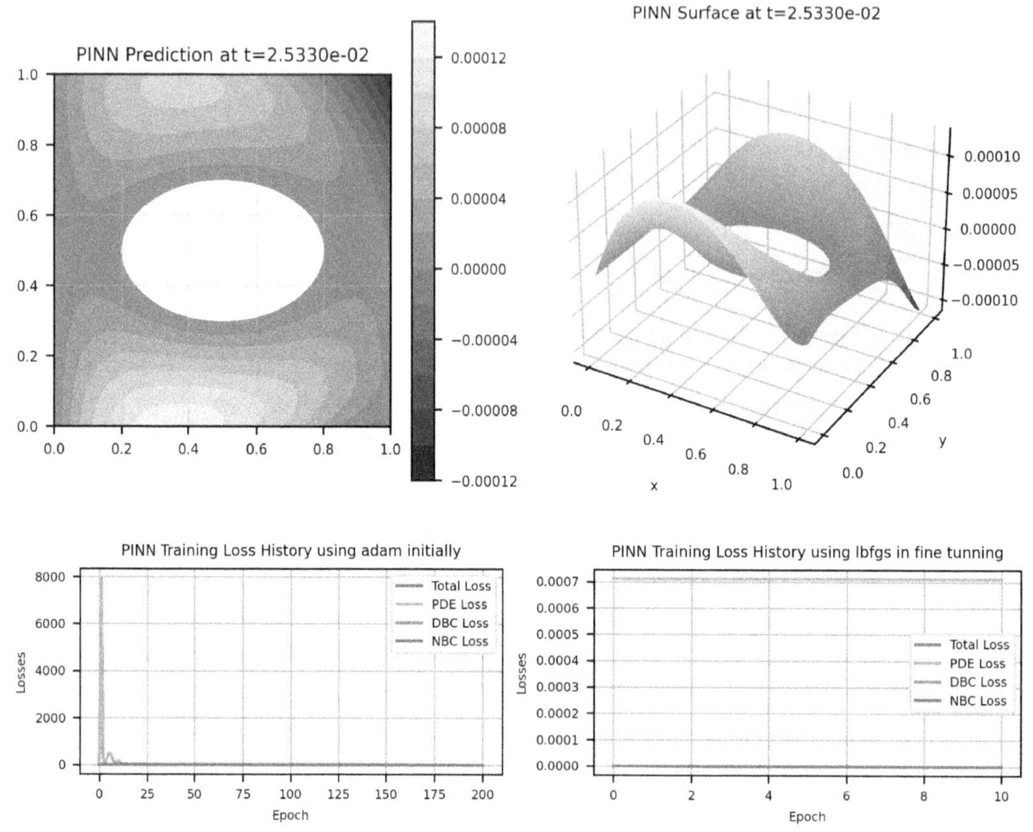

As shown in the surface plot, the solution function is quite complicated. From the distribution of the solution, we can gain some insight into its quality. Based on the setup of this problem, we know that the solution should be symmetric with respect to the line $y = 0.5$. However, the contour plots indicate that this symmetry is not well preserved. This is partly due to the fact that the training data points were randomly generated and thus not symmetrically distributed.

4.9.3 A Convergence Study

Note that the samples-neurons theorem , $m \geq P^*$, provides a lower bound for the number of sampling points needed to ensure that the solution is data-driven [12]. Improving the accuracy of the solution generally requires increasing the number of sampling points.

Therefore, another way to assess the reliability of the solution is to train a new model with more sampling points. If the solution from the coarse model is reasonably close (though not identical) to that of the refined model, we gain additional confidence in the trained PINN model. The following snippet trains a model using twice as many sampling points.

```
# Build and train a PINN with dense sampling data points:
# Generate data:
gr.set_seed_th(88)                                    # for reproducibility
```

```
x_Ω_th, x_DBC_th, u_DBC_th, x_NBC_th, normal_NBC,dudn_val=\
     sample_xyt_D_ellipseNBC(f_NBC, N_clctn=2*N_sample, N_DBC=2*N_DBC, \
             N_NBC=2*N_DBC, N_ellipse=4*N_DBC, x0=x0,y0=y0,ra=ra,rb=rb,
                                                  a=a,b=b,c=c,d=d,t_end=t_end)
print(f"Sampled {x_Ω_th.shape[0]} collocation points in the domain.")
print(f"Sampled {x_DBC_th.shape[0]} DBC boundary points.")
print(f"Sampled {x_NBC_th.shape[0]} NBC boundary points.")

# Create a new PINN model, DBC & IC are same function for this problem:
pinn1616 = PINN2D_DBC_IC(Nn, NL, f_DBC, f_IC, my_actvtn)

# Training excecution:
w_NBC = 100.                                            # weight for loss_NBC
Trainer  = trainerPDE_DBC_NBC
Residual = residual_heat
losses, totalTime = train_adam_lbfgs(pinn1616, Trainer, Residual, f_th,
       x_Ω_th, w_s_th, x_DBC_th, u_DBC_th, x_NBC_th, t_NBC_th, w_NBC_th,
                         I_NBC, epochs1=201, epochs2=11, use_lbfgs=True)
```

```
Sampled 1616 collocation points in the domain.
Sampled 114 DBC boundary points.
Sampled 570 NBC boundary points.
==================================================
201 epoches training with Adam:
Epoch    0, Loss:1.087e+01, PDE:1.087e+01, DBC:0.000e+00, NBC:9.697e-02
Epoch  100, Loss:1.269e-02, PDE:1.269e-02, DBC:0.000e+00, NBC:5.485e-05
Epoch  200, Loss:5.831e-03, PDE:5.831e-03, DBC:0.000e+00, NBC:2.314e-05
Time taken by adam: 22.233448 seconds

==================================================
11 epoches fine-tuning with LBFGS:
Epoch    0, Loss:3.847e-04, PDE:3.847e-04, DBC:0.000e+00, NBC:7.868e-07
Epoch   10, Loss:3.544e-04, PDE:3.544e-04, DBC:0.000e+00, NBC:6.383e-07

Training complete.
Time taken by lbfgs: 77.653305 seconds
Total time: 99.886752 seconds
Ratio (time_lbfgs/time_adam) of time taken per epoch:63.819957
```

```
# Plot the solution:
t_obs= 0.5* t_end                       # take a time observation slice at t_obs
X, Y, u_pred = data4plot_XYT(pinn1616, is_inside_ellipse,
  exact_u_xyt=None, x0=x0, y0=y0, ra=ra, rb=rb, N_plot=500, t_obs=t_obs)
plot2Dt_hole2(X, Y, u_pred, t_obs, vmin=-9.e-4, vmax=1.2e-3)
plot3loss2(losses, ('adam initially','lbfgs in fine tunning'),
                     PDE='PDE', C1='DBC', C2='NBC', CL='Loss',log=False)
```

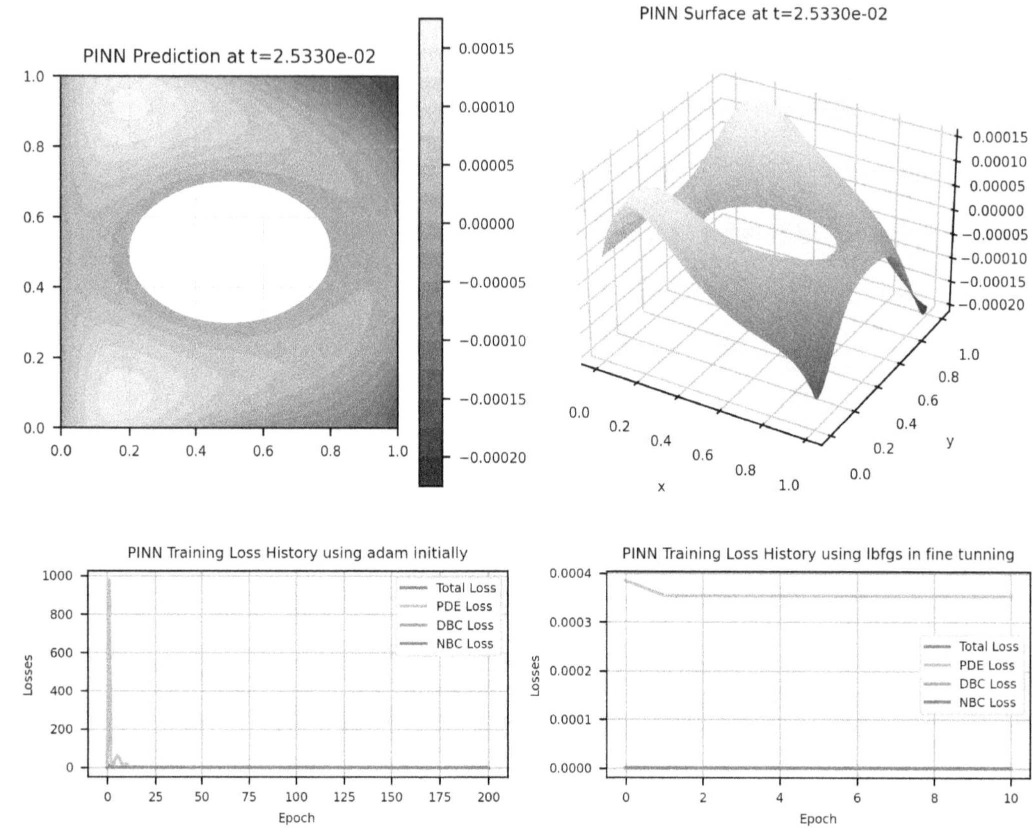

It is shown that the symmetry is significantly improved. The training took about 2 minutes, which is roughly double the time required for the previous coarse model.

Readers are encouraged to train a model with different number of sampling points using the provided code and compare the predicted results with those from the coarse model.

4.10 Remarks

This chapter presents various PINN models for parabolic PDEs, which govern diffusion-type problems, using heat transfer as a representative example. Leveraging the strong performance of collocation-based schemes shown in previous chapters, the training data is generated via collocation sampling guided by the samples-neurons theorem. The following key points are noted:

1. Both the Dirichlet boundary conditions (DBCs) and initial conditions (ICs) can be enforced as additional terms in the total composite loss; however, it is preferable to enforce them directly within the PINN class for improved efficiency and accuracy.

2. For the problem of the square plate, the trained PINN models produced the exact unique solution.

3. Problems involving domains with holes have been successfully modeled using PINNs, accounting for DBCs, ICs, and Neumann boundary conditions (NBCs). These example problems have unique solutions but unknown; the PINN models provide at least locally

optimal solutions.

The modeling strategies and techniques discussed in this chapter are, in principle, applicable to a wide range of diffusion-related problems, including but not limited to mass/population diffusion, viscous diffusion in fluids, magnetic diffusion in conductors, and neutron diffusion in nuclear reactors.

Chapter 5

PINN for Wave Equations

Let's import the necessary external modules or dependencies for later use in this chapter.

```
# Place cursor in this cell, and press Ctrl+Enter to import dependencies.
import os
import sys                                  # for accessing the computer system
sys.path.append('../grbin/')                   # Add in the path to your system

from commonImports import *              # Import dependencies from '../grbin/'
import grcodes as gr                          # Import the module of the author
importlib.reload(gr)                      # When grcodes is modified, reload it
import math
import torch

init_printing(use_unicode=True)               # For latex-like quality printing
np.set_printoptions(precision=4, suppress=True)   # Digits in print-outs
```

Wave propagation is a phenomenon encountered in many practical applications, including acoustic waves (Chapter 15 of Ref. [20]), elastic waves [22], seismic waves, water waves, electromagnetic waves (governed by Maxwell's equations), and quantum mechanics (governed by complex-valued wave equations), just to name a few. Wave equations are typical examples of **hyperbolic partial differential equations (PDEs)** and have been extensively studied using traditional numerical methods.

With the techniques developed in the previous chapters, this chapter focuses on building **Physics-Informed Neural Network (PINN)** models for **time-dependent wave equations**, governed by second-order PDEs involving both time and space.

5.1 Governing PDE for 2D Wave Equation

We begin by considering a basic scalar **2D wave equation**, which evolves over a one-dimensional time domain and a two-dimensional spatial domain — specifically, a unit square. The governing PDE is given as follows:

$$\frac{\partial^2 u}{\partial t^2} = c^2 \underbrace{\left(\frac{\partial^2 u}{\partial x^2} + \frac{\partial^2 u}{\partial y^2}\right)}_{\Delta u} + f(x,y,t), \quad (x,y,t) \in (0,1)^2 \times (0, t_{\text{end}}] \tag{5.1}$$

In which:

- The unknown function $u(x, y, t)$ represents the distribution of wave displacement.

- The force/source term $f(x, y, t)$ denotes a given excitation, typically defined over the space-time domain .

- The constant c is the wave speed, which is usually positive. For example, the longitudinal wave speed in aluminum is approximately 6350 m/s.

Equation (5.1) is a partial differential equation (PDE) involving spatial coordinates (x, y) defined over a domain Ω, and time $t \in (0, t_{\text{end}}]$, where t_{end} denotes the end time of the simulation. Therefore, the problem is inherently **three-dimensional** in terms of its independent variables.

Unlike the heat transfer equation discussed in the previous chapter — which involves only a first-order time derivative — this wave equation contains a **second-order time derivative**. However, the spatial operator remains the **Laplace operator**, as before.

Essentially, Eq.(5.1) asserts that the second time derivative of u (representing acceleration) is equal to the sum of its second spatial derivatives (associated with internal stresses), scaled by the square of the wave speed, plus the applied body force. This expression can be interpreted as a **generalized form of Newton's second law** applied at every point in the domain Ω.

Equation (5.1) is typically accompanied by the following conditions:

- **Dirichlet boundary condition (DBC):** the displacement u is prescribed on part of the boundary of the spatial domain;

- **Initial condition (IC):** $u(x, y, 0)$ specifies the initial displacement field at time $t = 0$;

- **Initial velocity (IV):** $\dot{u}(x, y, 0)$ provides the initial velocity distribution at time $t = 0$, where $\dot{u}$ is shorthand for $\frac{\partial u}{\partial t}$.

Note that there could also be Neumann boundary conditions (NBCs), as discussed in the previous chapters, because the spatial partial derivatives in the wave equation are of second-order. In this chapter, however, we will not repeat the discussion from the previous chapter on how to handle NBCs. The example problems studied here will be set without NBCs so that our focus can remain on the DBC, IC, and IV together with the PDE.

5.1.1 Exact Solution of Waves Induced by Initial Condition

The Dirichlet boundary condition (DBC) and initial condition (IC) are similar to those for the heat transfer problem. The initial velocity condition (IV) arises due to the second-order time derivative in the wave equation. Under the following conditions:

- **Zero Dirichlet boundary conditions** on all edges of the unit square domain:

$$u(x, y)\big|_{\text{on all edges}} = 0 \tag{5.2}$$

- **Zero force term** in the entire spatial-temporal domain:

$$f(x, y, t) = 0, \quad (x, y, t) \in (0, 1)^2 \times (0, t_{\text{end}}] \tag{5.3}$$

- **Initial displacement** at $t = 0$ given by:

$$f_{\text{IC}}(x, y) = u(x, y, 0) = \sin(\pi x)\sin(\pi y) \tag{5.4}$$

- **Zero initial velocity** throughout the domain:

$$f_{\text{IV}}(x, y) = \frac{\partial u}{\partial t}(x, y, 0) = 0 \tag{5.5}$$

the exact solution to Eq. (5.1) becomes:

$$u_{\text{exact}}(x, y, t) = \cos(\sqrt{2}\pi c\, t)\sin(\pi x)\sin(\pi y) \tag{5.6}$$

This can be readily confirmed by substituting Eq.(5.6) into the PDE and verifying that it satisfies the boundary condition (5.2), the initial condition (5.4), and the initial velocity condition (5.5). The Sympy snippet is as follows:

```
# Define symbols and functions in Sympy to be used in this code cell:
x, y, t = sp.symbols('x, y, t')

f_sp = 0                                              # define the surce term
IC_sp = sp.sin(sp.pi * x) * sp.sin(sp.pi * y)                 # define the IC
IV_sp = 0                                                     # define the IV
# Define the exact solution:
u_sp = sp.cos(sqrt(2)*sp.pi*t) * sp.sin(sp.pi*x) * sp.sin(sp.pi*y)

# Compute the wave equation:
waveEq = sp.diff(sp.diff(u_sp,x),x)+sp.diff(sp.diff(u_sp,y),y) + f_sp -\
                                               sp.diff(sp.diff(u_sp,t),t)
display(Math(f"\\text{{Is the wave Eq. satisfied? }}{waveEq==0}"))
display(Math(f"\\text{{Is IC = IC_sp? }}{u_sp.subs(t,0)==IC_sp}"))
display(Math(f"\\text{{Is IV = IV_sp? }}\
                                   {sp.diff(u_sp,t).subs(t,0)==IV_sp}"))
del x, y, t                       # release these variables for other use later
```

Is the wave Eq. satisfied? *True*

Is IC = IC_sp? *True*

Is IV = IV_sp? *True*

This setup of the wave propagation problem involves a wave field generated by the initial condition. In other words, the cause of the wave is the $f_{\text{IC}}(x, y)$. Since the amplitude of $f_{\text{IC}}(x, y)$ is already unity, normalization is not required for this problem.

With this setting, it is evident that the solution to the wave equation is **sinusoidal**, a characteristic feature of wave phenomena. Since c typically takes a large value in SI units (e.g.,

$c = 6350$ m/s for longitudinal waves in aluminum), the time variable t must be very small in practice—otherwise, the term $\cos(\sqrt{2}\pi c t)$ would undergo many oscillations .

To maintain physical realism and numerical stability in computations, we will later constrain t appropriately in relation to the spatial domain.

To simplify the formulation—and **without loss of generality**—we can perform a **dimensionless analysis** by setting $c = 1$. This normalization helps keep numerical values close to unity, which is ideal for training PINN models. For physical predictions involving an actual wave speed c, one can scale the time variable as $t \mapsto ct$ in the trained model, and the solution remains valid.

The following code snippet defines the exact solution, as well as the functions for the initial and boundary conditions in Numpy, for later study and comparison purposes.

```
# Define the known functions in Numpy for the 2D Wave Equation:

def exact_u_wave(x, y, t):
    '''Exact solution for the wave equation.'''
    return np.cos(np.sqrt(2)*np.pi*t) * np.sin(np.pi*x)*np.sin(np.pi*y)

def f_IC(x, y):
    '''Initial condition (IC) for the wave equation. '''
    return torch.sin(np.pi * x) * torch.sin(np.pi * y)

def f_DBC(x, y):
    '''Function for the Dirichlet boundary for the wave equation. '''
    return torch.sin(np.pi*x) * torch.sin(np.pi*y)
```

5.1.2 Time Domain Size

For the given spatial domain $(0,1)^2$, the value of t_{end} (which defines the time domain $(0, t_{\text{end}}]$) must be properly set when sampling data for PINN models. This requirement is a major distinction from static problems (e.g., Poisson's equation).

In a PINN model for Eq.(5.1), we treat the space-time domain as a **3D domain** and apply the **equal-number sampling** scheme. That is, we sample the same number of points along the x-, y-, and t-axes. This enables the construction of an $(m, 3)$ tensor that can flow efficiently through the neural network.

Based on the exact solution in Eq.(5.6), a characteristic upper bound for t_{end} can be derived as:

$$t_{\text{char}} \leq \sqrt{2} \times x_{\text{end}} = \frac{\sqrt{2}}{2} \approx 0.707 \tag{5.7}$$

Here, x_{end} denotes **half the spatial domain size**. This constraint ensures that, for $c = 1$, the wave emitted from the center is contained within the domain $(0,1)^2$ during the time interval $(0, t_{\text{char}}]$.

If the simulation end time is set to $t_{\text{end}} = 2t_{\text{char}}$, the wave will have traveled across the domain and reflected back once, thereby capturing wave-boundary interactions.

5.2 Procedure and Codes for PINN Models

5.2.1 Preparation for Coding

The following snippet imports the necessary Python classes and functions developed in previous chapters. These utilities will be used in this chapter for PINN model construction, training, and evaluation:

```
from grcodes import train_adam_lbfgs, printSummary, compute_NBC_loss,\
            sln_error, plot_sln_XY, PINN2D_DBC_IC, trainerPDE_DBC_NBC,\
            print_model_parameters, sample_xyt_D_ellipseNBC, plot3loss2
```

Readers may review the previous chapter for the functionality of these snippets. We will not repeat the details here.

The following snippet defines the device for training PINN models. It ensures that a CUDA GPU is used for fast computation if available; otherwise, the CPU is used.

```
# Device management: ensure CUDA is available if you plan to use GPU
device = torch.device('cuda' if torch.cuda.is_available() else 'cpu')
gr.device = device                           # inject device to gr module
torch.cuda.empty_cache()                     # Clear previous model state
print(f"Using device: {device}")

# user defined activation
def my_actvtn(x):
    return torch.tanh(x) #sin(x) #
```

```
Using device: cpu
```

We are now ready to build and train PINN models.

5.2.2 A Plain PINN Class for Wave Equation

The following snippet defines a plain PINN class for solving 2D wave equations. It closely follows the design introduced in Chapter 2 but is implemented in a slightly more concise form. The primary distinction lies in the extension of the domain to a 3D space-time setting, requiring the input dataset to have shape $[m, 3]$, corresponding to the triplet (x, y, t).

```
class PINN_Wave2D(nn.Module):
    '''A plain PINN class for wave equation in torch. It uses tanh as
       the activation function.'''
    def __init__(self, Nn, NL):
        super().__init__()
        layers = []
        for _ in range(NL):
            layers.append(nn.Linear(3 if _== 0 else Nn, Nn))
            layers.append(nn.Tanh())          # or use nn.Sin() if desired
        layers.append(nn.Linear(Nn, 1))
        self.model = nn.Sequential(*layers)

    def forward(self, xyt):
```

```
        return self.model(xyt)
```

A PINN model can use various activation functions. The following class uses cosine function as the activation function, which will be used in later examples. Readers can also use it for any PINN model with a user-defined function for the same class.

```
class Cosine(nn.Module):
    def forward(self, x):
        return torch.cos(x)                            # tanh(x), sin(x)...

class PINN_Wave2Dcos(nn.Module):
    '''A plain PINN class for the 2D wave equation using sine
    activations in PyTorch.'''
    def __init__(self, Nn, NL):
        super().__init__()
        layers = []
        for i in range(NL):
            layers.append(nn.Linear(3 if i == 0 else Nn, Nn))
            layers.append(Cosine())    # use sine activation via subcalss
        layers.append(nn.Linear(Nn, 1))
        self.model = nn.Sequential(*layers)

    def forward(self, xyt):
        return self.model(xyt)
```

5.2.3 Residual Computation of Wave Equations

The following snippet defines the computation of the residual for the wave equation. This residual quantifies the discrepancy between the current PINN model prediction and the governing PDE. The implementation in Python is structurally similar to that used for the heat transfer problem, with one key difference: the wave equation involves a **second-order derivative with respect to time**, whereas the heat transfer equation involves only a **first-order time derivative**.

Although this may appear to be a minor modification, it fundamentally changes the nature of the PDE and leads to drastically different solution behaviors, as evident from the exact solution discussed earlier.

```
def residual_wave(pinn, f_th, xyt, w_s_th=None):
    '''
    Compute the residual in the space-time domain for the wave equation.
    w_s_th: dummy weight, not used in a collocation sheme.
    '''
    xyt.requires_grad_(True)
    x1 = xyt[:, 0:1]; x2 = xyt[:, 1:2]              # get x and y coordinates
    t  = xyt[:, 2:3]                                     # get t coordinates

    x1_device = x1.to(device)
    x2_device = x2.to(device)
    t_device  = t.to(device)
```

```
    u = pinn(xyt)
    grads = torch.autograd.grad(u, xyt, grad_outputs=torch.ones_like(u),
                                            create_graph=True)[0]
    u_x, u_y, u_t = grads[:, 0:1], grads[:, 1:2], grads[:, 2:3]

    u_xx= torch.autograd.grad(u_x,xyt,grad_outputs=torch.ones_like(u_x),
                                        create_graph=True)[0][:, 0:1]

    u_yy= torch.autograd.grad(u_y,xyt,grad_outputs=torch.ones_like(u_y),
                                        create_graph=True)[0][:, 1:2]

    u_tt= torch.autograd.grad(u_t,xyt,grad_outputs=torch.ones_like(u_t),
                                       create_graph=True)[0][:, 2:2+1]

    f_val = f_th(x1_device, x2_device, t_device)
    residual = (u_tt - u_xx - u_yy - f_val)**2            # PDE residaul

    return residual.mean(), residual                     # scalar, tensor
```

5.2.4 Trainer for PINN of Wave Equation

Since the wave equation involves a second-order time derivative, the associated PINN trainer must handle two types of initial conditions: one for the displacement field $u(x, y, 0)$ and another for the velocity field $\dot{u}(x, y, 0)$. These must be satisfied by the solution along with the Dirichlet boundary conditions (DBCs) and the governing PDE itself.

Accordingly, the trainer implements a **composite loss function** , as defined in Eq.(2.6) in Chapter 2, which consists of the following components:

- PDE residual loss (`loss_PDE`)
- Initial condition loss (`loss_IC`)
- Initial velocity loss (`loss_IV`)
- Boundary condition loss (`loss_DBC`)

To enable **adaptive training**, a few additional lines have been included in the trainer function. Adaptive training will be discussed in more detail later and demonstrated through a specific example.

Let us now define a function, `trainerPDE_ICs_DBC`, that computes and logs each individual loss term, the total composite loss, and its gradient with respect to the trainable parameters of the PINN model:

```
def trainerPDE_ICs_DBC(pinn, residual, optimizer, f_th, xyt_clctn,w_s_th,
                xyt_DBC, u_DBC, xyt_IC, u_IC, xyt_IV, u_IV, epochs=201,
                                                       adapt_every=0):
    """
```

```
    Train a PINN using loss_PDE + loss_DBC + loss_IC + loss_IV.
    Args: pinn:       the PINN model (nn.Module)
          residual:   function to compute residual loss
          optimizer:  PyTorch optimizer (e.g. Adam, LBFGS)
          epochs:     number of training iterations for PINN
          f_th:       the force/source function~ f(x, y)
          w_s_th:     weights (dummy for collocation), a tensor on device
          xyt_clctn: points (tensor, requires_grad=True, on device)
          xyt_DBC:    cooresponding boundary points (tensor,
                      requires_grad=True, on device)
          xyt_IC, xyt_IV: points for initial conditions (tensor,
                        requires_grad=True, on device)
          u_DBC, u_IC, u_IV: target values on bondary (tensor,on device)
          adapt_every: if nonezero, adaptive smapling points are added.
    """
    loss_history={"total": [], "PDE": [], "DBC": [], "IC": [], "IV": []}

    if epochs <= 301: n_prnt = 100
    else:             n_prnt = max(1, round(epochs / 3, -2))

    for epoch in range(epochs):
        # Initialize loss variables for the current epoch
        loss_PDE = None; loss_DBC = None; loss_IC = None
        loss_IV  = None; total_loss = None

        def closure():
            nonlocal loss_PDE, loss_DBC, loss_IC, loss_IV, total_loss
            # Allow modifying variables from outer scope
            optimizer.zero_grad()
            loss_PDE, _ = residual(pinn, f_th, xyt_clctn, w_s_th)
            u_pred_DBC = pinn(xyt_DBC)
            loss_DBC = torch.mean((u_pred_DBC - u_DBC)**2)
            loss_IC = torch.mean((pinn(xyt_IC) - u_IC)**2)    # IC for u

            xyt_IV.requires_grad_(True) # IV: du_dt, velocity condition
            u_pred = pinn(xyt_IV)
            v_pred = torch.autograd.grad(u_pred, xyt_IV,
                   torch.ones_like(u_pred), create_graph=True)[0][:,2:3]
            loss_IV = torch.mean((v_pred - u_IV)**2)
            total_loss = loss_PDE + loss_DBC + loss_IC + loss_IV

            total_loss.backward()                 # Perform backward pass

            return total_loss  # Return the total loss for the optimizer

        if isinstance(optimizer, torch.optim.LBFGS):
            optimizer.step(closure)
            # losses were updated by the last closure call:
            total_loss_val = total_loss.item()
            loss_PDE_val = loss_PDE.item()
```

```
        loss_DBC_val = loss_DBC.item()
        loss_IC_val = loss_IC.item()
        loss_IV_val = loss_IV.item()

    else: # Adam or other optimizers
        total_loss_th = closure()
        optimizer.step()
        total_loss_val = total_loss_th.item()
        loss_PDE_val = loss_PDE.item()         # set inside closure
        loss_DBC_val = loss_DBC.item()
        loss_IC_val = loss_IC.item()
        loss_IV_val = loss_IV.item()

    # Logging:
    with torch.no_grad():
        loss_history["total"].append(total_loss_val)
        loss_history["PDE"  ].append(loss_PDE_val)
        loss_history["DBC"  ].append(loss_DBC_val)
        loss_history["IC"   ].append(loss_IC_val)
        loss_history["IV"   ].append(loss_IV_val)

    if epoch % n_prnt == 0 or epoch == epochs - 1:
        print(f"Epoch{epoch:5d}, Loss:{total_loss_val:.2e}, "
              f"PDE:{loss_PDE_val:.2e}, DBC:{loss_DBC_val:.2e},"
              f" IC:{loss_IC_val:.2e}, IV:{loss_IV_val:.2e}")

    if adapt_every!=0 and epoch % adapt_every == 0 and epoch > 0:
        points_to_add = adaptive_sample(pinn, N_new=2500)
        xyt_clctn = torch.cat([xyt_clctn, points_to_add], dim=0)
        print(f"New collocation set size: {xyt_clctn.shape}")

return loss_history
```

5.2.5 Data Sampling and Preparation

We need one more snippet to generate the sampling dataset for the wave equation. This process is similar to that used for the heat transfer problem. The key differences are:

- The wave equation involves **second-order time derivatives**, so the time axis must be handled carefully.
- An additional **initial velocity condition** must be enforced, in addition to the initial and boundary conditions.

The following snippet prepares the dataset, including collocation points in the 3D space-time domain, and ensures correct handling of both the initial condition $u(x, y, 0)$ and the initial velocity $\dot{u}(x, y, 0)$.

```
def sampleData(N_clctn=2500, N_DBC=50, N_IC=50, x_end=1.0, y_end=1.0,
                                                        t_end=0.7):
    '''
```

```
    Sample the data points inside and on boundaries of a rectangular
    domain (a,b)×(c,d) for a PINN model of the wave equation.
    `f_DBC` and `f_IC` are used from the surrounding scope, and hence
    must be defined before invoking `sampleData`.
    '''
    # Inside the space-time domain:
    #xyt_clctn = np.random.rand(N_clctn, 3)  # if x,y,t are all in (0,1)
    #scaled = start + (end - start)*np.random.rand(N_clct)        # scaling
    xyt_clctn = np.zeros((N_clctn, 3))
    xyt_clctn[:, 0] = np.random.rand(N_clctn)*x_end      # x in (0, x_end)
    xyt_clctn[:, 1] = np.random.rand(N_clctn)*y_end      # y in (0, y_end)
    xyt_clctn[:, 2] = np.random.rand(N_clctn)*t_end      # t in (0, t_end)

    # For DBCs o n the four edges:
    x = np.random.rand(N_DBC)
    y = np.random.rand(N_DBC)
    t = np.random.rand(N_DBC)*t_end

    xyt_DBC = np.vstack([
        np.stack([x,                 np.zeros(N_DBC), t], axis=1),
        np.stack([x,           y_end*np.ones (N_DBC), t], axis=1),
        np.stack([np.zeros(N_DBC),        y,          t], axis=1),
        np.stack([x_end*np.ones(N_DBC),  y,           t], axis=1)])

    xyt_DBC_th = torch.tensor(xyt_DBC, dtype=torch.float32,
                                                 device=device)
    u_DBC = f_DBC(xyt_DBC_th[:, 0:1], xyt_DBC_th[:, 1:2])
                                     #xyt_DBC_th[:, 2:3])      not needed

    # For ICs and IVs in the domain:
    x = np.random.rand(N_IC)
    y = np.random.rand(N_IC)
    t0 = np.zeros_like(x)

    xyt_IC = np.stack([x, y, t0], axis=1)
    xyt_IC_th=torch.tensor(xyt_IC,dtype=torch.float32,device=device)
    u_IC = f_IC(xyt_IC_th[:, 0:1], xyt_IC_th[:, 1:2])

    xyt_IV_th = xyt_IC_th.clone()
    u_IV = torch.zeros_like(u_IC)

    return (torch.tensor(xyt_clctn, dtype=torch.float32, device=device),
                  xyt_DBC_th, u_DBC, xyt_IC_th, u_IC, xyt_IV_th, u_IV)
```

Accordingly, we define the following function to manage the training process. It supports the use of both the first-order optimizer `adam` for initial training and the quasi-second-order optimizer `lbfgs` for fine tuning, if desired. The function invokes the trainer defined earlier and also accommodates the use of adaptive sampling datasets.

```
def train_t_adam_lbfgs(pinn, trainer, residual, f_th, xyt_clctn, w_s_th,
                xyt_DBC, u_DBC, xyt_IC, u_IC, xyt_IV, u_IV, epochs1=201,
              epochs2=21, use_lbfgs=True, adapt_every=50, lr_adam=1e-3):
    '''
    Train a PINN model using both adam and lbfgs, by shoice.
    adam trains initial `epochs1`, and lbfgs fine tuns `epochs2` ecpochs
    '''
    #pinn.to(device) # Move model to device BEFORE creating optimizers.
    import time
    start_time = time.perf_counter()
    # 1. Train with Adam
    adam = torch.optim.Adam(pinn.parameters(), lr=lr_adam)
    print("="*50)
    print(f"{epochs1} epoches training with Adam:")

    history_adam = trainer(pinn, residual, adam, f_th, xyt_clctn, w_s_th,
              xyt_DBC, u_DBC, xyt_IC, u_IC, xyt_IV, u_IV, epochs=epochs1,
                                              adapt_every=adapt_every)
    end_time = time.perf_counter()
    time_adam = end_time - start_time
    print(f"Time taken by adam: {time_adam:.6f} seconds")

    if use_lbfgs:                                 # Use the renamed flag
        start_time = time.perf_counter()
        # 2. Fine-tune with LBFGS. # Create NEW LBFGS optimizer instance
        lbfgs_optimizer = torch.optim.LBFGS(pinn.parameters(), lr=1.0,
              max_iter=1500, tolerance_grad=1e-9, tolerance_change=1e-9,
              history_size=100, line_search_fn='strong_wolfe')
        print("\n" + "="*50)                    # standard
        print(f"{epochs2} epoches fine-tuning with LBFGS:")

        # Pass the SAME tensors that were used for Adam training
        history_lbfgs = trainer(pinn, residual, lbfgs_optimizer, f_th,
           xyt_clctn, w_s_th, xyt_DBC, u_DBC, xyt_IC, u_IC, xyt_IV, u_IV,
                                      epochs=epochs2, adapt_every=0)
        print(f"\nTraining complete.")
        end_time = time.perf_counter()
        time_lbfgs = end_time - start_time
        totalTime = time_adam + time_lbfgs
        print(f"Time taken by lbfgs: {time_lbfgs:.6f} seconds")
        print(f"Total time: {totalTime:.6f} seconds")
        print(f"Ratio (time_lbfgs/time_adam) of time taken per epoch:"
              f"{((time_lbfgs/epochs2)/(time_adam/epochs1)):.6f}")

        return (history_adam, history_lbfgs), totalTime
    else:
        time_lbfgs = 0.
        print(f"\nTraining complete.")
        return history_adam, time_adam       # Only Adam history and time
```

5.3 Example: Waves Induced by Initial Condition

We now build and train PINN models for the fundamental wave solutions in a square plate with fixed edges, as defined in detail in Section 5.1.

5.3.1 Training of Plain PINN

The procedure for training a PINN model for the wave equation is largely similar to that used for the heat transfer problem. The following snippet, with accompanying comments, outlines the training process:

```
gr.set_seed_th(88)                                    # for reproducibility
# Initialization lists for record-keeping:
pinns = []                                       # for logging key variables
timeAdam  = []; timeLbfgs = []
Trainers  = []; Residuals = []
totalTimes= []
error_L2s = []; errorVIs = []                                # log the error
LossTerms = []                               # log terms in the composite loss

# Use exact functions for DBC and IC:
def f_DBC(x, y): return torch.sin(np.pi*x) * torch.sin(np.pi*y)
def f_IC (x, y): return torch.sin(np.pi*x) * torch.sin(np.pi*y)

p = 3                                                 # 3 features, x, y, t
Nn, NL = 100, 2                                              # a new model
pinn = PINN_Wave2D(Nn, NL).to(device)           # use the plain pinn class
P_star, P_total = print_model_parameters(pinn, print_table=False)

N_sample = 1*P_star            # minimum needed training samples for pinn
N_DBC = 2*int(sqrt(N_sample))+1
N_IC = N_sample

t_end = 1.0             # the characteristic time~ is 0.707 ≈ 1/np.sqrt(2)
xyt_clctn, xyt_DBC, u_DBC, xyt_IC, u_IC, xyt_IV, u_IV = sampleData(
                              N_clctn=N_sample, N_DBC=N_DBC, N_IC=N_IC,
                                        x_end=1., y_end=1., t_end=t_end)
w_s_th = 1. # dummy, not used here
print(f"Sampled {xyt_clctn.shape[0]} collocation points in the domain.")
print(f"Sampled {xyt_DBC.shape[0]} boundary points.")

def f_th(x, y, t):
    '''Source Term Function for the Poisson's equation in torch'''
    return 0.*x + 0.*y + 0.*t # one may replace the dummy with actual f

x_NBC_th, t_NBC_th, w_NBC_th = None,  None,  None
```

```
P* (Total pseudo-dimension~ of the affine space~) = 206
Total trainable parameters: 10601
Sampled 206 collocation points in the domain.
Sampled 116 boundary points.
```

For this example wave problem, we doubled the capacity of the PINNs built for the Poisson problems in Chapter 2, using the same configuration as in the heat transfer problem in Chapter 4. This `pinn` is built with two hidden layers , each with 100 neurons. It has $P^* = 206$, which is the total pseudo-dimension of the affine space of the model.

When a collocation method is used, the number of sampling points must be equal to or greater than the P^* value, based on the samples-neurons theorem [12]. In this example, we use 206 collocation sampling points, which is the minimum requirement.

The following snippet trains the model.

```
gr.set_seed_th(88)
DI_rule = ''               # collocation do not use Domain Integration rule.
LossTerm = 'DBC+IC+IV'  # use loss_DBC+loss_IC+loss_IV+loss_PDE (always)
Trainer  = trainerPDE_ICs_DBC
Residual = residual_wave
losses, totalTime = train_t_adam_lbfgs(pinn, Trainer, Residual, f_th,
          xyt_clctn, w_s_th, xyt_DBC, u_DBC, xyt_IC, u_IC, xyt_IV, u_IV,
               epochs1=201, epochs2=11, use_lbfgs=True, adapt_every=0)
```

```
==================================================
201 epoches training with Adam:
Epoch    0, Loss:1.92e-01, PDE:5.01e-03, DBC:6.95e-03, IC:1.76e-01, IV:3.
  91e-03
Epoch  100, Loss:1.17e-01, PDE:1.10e-03, DBC:2.28e-02, IC:9.04e-02, IV:2.
  72e-03
Epoch  200, Loss:6.10e-02, PDE:1.48e-03, DBC:1.89e-02, IC:3.92e-02, IV:1.
  42e-03
Time taken by adam: 1.821263 seconds

==================================================
11 epoches fine-tuning~ with LBFGS:
Epoch    0, Loss:1.68e-04, PDE:4.55e-05, DBC:9.38e-05, IC:1.35e-05, IV:1.
  55e-05
Epoch   10, Loss:7.75e-06, PDE:2.88e-06, DBC:2.55e-06, IC:6.11e-07, IV:1.
  71e-06

Training complete.
Time taken by lbfgs: 65.037590 seconds
Total time: 66.858853 seconds
Ratio (time_lbfgs/time_adam) of time taken per epoch:652.522128
```

The PINN model was trained in a few minutes on the author's laptop, reducing the total L2 loss from the 10^{-1} level to the 10^{-5} level. In this setup, the first-order optimizer `adam` was run for 200 epochs and completed in just a few seconds, demonstrating its efficiency. The quasi-second-order optimizer `lbfgs` was then used for 10 epochs; this phase took most of the training time but successfully drove the losses down by approximately three orders of magnitude.

Since our PINN model is the simplest plain form, all loss terms are evaluated separately, as shown in the printed outputs. These individual terms are then summed to form the total loss. The optimizer minimizes this composite loss, although each component may decrease at a different rate. Consequently, the training process is relatively slow, consistent with the discussions in Chapter 2.

Among the loss components:

- loss_PDE was reduced by 2 to 3 orders,
- loss_DBC by about 3 orders,
- loss_IC by approximately 5 orders, and
- loss_IV by around 3 orders.

Based on this behavior, we expect the solution accuracy to lie in the range of 10^{-2} to 10^{-3}. The terms loss_DBC and loss_IV largely control the final solution accuracy, as these represent the slowest-converging (worst-case) loss terms. Let us now assess this accuracy quantitatively.

5.3.2 Results Accuracy Evaluation

The following snippet evaluates the solution accuracy for this time-dependent problem at a specific observation time slice t_{obs}.

```
# Evaluate the results using independent data points:
N_plot = 100             # density of the grid used for evaluation purpose
# Sampling over rectangular domain (a, b) × (c, d):
a = 0.0; b = 1.0; c = 0.0; d = 1.0

x = np.linspace(a, b, N_plot)          # Create meshgrid over (a,b)x(c,d)
y = np.linspace(c, d, N_plot)

X, Y = np.meshgrid(x, y)
t_obs = 0.5*t_end
print(f"t_obs = {t_obs}")
T = t_obs * np.ones_like(X)                                # a section on time

XYT = np.stack([X.flatten(), Y.flatten(), T.flatten()], axis=1)
XYT_th = torch.tensor(XYT, dtype=torch.float32)

u_pred = pinn(XYT_th).detach().numpy().reshape(N_plot, N_plot)
u_exact = exact_u_wave(X, Y, t_obs)

error_L2, errorVI, u_diff = sln_error(u_pred, u_exact)
```

```
t_obs = 0.5
Maximum Error: 3.8389e-03
Mean Absolute Error: 1.1999e-03
Error Variation Index (ratio of Max/Mean): 3.1993e+00
L2 Relative Error: 4.8191e-03
```

The relative error is about the order of 10^{-3}, as expected. This level of accuracy is typical for a plain PINN model, consistent with observations in the previous chapters.

5.3.3 Contour Plots of Wave Solution

The following snippet computes and plots the contours of the PINN solution and the exact solution, along with the error distribution at the same time slice. The Python functions for plotting are from previous chapters.

```
# Plot the contours of both PINN solution and the exact one:
plot_sln_XY(X, Y, u_pred, u_exact, vmin=-0.6, vmax=0.)
```

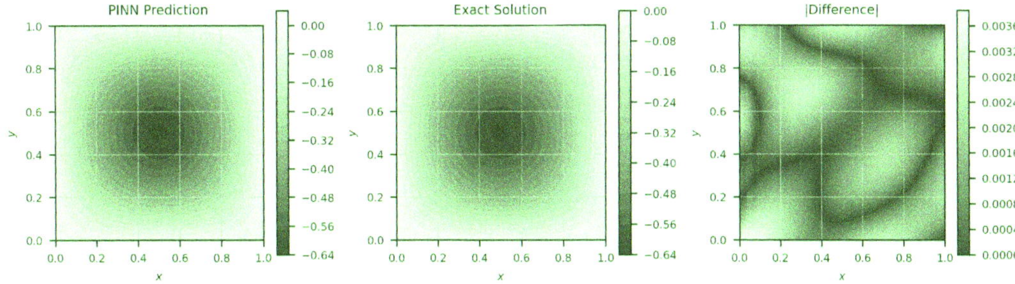

The distribution of the error is smooth due to the nature of the neural network, which is constructed from smooth elementary functions as the activation function. Since the error is measured as the absolute difference between the PINN prediction and the exact solution, the valleys (dark-colored regions) correspond to the lines where the error is zero.

Because the sampling points for training are randomly generated, the error distribution appears random as well. Generally, it will not be symmetric, even though the problem and the exact solution themselves are symmetric, due to the randomness in sampling. The error Variation Index (`errorVI`), defined as the ratio of the maximum to mean absolute errors, is about 3.2.

5.3.4 Plot of Loss History

Since our model is trained consecutively with both `adam` and `lbfgs` optimizers, we have two sets of loss function histories. We use the function `gr.plot3loss2()` to generate the plots, as introduced in Chapter 2.

```
plot3loss2(losses, ('adam initially','lbfgs in fine tunning'),
           PDE='PDE', C1='DBC', C2='IC', C3='IV', CL='Loss',log=False)
```

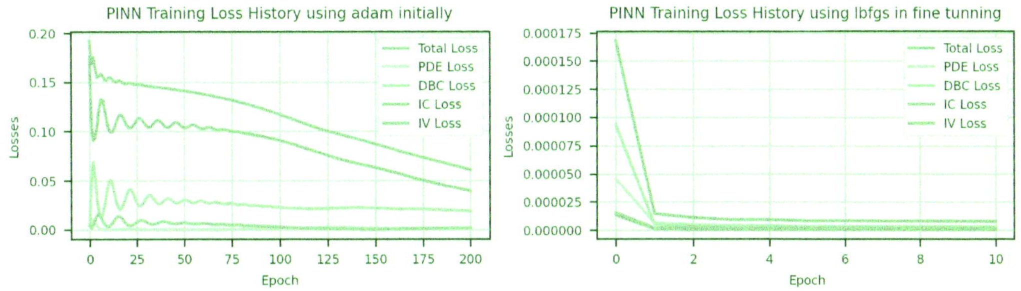

The foregoing history plots of losses show that the `adam` optimizer provides some initial parameter tuning, but the heavy lifting of driving the loss down is done by `lbfgs`. For this example problem, all the losses are reduced effectively.

Readers are encouraged to train the same PINN model while switching off the `lbfgs` optimizer and then plot the loss history of `adam` alone. The oscillatory behavior in the loss, as discussed in previous chapters, will be observed. This reiterates the importance of using a second-order optimizer like `lbfgs` for efficient and stable convergence.

5.3.5 Performance Records

The following snippets maintain records of the trained PINN models for future reference and comparison.

```
def recordSheet():
    '''
    Log the key performance indicators of PINN models for comparison
    study. '''
    new_error_L2_str = f"{error_L2:.3e}"

    if new_error_L2_str not in error_L2s:
        pinns.append(pinn.__class__.__name__)           # log pinn models
        Residuals.append(Residual.__name__ + DI_rule)     # log residuals
        #Trainers.append(Trainer.__name__)                 # log trainers
        LossTerms.append(LossTerm)  # additional terms in composite loss
        totalTimes.append(f"{totalTime:.3e}")                 # log time
        error_L2s.append(new_error_L2_str)                  # log the error
        errorVIs.append(f"{errorVI:.3e}")         # error variation index
    else:
        print(f"Error_L2 value '{new_error_L2_str}'"\
              f" already exists in error_L2s. Skipping update.")

recordItems=[pinns, LossTerms, Residuals,totalTimes,error_L2s,errorVIs]
```

```
# Create record sheet (exercute this only once!):
recordSheet()
```

The following snippet clears the entire record sheet or deletes specific records from it, if needed.

```
# for item in recordItems: item.clear()      # to clear the entire record
# for item in recordItems: item.pop(-1)          # to delete one record
```

```
printSummary(pinns, LossTerms, Residuals,totalTimes, error_L2s,errorVIs)
```

```
#  PINN model        Loss_terms residual    total_time error_L2s  errorVIs
========================================================================
1  _Wave2D           DBC+IC+IV  _wave       6.686e+01  4.819e-03  3.199e+00
```

Note that the time logging in the case studies will vary depending on the status of the computer, and is therefore only indicative.

The 2D wave equation can be modeled by PINNs more effectively through improved model construction techniques, as discussed previously. For the wave equation, an effective approach is to enforce both the initial displacement condition and the initial velocity condition exactly, in a hard way, by using a **trial solution** form.

5.4 Trial Solution Formula for Initial Conditions

To enforce the initial displacement condition

$$u(x, y, 0) = f_{\mathrm{IC}}(x, y) \tag{5.8}$$

exactly, the following **trial solution** form can be used:

$$u_{\mathrm{trial}}(x, y, t) = (1-t) f_{\mathrm{IC}}(x, y) + t \cdot \mathcal{N}(x, y, t)_{\hat{\mathbf{w}}} \tag{5.9}$$

This construction clearly satisfies the initial displacement condition at $t = 0$:

$$u_{\mathrm{trial}}(x, y, 0) = f_{\mathrm{IC}}(x, y). \tag{5.10}$$

For our problem, $f_{\mathrm{IC}}(x, y)$ is given in Eq.(5.4).

Hence, the neural network only needs to train the function $\mathcal{N}_{\hat{\mathbf{w}}}(x, y, t)$, while the initial condition is satisfied exactly by construction. This approach is expected to simplify training.

To implement this, one creates a PINN instance using the `PINN2D_DBC_IC` class template defined in Chapter 4 and imported from the `gr` module.

5.4.1 Trial Solution Form for IC and DBC

Similar to the heat transfer problems, to enforce zero Dirichlet boundary conditions (DBC) on all four edges of the square domain $[0, 1] \times [0, 1]$, we multiply the neural network output by a polynomial function:

$$f_{\mathrm{DBC}}(x, y) = x(1-x)y(1-y) \tag{5.11}$$

Combining these trial solutions, we obtain

$$u_{\mathrm{trial}}(x, y, t) = (1-t)u(x, y, 0) + tx(1-x)y(1-y) \cdot \mathcal{N}(x, y, t)_{\hat{\mathbf{w}}} \tag{5.12}$$

where the function $x(1-x)y(1-y)$ ensures that $u = 0$ on all boundary edges.

This polynomial multiplier approach was used in the previous chapter. Here, we also consider an alternative multiplier:

$$f_{\mathrm{DBC}}(x, y) = \sin(\pi x) \sin(\pi y) \tag{5.13}$$

The PINN class with exact IC and DBC enforcement is also `PINN2D_DBC_IC`.

5.4.2 Example: PINN Model with IC Auto-Enforced

In this example, we create a PINN model using the class PINN2D_DBC_IC, but intentionally pass a dummy f_{DBC} function that is constant 1, effectively disabling the automatic enforcement of boundary conditions at the class level. Thus, we need to handle a composite loss function consisting of loss_PDE and loss_DBC, which allows us to examine the effects of the composite loss.

The following snippet illustrates this setup:

```
# Create a new PINN model, only IC is enforced internally:
gr.set_seed_th(88)                                  # for reproducibility

def f_DBC(x, y): return torch.ones_like(x)*torch.ones_like(y) #dummy DBC
def  f_IC(x, y): return torch.sin(np.pi*x)*torch.sin(np.pi*y)  # true IC

pinn = PINN2D_DBC_IC(Nn, NL, f_DBC, f_IC, my_actvtn)     # IC is enforced

def f_DBC(x, y): return torch.sin(np.pi*x)*torch.sin(np.pi*y) # true DBC
xyt_clctn, xyt_DBC, u_DBC, xyt_IC, u_IC, xyt_IV, u_IV = sampleData(
                              N_clctn=N_sample, N_DBC=N_DBC, N_IC=N_IC,
                                     x_end=1., y_end=1., t_end=t_end)

LossTerm = 'DBC+IV'         # use loss_DBC+ loss_IV+loss_PDE (always used)
Trainer  = trainerPDE_ICs_DBC
Residual = residual_wave
losses, totalTime = train_t_adam_lbfgs(pinn, Trainer, Residual, f_th,
          xyt_clctn, w_s_th, xyt_DBC, u_DBC, xyt_IC, u_IC, xyt_IV, u_IV,
               epochs1=201, epochs2=11, use_lbfgs=True, adapt_every=0)
```

```
==================================================
201 epoches training with Adam:
Epoch   0, Loss:2.43e+01, PDE:2.41e+01, DBC:2.29e-03, IC:0.0, IV:1.76e-01
Epoch 100, Loss:1.15e+01, PDE:1.09e+01, DBC:3.69e-01, IC:0.0, IV:2.18e-01
Epoch 200, Loss:7.33e+00, PDE:6.66e+00, DBC:3.96e-01, IC:0.0, IV:2.68e-01
Time taken by adam: 2.941405 seconds

==================================================
11 epoches fine-tuning with LBFGS:
Epoch   0, Loss:9.78e-04, PDE:3.68e-04, DBC:4.59e-04, IC:0.0, IV:1.51e-04
Epoch  10, Loss:6.54e-05, PDE:2.58e-05, DBC:2.59e-05, IC:0.0, IV:1.36e-05

Training complete.
Time taken by lbfgs: 130.767803 seconds
Total time: 133.709208 seconds
Ratio (time_lbfgs/time_adam) of time taken per epoch:812.361493
```

Since the initial condition (IC) is now enforced inside the PINN class, the loss_IC term becomes zero. However, because the Dirichlet boundary condition (DBC) is disabled at the class level, the loss_DBC is nonzero and must be driven down through the composite loss function during training.

Let us now assess the performance of this setup.

```
# Evaluate the accuracy of the results:
u_pred = pinn(XYT_th).detach().numpy().reshape(N_plot, N_plot)
u_exact = exact_u_wave(X, Y, t_obs)

error_L2, errorVI, u_diff = sln_error(u_pred, u_exact)
```

```
Maximum Error: 1.3115e-02
Mean Absolute Error: 2.6323e-03
Error Variation Index (ratio of Max/Mean): 4.9825e+00
L2 Relative Error: 1.2261e-02
```

```
# Plot the contours of both PINN solution and the exact one:
plot_sln_XY(X, Y, u_pred, u_exact, vmin=-0.6, vmax=0.)

plot3loss2(losses, ('adam initially','lbfgs in fine tunning'),
           PDE='PDE', C1='DBC', C2='IC', C3='IV', CL='Loss',log=False)
```

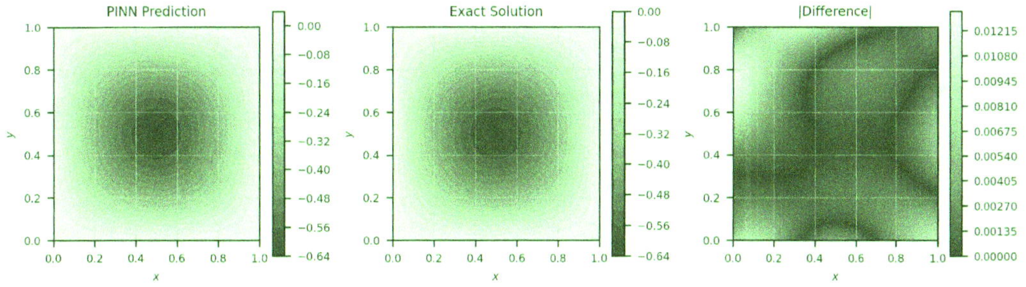

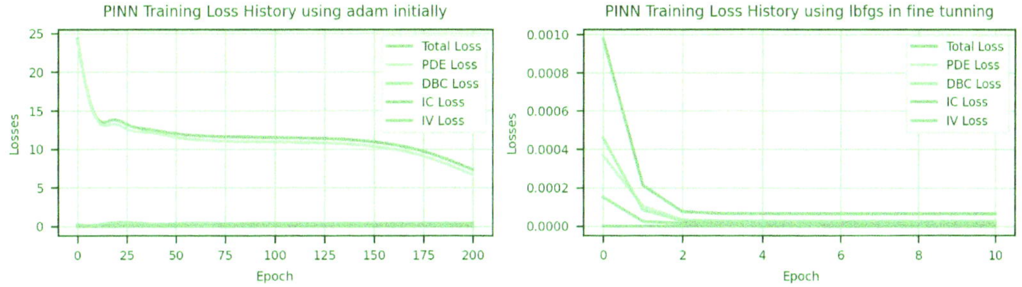


```
recordSheet()                                  # Update record sheet
```

```
printSummary(pinns, LossTerms, Residuals, totalTimes,error_L2s,errorVIs)
```

```
#  PINN model        Loss_terms residual    total_time error_L2s  errorVIs
===========================================================================
1  _Wave2D           DBC+IC+IV  _wave       6.686e+01  4.819e-03  3.199e+00
2  2D_DBC_IC         DBC+IV     _wave       1.337e+02  1.226e-02  4.982e+00
```

The relative accuracy for model-2 is higher than that of model-1. Since the initial velocity (IV) condition is still enforced outside the class, and the Dirichlet boundary condition (DBC) is also enforced externally, these two terms dominate the error. As a result, model-2 performs worse than model-1 on all assessment metrics.

5.4.3 Example: PINN with IC and DBC Hard Enforced

In this example, we create a PINN model using the class `PINN2D_DBC_IC` and provide the actual `f_DBC` function, so that the DBC is also enforced automatically at the class level. In this setup, only `loss_IV` remains as an external loss term, in addition to `loss_PDE` and `loss_DBC`. The training snippet is as follows.

```
#Create a new PINN model of same size, DBC & IC are enforced internally:
gr.set_seed_th(88)                                  # for reproducibility

def f_DBC(x, y): return torch.sin(np.pi*x)*torch.sin(np.pi*y) # true DBC
def  f_IC(x, y): return torch.sin(np.pi*x)*torch.sin(np.pi*y)  # true IC
pinn = PINN2D_DBC_IC(Nn, NL, f_DBC, f_IC, my_actvtn)   # DBC&IC enforced

xyt_clctn, xyt_DBC, u_DBC, xyt_IC, u_IC, xyt_IV, u_IV = sampleData(
                           N_clctn=N_sample, N_DBC=N_DBC, N_IC=N_IC,
                                  x_end=1., y_end=1., t_end=t_end)

LossTerm = 'IV'                     # use loss_IV + loss_PDE (always used)
Trainer  = trainerPDE_ICs_DBC
Residual = residual_wave
losses, totalTime = train_t_adam_lbfgs(pinn, Trainer, Residual, f_th,
          xyt_clctn, w_s_th, xyt_DBC, u_DBC, xyt_IC, u_IC, xyt_IV, u_IV,
               epochs1=201, epochs2=11, use_lbfgs=True, adapt_every=0)
```

```
==================================================
201 epoches training with Adam:
Epoch   0, Loss:2.66e+01, PDE:2.64e+01, DBC:5.91e-16, IC:0.0, IV:1.97e-01
Epoch 100, Loss:1.39e+01, PDE:1.17e+01, DBC:2.42e-15, IC:0.0, IV:2.14e+00
Epoch 200, Loss:2.54e+00, PDE:1.90e+00, DBC:3.99e-15, IC:0.0, IV:6.40e-01
Time taken by adam: 3.425753 seconds

==================================================
11 epoches fine-tuning with LBFGS:
Epoch   0, Loss:4.67e-06, PDE:4.55e-06, DBC:4.35e-15, IC:0.0, IV:1.20e-07
Epoch  10, Loss:3.51e-06, PDE:3.40e-06, DBC:4.35e-15, IC:0.0, IV:1.07e-07

Training complete.
Time taken by lbfgs: 11.304125 seconds
Total time: 14.729878 seconds
Ratio (time_lbfgs/time_adam) of time taken per epoch:60.295417
```

This time, both `Loss_DBC` and `Loss_IC` are automatically zero due to the trial solution construction within the PINN class. However, `Loss_IV` (initial velocity loss) is still nonzero and must be minimized as part of the composite loss.

Let's examine the model's performance:

```
# Evaluate the results:
u_pred = pinn(XYT_th).detach().numpy().reshape(N_plot, N_plot)
u_exact = exact_u_wave(X, Y, t_obs)
```

```
error_L2, errorVI, u_diff = sln_error(u_pred, u_exact)
```

```
Maximum Error: 1.5336e-04
Mean Absolute Error: 3.7007e-05
Error Variation Index (ratio of Max/Mean): 4.1442e+00
L2 Relative Error: 1.8186e-04
```

Readers may uncomment the following snippet to plot the results.

```
# # Plot the contours of both PINN solution and the exact one:
# plot_sln_XY(X, Y, u_pred, u_exact, vmin=-0.6, vmax=0.)

# plot3loss2(losses, ('adam initially','lbfgs in fine tunning'),
#              PDE='PDE', C1='DBC', C2='IC', C3='IV', CL='Loss',log=False)
```

```
recordSheet()                                         # Create record sheet
```

```
printSummary(pinns, LossTerms, Residuals, totalTimes,error_L2s,errorVIs)
```

```
#  PINN model        Loss_terms residual    total_time error_L2s  errorVIs
==========================================================================
1  _Wave2D           DBC+IC+IV  _wave       6.686e+01  4.819e-03  3.199e+00
2  2D_DBC_IC         DBC+IV     _wave       1.337e+02  1.226e-02  4.982e+00
3  2D_DBC_IC         IV         _wave       1.473e+01  1.819e-04  4.144e+00
```

Model-3 performed significantly better than the previous two models on accuracy and speed.

Let's construct a PINN model in such a way that all the conditions are enforced inside the class.

5.5 Trial Solution Form for All Conditions

To enforce both the initial displacement condition

$$u(x, y, 0) = f_{\text{IC}}(x, y) \tag{5.14}$$

and the initial velocity condition

$$\frac{\partial u}{\partial t}(x, y, 0) = 0 \tag{5.15}$$

exactly, the following **trial solution** form can be used:

$$u_{trial}(x, y, t) = (1 - t^2) f_{\text{IC}}(x, y) + t^2 \cdot \mathcal{N}_{\hat{\mathbf{w}}}(x, y, t) \tag{5.16}$$

This construction clearly satisfies:

- Initial displacement: $u(x, y, 0) = f\_\{\text{IC}\}(x, y)$,
- Initial velocity: $\dot{u}(x, y, 0) = 0$.

Therefore, we only need to train the neural network $\mathcal{N}_{\hat{\mathbf{w}}}(x, y, t)$ with learning parameters $\hat{\mathbf{w}}$. These IC and IV conditions satisfied exactly by construction.

Further, to enforce zero Dirichlet boundary conditions (DBC) on all four edges of the square domain, we multiply by the polynomial

$$x(1-x)y(1-y) \tag{5.17}$$

or alternatively the sinusoidal form in Eq.(5.13).

Combining all these, the full trial solution reads:

$$u_{trial}(x,y,t) = (1-t^2)u(x,y,0) + t^2x(1-x)y(1-y) \cdot \mathcal{N}_{\hat{\mathbf{w}}}(x,y,t) \tag{5.18}$$

This guarantees $u = 0$ on all boundaries, and exactly enforces IC and IV conditions.

5.5.1 PINN Class with All Conditions Enforced

Next, we define a PINN class that incorporates DBC, IC, and IV conditions all enforced automatically via the trial solution inside the model.

```
class PINN_Wave2D_DBC_ICIV(nn.Module):
    '''
    PINN class with DBC, IC, and IV conditions all enforceable.
    '''
    def __init__(self, Nn, NL, f_DBC, f_IC, my_actvtn):
        super().__init__()                                  # 3D: x, y, t
        self.f_DBC = f_DBC                                  # Save to self
        self.f_IC  = f_IC                                   # Save to self
        self.my_actvtn = my_actvtn                          # Save to self

        self.hidden = nn.ModuleList([nn.Linear(3, Nn)] +
                        [nn.Linear(Nn, Nn) for _ in range(NL - 1)])
        self.out = nn.Linear(Nn, 1)

        for layer in self.hidden:           # Apply Xavier initialization
            nn.init.xavier_normal_(layer.weight)
        nn.init.xavier_normal_(self.out.weight)

    def forward(self, xyt):
        x = xyt[:, 0:1]
        y = xyt[:, 1:2]
        t = xyt[:, 2:3]

        z = xyt
        for layer in self.hidden:
            z = my_actvtn(layer(z))                       # use self.my_actvtn

        N = self.out(z)

        u_IC  = self.f_IC(x, y)                               # use self.f_IC
        uf_DBC = self.f_DBC(x, y)     # function for DBCs, use self.f_DBC
```

```
        return (1-t**2) * u_IC + t**2 * (uf_DBC * N)  # for DBC, IC & IV
```

5.5.2 Example: PINN with All Conditions Auto-Enforced

The following snippet trains a PINN model in which all physical conditions — Dirichlet boundary conditions (DBC), initial displacement condition (IC), and initial velocity condition (IV) — are enforced by construction within the class.

We use the PINN_Wave2D_DBC_ICIV class defined in the foregoing code cell, which internally applies the full trial solution form given in Eq.(5.18). This class automatically embeds the conditions into the network output through appropriate multiplicative functions. As a result, only the PDE residual contributes to the loss function, simplifying the training process significantly.

The code snippet to train the model is as follows:

```
# A new PINN model of same size, DBC, IC & IV are enforced in class:
gr.set_seed_th(88)                                 # for reproducibility

def f_DBC(x, y): return torch.sin(np.pi*x)*torch.sin(np.pi*y) # true DBC
def  f_IC(x, y): return torch.sin(np.pi*x)*torch.sin(np.pi*y)  # true IC

# DBC, IC & IV are all enforced in class:
pinn = PINN_Wave2D_DBC_ICIV(Nn, NL, f_DBC, f_IC, my_actvtn)

xyt_clctn, xyt_DBC, u_DBC, xyt_IC, u_IC, xyt_IV, u_IV = sampleData(
                              N_clctn=N_sample, N_DBC=N_DBC, N_IC=N_IC,
                                     x_end=1., y_end=1., t_end=t_end)
LossTerm = 'PDE_only'
Trainer  = trainerPDE_ICs_DBC
Residual = residual_wave
losses, totalTime = train_t_adam_lbfgs(pinn, Trainer, Residual, f_th,
          xyt_clctn, w_s_th, xyt_DBC, u_DBC, xyt_IC, u_IC, xyt_IV, u_IV,
                epochs1=201, epochs2=11, use_lbfgs=True, adapt_every=0)
```

```
==================================================
201 epoches training with Adam:
Epoch   0, Loss:2.58e+01, PDE:2.58e+01, DBC:7.43e-16, IC:0.0, IV:0.00e+00
Epoch 100, Loss:2.93e+00, PDE:2.93e+00, DBC:3.92e-15, IC:0.0, IV:0.00e+00
Epoch 200, Loss:1.62e+00, PDE:1.62e+00, DBC:3.94e-15, IC:0.0, IV:0.00e+00
Time taken by adam: 4.080018 seconds

==================================================
11 epoches fine-tuning with LBFGS:
Epoch   0, Loss:3.42e-05, PDE:3.42e-05, DBC:4.35e-15, IC:0.0, IV:0.00e+00
Epoch  10, Loss:1.28e-05, PDE:1.28e-05, DBC:4.35e-15, IC:0.0, IV:0.00e+00

Training complete.
Time taken by lbfgs: 8.500244 seconds
Total time: 12.580262 seconds
Ratio (time_lbfgs/time_adam) of time taken per epoch:38.069103
```

This time, Loss_DBC, Loss_IC, and Loss_IV are all automatically zero, and only a single loss term, Loss_PDE, remains active during training. As a result, the accuracy is higher, as reflected in the output of the following snippet.

```
# Evaluate the results:
u_pred = pinn(XYT_th).detach().numpy().reshape(N_plot, N_plot)
u_exact = exact_u_wave(X, Y, t_obs)

error_L2, errorVI, u_diff = sln_error(u_pred, u_exact)
```

```
Maximum Error: 8.5125e-05
Mean Absolute Error: 2.9427e-05
Error Variation Index (ratio of Max/Mean): 2.8927e+00
L2 Relative Error: 1.2666e-04
```

```
# Plot the contours of both PINN solution and the exact one:
plot_sln_XY(X, Y, u_pred, u_exact, vmin=-0.6, vmax=0.)

plot3loss2(losses, ('adam initially','lbfgs in fine tunning'),
           PDE='PDE', C1='DBC', C2='IC', C3='IV', CL='Loss',log=False)
```

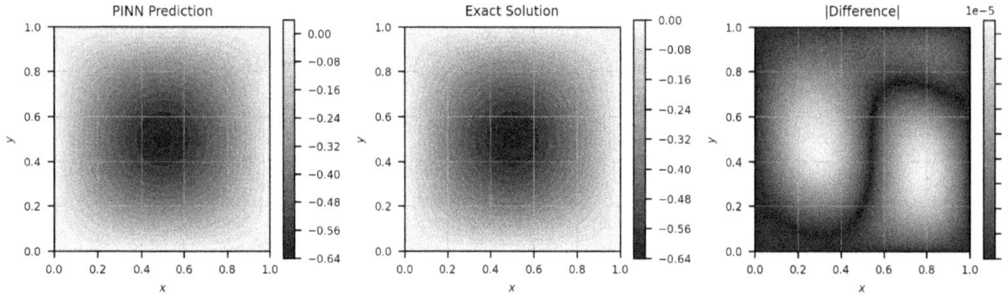

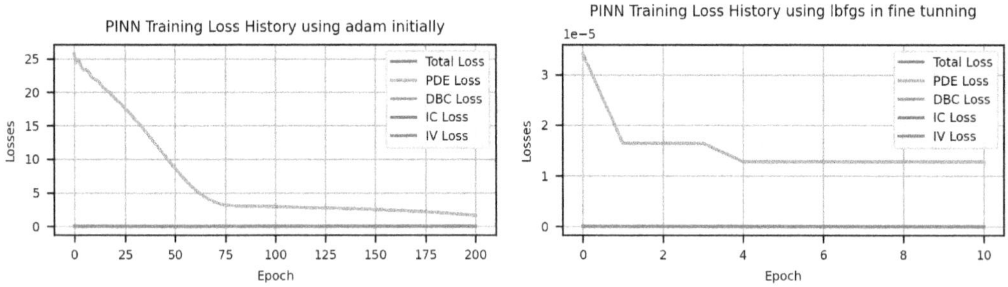


```
recordSheet()                                    # update record sheet
```

```
printSummary(pinns, LossTerms, Residuals, totalTimes,error_L2s,errorVIs)
```

```
#  PINN model        Loss_terms residual    total_time error_L2s  errorVIs
==========================================================================
1  _Wave2D           DBC+IC+IV  _wave       6.686e+01  4.819e-03  3.199e+00
2  2D_DBC_IC         DBC+IV     _wave       1.337e+02  1.226e-02  4.982e+00
```

```
3  2D_DBC_IC          IV          _wave        1.473e+01  1.819e-04  4.144e+00
4  _Wave2D_DBC_ICIV   PDE_only    _wave        1.258e+01  1.267e-04  2.893e+00
```

Compared to model-3, model-4 is more accurate and faster. Its solution error variation is also the lowest. Model-4 is the best model trained so far for this problem.

5.6 PINN with Adaptive Sampling

5.6.1 Adaptation Strategy and Code

Random sampling can sometimes result in a highly biased distribution of points. In Chapters 2 and 3, we introduced integral approaches to address this issue. Here, we use the so-called **adaptive sampling** technique to tackle it more dynamically.

The essential idea is to use the current neural network to compute the PDE residuals at a new set of randomly sampled points. These residuals reveal how well the current model satisfies the PDE at those points. We then identify the subset of points with the highest residual values — for example, the top 10%. These corresponding sampling points are added to the existing training set to refine the model. This process of adding new sampling points can be repeated periodically over several training epochs.

The following function identifies the adaptation points, which will be used in `trainerPDE_ICs_DBC`:

```python
def adaptive_sample(pinn, N_new=2500):
    '''Periodically resamples the domain using the PDE residual
       to focus on high-error regions.'''
    xyt = np.random.rand(N_new, 3).astype(np.float32)
    xyt_th = torch.tensor(xyt, device=device)

    _, res = residual_wave(pinn, f_th, xyt_th, w_s_th)
    res_np = res.detach().cpu().numpy().flatten()

    idx = np.argsort(-res_np)[:int(0.01 * N_new)]             # Top 1%
    points_to_add = torch.tensor(xyt[idx], dtype=torch.float32,
                                                  device=device)
    return points_to_add
```

5.6.2 Procedure for Adaptive Sampling

Let us test the adaptive sampling technique using the same example. The adaptive strategy employed here is as follows:

1. For simplicity, we adapt only the collocation points inside the problem domain. The points on the boundary remain unchanged throughout the process.

2. Adaptation is performed every 50 epochs during the initial training using `adam`. Since we plan to run `adam` for 200 epochs, there will be a total of four adaptation steps.

3. Each time, we generate a set of N_{adapt} random collocation points and use the current PINN model to compute the PDE residuals at these points. The top 1% of the points with the highest residuals are then added to the current set of collocation points. After

all adaptations, a total of $0.04N_{\text{adapt}}$ points will have been added to the initial training set of collocation points.

4. Upon completion of adaptation, we expect a more effective and well-distributed set of points inside the domain. The quasi-second-order optimizer `lbfgs` is then used to finalize the training.

We also need to make a small modification to the trainer function `trainerPDE_ICs_DBC()` provided earlier. Since the only change is in the collocation points, we make `xyt_clctn` a global variable so that it can be accessed and modified within the trainer. Additionally, the function `adaptive_sample()` is invoked at the end of each adaptation cycle inside the trainer. Everything else remains unchanged, so the modification to the code is minimal.

5.6.3 Example: PINN Model Trained via Adaptation

Let us generate a fresh set of data and create a new model to perform adaptive training:

```
gr.set_seed_th(88)                                   # for reproducibility

#Nn, NL = 100, 2                          # a new model with the same parameters
pinn = PINN_Wave2D(Nn, NL).to(device)
P_star, P_total = print_model_parameters(pinn, print_table=False)

def f_DBC(x, y): return torch.sin(np.pi*x)*torch.sin(np.pi*y) # True DBC
def  f_IC(x, y): return torch.sin(np.pi*x)*torch.sin(np.pi*y)  # True IC

# Generate the data:
xyt_clctn, xyt_DBC, u_DBC, xyt_IC, u_IC, xyt_IV, u_IV = sampleData(
                              N_clctn=N_sample, N_DBC=N_DBC, N_IC=N_IC,
                                      x_end=1., y_end=1., t_end=t_end)

print(f"Sampled {xyt_clctn.shape[0]} collocation points in the domain.")
print(f"Sampled {xyt_DBC.shape[0]} boundary points.")

def f_th(x, y, t):
    '''Source Term Function for the Poisson's equation in torch'''
    return 0.*x + 0.*y + 0.*t # one may replace the dummy with actual f

# A new PINN model of same size, DBC, IC & IV are enforced in the class:
pinn = PINN_Wave2D_DBC_ICIV(Nn, NL, f_DBC, f_IC, my_actvtn)

DI_rule  = '+adap'
LossTerm = 'PDE_only'
Trainer  = trainerPDE_ICs_DBC
Residual = residual_wave
losses, totalTime = train_t_adam_lbfgs(pinn, Trainer, Residual, f_th,
           xyt_clctn, w_s_th, xyt_DBC, u_DBC, xyt_IC, u_IC, xyt_IV, u_IV,
               epochs1=201, epochs2=11, use_lbfgs=True, adapt_every=50)
```

```
P* (Total pseudo-dimension of the affine space) = 206
Total trainable parameters: 10601
Sampled 206 collocation points in the domain.
Sampled 116 boundary points.
==================================================
201 epoches training with Adam:
Epoch   0, Loss:6.60e+01, PDE:6.60e+01, DBC:3.19e-17, IC:0.0, IV:0.00e+00
New collocation set size: torch.Size([231, 3])
Epoch 100, Loss:6.99e+00, PDE:6.99e+00, DBC:3.65e-15, IC:0.0, IV:0.00e+00
New collocation set size: torch.Size([256, 3])
New collocation set size: torch.Size([281, 3])
Epoch 200, Loss:6.81e+00, PDE:6.81e+00, DBC:4.82e-15, IC:0.0, IV:0.00e+00
New collocation set size: torch.Size([306, 3])
Time taken by adam: 3.945713 seconds

==================================================
11 epoches fine-tuning with LBFGS:
Epoch   0, Loss:8.32e-06, PDE:8.32e-06, DBC:4.35e-15, IC:0.0, IV:0.00e+00
Epoch  10, Loss:4.10e-06, PDE:4.10e-06, DBC:4.35e-15, IC:0.0, IV:0.00e+00

Training complete.
Time taken by lbfgs: 18.601260 seconds
Total time: 22.546974 seconds
Ratio (time_lbfgs/time_adam) of time taken per epoch:86.143052
```

It is observed that four adaptations were performed, with each step adding 25 points to the training set of collocation points. At the end of the adaptation process, the total number of collocation points inside the domain increased from the initial 206 to 306. This increase in collocation points typically improves the solution accuracy, although it is still subject to the inherent randomness of the PINN model. The effect on solution accuracy is assessed using the following snippet:

```python
# Evaluate the results:
u_pred = pinn(XYT_th).detach().numpy().reshape(N_plot, N_plot)
u_exact = exact_u_wave(X, Y, t_obs)

error_L2, errorVI, u_diff = sln_error(u_pred, u_exact)
```

```
Maximum Error: 8.7636e-05
Mean Absolute Error: 2.5043e-05
Error Variation Index (ratio of Max/Mean): 3.4994e+00
L2 Relative Error: 1.0933e-04
```

Let us take a look at the history of the losses using the following snippet:

```python
# Plot the contours of both PINN solution and the exact one (optional):
#plot_sln_XY(X, Y, u_pred, u_exact, vmin=-0.6, vmax=0., shrink=0.6)

plot3loss2(losses, ('adam initially','lbfgs in fine tunning'),
            PDE='PDE', C1='DBC', C2='IC', C3='IV', CL='Loss',log=False)
```

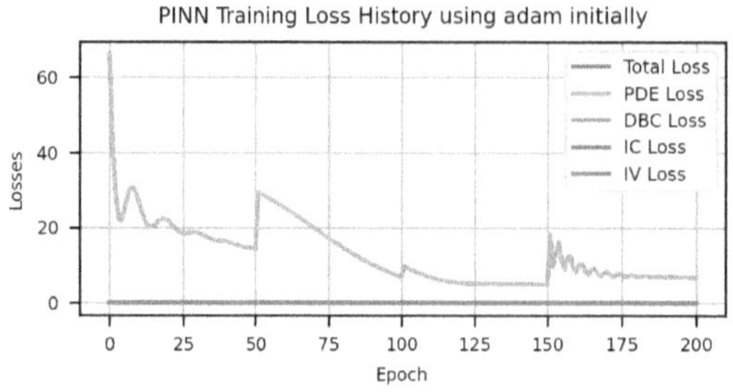

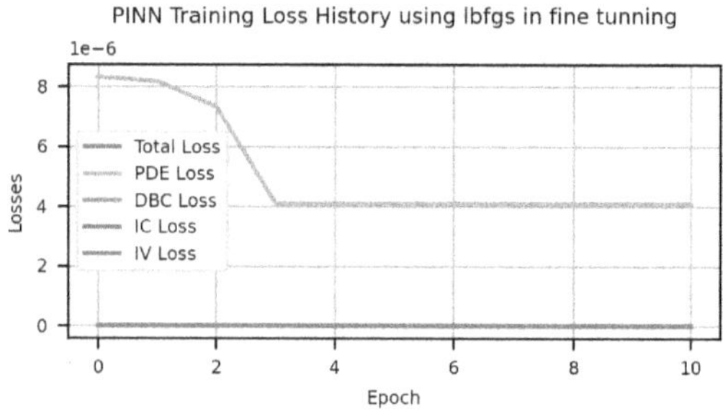

It is observed that the loss exhibits noticeable jumps whenever an adaptation step occurs during training with `adam`. Once `lbfgs` takes over, it effectively drives the loss down to the order of 10^{-5}.

```
recordSheet()
```

```
printSummary(pinns, LossTerms, Residuals, totalTimes,error_L2s,errorVIs)
```

```
#  PINN model         Loss_terms residual    total_time error_L2s  errorVIs
============================================================================
1  _Wave2D            DBC+IC+IV  _wave       6.686e+01  4.819e-03  3.199e+00
2  2D_DBC_IC          DBC+IV     _wave       1.337e+02  1.226e-02  4.982e+00
3  2D_DBC_IC          IV         _wave       1.473e+01  1.819e-04  4.144e+00
4  _Wave2D_DBC_ICIV PDE_only     _wave       1.258e+01  1.267e-04  2.893e+00
5  _Wave2D_DBC_ICIV PDE_only     _wave+adap  2.255e+01  1.093e-04  3.499e+00
```

As observed with this adaptive sampling technique, the accuracy is the highest among all these models. However, the training takes more time and the error distribution becomes less uniform than in model-4. This result suggests that adaptation can conceptually have a good chance to lead to improved accuracy—though not guaranteed—depending on the nature of the problem. Various factors may influence the outcome, such as the representational capacity of the original PINN model.

Many successful cases of using the adaptive sampling technique have been reported, such as in Refs. [6, 23, 27]. Since adaptive sampling is relatively easy to implement, it is a valuable technique to include in the toolbox.

5.7 Case Study: Time History of Solution

This section investigates the time history of solutions obtained from PINN models. According to Eq.(5.6), the solution varies with x (and y) at a frequency of π, while it varies with t at a frequency of $\sqrt{2}\pi$ (assuming $c = 1$). Therefore, t_{end} should ideally be around the characteristic time $t_{\text{char}} = \frac{1}{\sqrt{2}} \approx 0.707$ when using equal-number sampling in time.

Let us extend t_{end} to approximately 2, and examine how the PINN performs in predicting the time history of the solution . In this study, we use the plain PINN class. The time domain is extended to $(0, 2]$, while the spatial domain remains $(0, 1)^2$.

5.7.1 Train the PINN Model

The following snippet trains the PINN model:

```
gr.set_seed_th(88)                                    # for reproducibility

# Sample the data and define necessary functions:
t_end=2.                   # the characteristic time = 0.707 ≈ 1/np.sqrt(2)

xyt_clctn, xyt_DBC, u_DBC, xyt_IC, u_IC, xyt_IV, u_IV = sampleData(
                                N_clctn=N_sample, N_DBC=N_DBC, N_IC=N_IC,
                                         x_end=1., y_end=1., t_end=t_end)
print(f"Sampled {xyt_clctn.shape[0]} collocation points in the domain.")
print(f"Sampled {xyt_DBC.shape[0]} boundary points.")

x_NBC_th, t_NBC_th, w_NBC_th = None,  None,  None

def f_th(x, y, t):
    '''Source Term Function for the Poisson's equation in torch'''
    return 0.*x + 0.*y + 0.*t

# Build the model:
pinn = PINN_Wave2D(Nn, NL).to(device)
P_star, P_total = print_model_parameters(pinn, print_table=False)

LossTerm = 'DBC+IC+IV'        # use loss_DBC+loss_IC+loss_IV and loss_PDE
Trainer  = trainerPDE_ICs_DBC
Residual = residual_wave
losses, totalTime = train_t_adam_lbfgs(pinn, Trainer, Residual, f_th,
          xyt_clctn, w_s_th, xyt_DBC, u_DBC, xyt_IC, u_IC, xyt_IV, u_IV,
                 epochs1=201, epochs2=11, use_lbfgs=True, adapt_every=0)
```

```
Sampled 206 collocation points in the domain.
Sampled 116 boundary points.
P* (Total pseudo-dimension of the affine space) = 206
Total trainable parameters: 10601
==================================================
201 epoches training with Adam:
Epoch    0, Loss:1.89e-01, PDE:3.99e-03, DBC:5.25e-03, IC:1.76e-01, IV:3.
 ↪91e-03
Epoch  100, Loss:1.09e-01, PDE:5.38e-04, DBC:2.41e-02, IC:8.37e-02, IV:7.
 ↪13e-04
Epoch  200, Loss:8.20e-02, PDE:1.13e-03, DBC:1.56e-02, IC:6.44e-02, IV:8.
 ↪96e-04
Time taken by adam: 2.341872 seconds

==================================================
11 epoches fine-tuning with LBFGS:
Epoch    0, Loss:1.72e-03, PDE:2.27e-04, DBC:1.40e-03, IC:6.66e-05, IV:2.
 ↪58e-05
Epoch   10, Loss:1.20e-05, PDE:2.72e-06, DBC:6.10e-06, IC:2.00e-06, IV:1.
 ↪23e-06

Training complete.
```

```
Time taken by lbfgs: 96.400752 seconds
Total time: 98.742624 seconds
Ratio (time_lbfgs/time_adam) of time taken per epoch:752.178091
```

The training is successful.

5.7.2 Results Accuracy Evaluation

The following snippet computes the solution accuracy at a specific observation time slice, t_{obs}.

```
# Evaluate the results:
N_plot = 100                                                  # Grid density
# Sampling over rectangular domain (a, b)×(c, d):
a = 0.0; b = 1.0; c = 0.0; d = 1.0

x = np.linspace(a, b, N_plot)          # Create meshgrid over (a,b)x(c,d)
y = np.linspace(c, d, N_plot)

X, Y = np.meshgrid(x, y)
t_obs = 0.9*t_end                                             # a slice in time
print(f"t_obs = {t_obs}")
T = t_obs * np.ones_like(X)

XYT = np.stack([X.flatten(), Y.flatten(), T.flatten()], axis=1)
XYT_th = torch.tensor(XYT, dtype=torch.float32)

u_pred = pinn(XYT_th).detach().numpy().reshape(N_plot, N_plot)
u_exact = exact_u_wave(X, Y, t_obs)

error_L2, errorVI, u_diff = sln_error(u_pred, u_exact)
```

```
t_obs = 1.8
Maximum Error: 7.9435e-02
Mean Absolute Error: 5.5462e-03
Error Variation Index (ratio of Max/Mean): 1.4323e+01
L2 Relative Error: 1.1181e-01
```

The error is very large. Let us plot the time history results and find out what is happening.

5.7.3 Plot Time History of the Solution

We use the following code to plot the time history of the solution predicted by the trained PINN model, along with that of the exact solution. The time history curve covers the range $[0, 2]$ for the point at $x = 0.5$ and $y = 0.5$, which is the center of the unit square.

```
def plot_time_history(model, u_exact_f, x_obs, y_obs, N_plot=200,
            dpi=300, t_min=0.0, t_max=1.0, device=device, f_name='tmp'):
    """
    Plots the time history of u(x, y, t) at a fixed (x_obs, y_obs)
    using a trained PINN model and compares it with the exact solution.
    Parameters:
    - model          : Trained PINN model (shape [N, 3] of [x, y, t])
```

```
    - u_exact_f     : Callable, exact solution u(x, y, t)
    - x_obs, y_obs  : Fixed spatial coordinates
    - t_min, t_max  : Time range
    - N_plot        : Number of time samples for plot
    - device        : 'cpu' or 'cuda'        """
    t_vals = np.linspace(t_min, t_max, N_plot)          # Time samples
    xtyt=np.array([[x_obs, y_obs, t] for t in t_vals], dtype=np.float32)
    xtyt_tensor = torch.tensor(xtyt).to(device)

    model.eval()                                       # Model prediction
    with torch.no_grad():
        u_pred = model(xtyt_tensor).cpu().numpy().flatten()

    u_exact_vals = np.array([u_exact_f(x_obs,y_obs,t) for t in t_vals])

    plt.figure(figsize=(5, 3), dpi=dpi)                            # Plot
    plt.plot(t_vals, u_exact_vals, 'g-', zorder=1, \
                                    label='Exact Solution $u_{exact}$')
    plt.plot(t_vals, u_pred, 'r--', zorder=2, \
                                    label='PINN Prediction $u_{pred}$')
    plt.xlabel('Time $t$');     plt.ylabel('$u(x, y, t)$')
    plt.title(f'Time History at $(x={x_obs}, y={y_obs})$')
    plt.legend();     plt.grid(True);     plt.tight_layout()
    plt.savefig(f_name+'.png', dpi=500);      plt.show()
```

```
# Plot the time history of solutions of PINN and exact:
x_obs, y_obs = 0.5, 0.5
plot_time_history(pinn, exact_u_wave, x_obs, y_obs, N_plot=200, dpi=300,
         t_min=0.0, t_max=2.0, device=device, f_name='imagesPN/tmp.png')
```

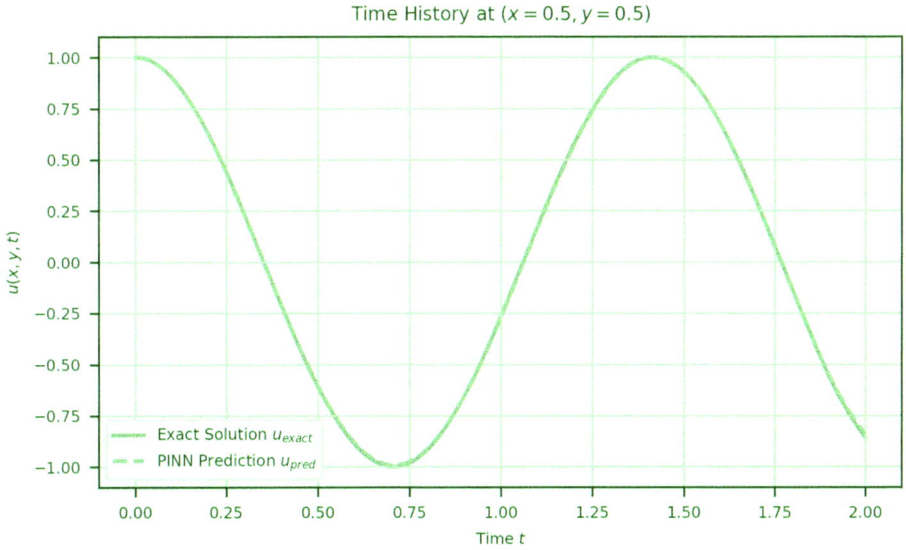

The curve clearly shows that the error becomes noticeable near t_{end}.

5.7.4 Improvement via Alternative Activations

In the previously trained PINN models, tanh() was used as the activation function. As demonstrated in the previous chapter, other types of activation functions can also be used. The following snippet trains a PINN model using the cos() function as the activation function. To do this, we use the PINN_Wave2Dcos class defined earlier. All other settings remain unchanged. The snippet is as follows.

```
# Build the model using cosine activation:
Nn, NL = 50, 1                                              # a new model
pinn = PINN_Wave2Dcos(Nn, NL).to(device)
gr.print_model_parameters(pinn, print_table=False)
print(f"Sampling poionts inside the domain: {len(xyt_clctn)}")

LossTerm = 'DBC+IC+IV'  # use loss_DBC+loss_IC+loss_IV+loss_PDE (always)
Trainer  = trainerPDE_ICs_DBC
Residual = residual_wave
losses, totalTime = train_t_adam_lbfgs(pinn, Trainer, Residual, f_th,
          xyt_clctn, w_s_th, xyt_DBC, u_DBC, xyt_IC, u_IC, xyt_IV, u_IV,
                epochs1=201, epochs2=11, use_lbfgs=True, adapt_every=0)
```

```
P* (Total pseudo-dimension of the affine space) = 55
Total trainable parameters: 251
Sampling poionts inside the domain: 206
==================================================
201 epoches training with Adam:
Epoch    0, Loss:5.20e-01, PDE:7.85e-04, DBC:1.40e-02, IC:4.87e-01, IV:1.
  ↪81e-02
Epoch  100, Loss:1.31e-01, PDE:2.14e-04, DBC:2.48e-02, IC:1.01e-01, IV:4.
  ↪35e-03
Epoch  200, Loss:1.19e-01, PDE:1.80e-04, DBC:2.31e-02, IC:9.27e-02, IV:3.
  ↪32e-03
Time taken by adam: 0.872778 seconds

==================================================
11 epoches fine-tuning with LBFGS:
Epoch    0, Loss:6.60e-07, PDE:2.80e-07, DBC:2.70e-07, IC:3.02e-08, IV:8.
  ↪01e-08
Epoch   10, Loss:6.57e-07, PDE:2.75e-07, DBC:2.68e-07, IC:3.11e-08, IV:8.
  ↪28e-08

Training complete.
Time taken by lbfgs: 7.788174 seconds
Total time: 8.660952 seconds
Ratio (time_lbfgs/time_adam) of time taken per epoch:163.055420
```

The training is much faster, and the composite loss has dropped by five orders of magnitude for this problem. Let us now check the solution accuracy.

5.7.5 Results Accuracy Evaluation

```
# Evaluate the results:
u_pred = pinn(XYT_th).detach().numpy().reshape(N_plot, N_plot)
u_exact = exact_u_wave(X, Y, t_obs)

error_L2, errorVI, u_diff = sln_error(u_pred, u_exact)
```

```
Maximum Error: 1.5379e-03
Mean Absolute Error: 4.8162e-04
Error Variation Index (ratio of Max/Mean): 3.1933e+00
L2 Relative Error: 8.3228e-03
```

Compared to `tanh()`, using `cos()` as the activation function has improved the L2 error by more than one order of magnitude.

5.7.6 Plot Time History of the Solution

We use the following code to plot the time history of the solution predicted by the trained PINN model with `cos()` activation, along with that of the exact solution for the point at the center of the unit square.

```
# Plot the time history of solutions of PINN and exact:
x_obs, y_obs = 0.5, 0.5
plot_time_history(pinn, exact_u_wave, x_obs, y_obs, N_plot=200, dpi=300,
         t_min=0.0, t_max=2.0, device=device, f_name='imagesPN/tmp.png')
```

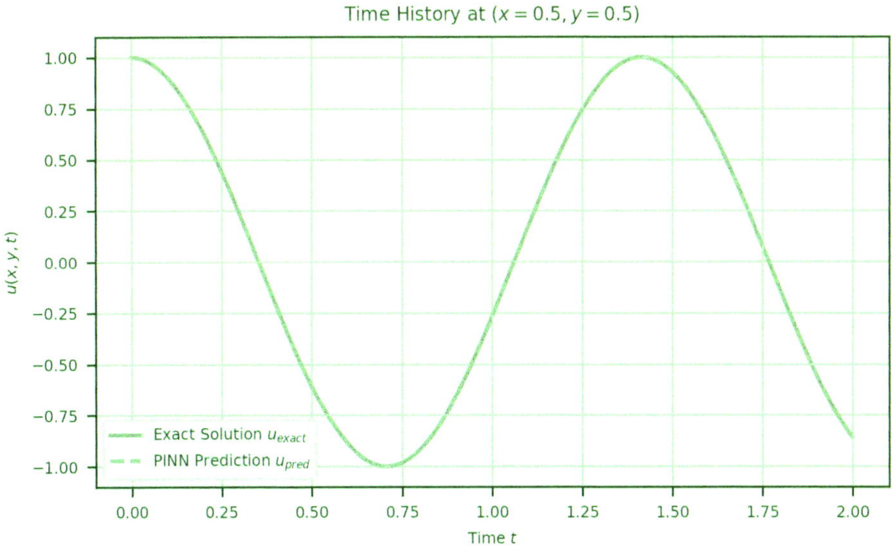

The preceding plot shows the time history curve of u at the center of the unit square, obtained using the PINN model trained with cosine activation. Compared to the time-history curve obtained earlier using `tanh()` activation, the curve obtained using `cos()` activation agrees better with the exact solution, especially near the t_{end}. This improvement is due to the fact that the exact solution for this wave problem includes a cosine function of t, as shown in

Eq.(5.6). We note the following:

> If we know the functional form of the solution, using that function as the activation function can have a good chance to improve the accuracy of the trained PINN model.

5.8 Case Study: Wave Excited by a Pulse-Like Force

In this section, we investigate wave propagation within a unit square plate subjected to a prescribed pulse-like body force. A physics-informed neural network (PINN) model is employed based on a trial solution formulation, which inherently satisfies the Dirichlet boundary conditions (DBC), initial conditions (IC), and initial velocity conditions (IV). This approach enables a focused examination of the effects induced by the spatial and temporal localization of the applied force.

The geometric and physical setting of the problem remains consistent with that of the square plate considered in preceding sections. The distinguishing feature of the current study is as follows:

> The external force is spatially localized and temporally transient , exhibiting rapid variation that emulates wave-like dynamics.

> The initial condition is identically zero, implying that the applied body force solely governs the causality of the resulting wave field.

This problem poses a greater level of difficulty for the PINN framework due to the emergence of sharp wave fronts and potential multiple reflections within the domain.

Most of the Python functions developed previously are directly reusable. In particular, the PINN model class `trainerPDE_ICs_DBC`, which incorporates the trial solution methodology, and the residual evaluation routine `residual_wave()` can be applied without modification. The principal adjustments required for this case study include:

- Designing an appropriate data sampling strategy to resolve the localized and rapidly varying features of both the applied body force and the resulting solution,
- Normalizing the residual with respect to the magnitude of the forcing function to maintain the causality level around unity, and
- Modifying the visualization tools to effectively capture and illustrate the wave propagation dynamics within the plate.

5.8.1 Pulse Force Function Construction

In this example, the force term in Eq.(5.1) is nonzero and has a pulse-like form to generate waves. The force function is designed to exhibit both **spatial concentration** and a **short time duration**, resembling an impulse. There are various ways to construct such a function, provided it remains smooth to ensure compatibility with the PINN framework. Here, we employ a Gaussian function, formulated as follows:

$$f(x,y,t) = C\,\underbrace{e^{-\tau^2(t-\tau_0)^2}}_{f_t}\; e^{-\alpha^2((x-\xi_0)^2+(y-\eta_0)^2)} \tag{5.19}$$

Here:

- C is a scaling constant, which will be determined later through a residual normalization process.
- τ controls the decay rate of the force in time.
- τ_0 specifies the peak time at which the force reaches its maximum.
- α controls the rate of spatial decay (i.e., how rapidly the force fades away from its center).
- (ξ_0, η_0) denotes the spatial location where the force is centered and peaks.

The following snippet constructs this Gaussian force function with the constant $C = 1$ and plots its time-dependent component.

```
# Compute and plot time variation term in the force function, f_t:
t = np.linspace(0, 1, 500)
τ0= 0.07                                                   # shift back
τ = 30.0                                         # for high concentration
def f_t(t, τ0= 0.07, τ = 30.0):
    return np.exp(-(τ*(t-τ0))**2)

fig = plt.figure(figsize=(4, 2.5), dpi=300)              # Plot the surface
plt.plot(t, f_t(t), label='$f_t = exp(-(τ(t-τ_0))^2)$', color='blue')
plt.xlabel('t')
plt.ylabel('$f_t$')
plt.title('Time variation term in the force function')
plt.legend(); plt.grid(True); plt.show()
```

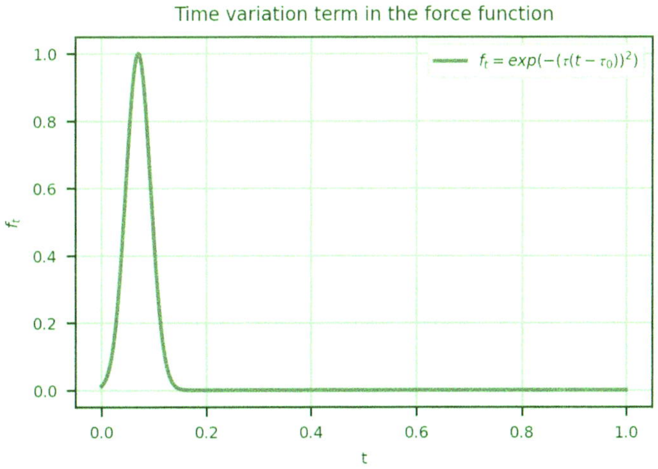

As shown in the preceding plot, the force varies sharply in time. Its significant values lie in the range $t \in [0, 0.14]$, with a peak at $t = 0.07$. This pulse generates a wave in the problem domain (a unit square) that continues to propagate indefinitely inside the domain, even though the force ceases at $t = 0.07$.

The force is also highly concentrated in space, reaching its maximum at the center of the unit square, at $x = 0.5$ and $y = 0.5$. The following snippet computes and plots the spatial variation of the force term.

```
from mpl_toolkits.mplot3d import Axes3D
a=0; b=1; c=0; d=1                                              # domain
# Compute and plot spatial variation term in the force function, f_xy:
def f_xy(x, y, ξ0=0.15, η0=0.15, α = 15.0):
    return np.exp(-(α*(x-ξ0))**2-(α*(y-η0))**2)

x = np.linspace(a, b, 300)                                 # Define the grid
y = np.linspace(c, d, 300)
X, Y = np.meshgrid(x, y)
ξ0 = η0 = 0.5                                          # center of the pulse
α = 15.0                                            # for high concentration
t_abs = 0.
Z = f_xy(X, Y, ξ0=ξ0, η0=η0, α=α)*f_t(t_abs)           # concentrated pulse

fig = plt.figure(figsize=(6, 4), dpi=300)                # Plot the surface
ax = fig.add_subplot(111, projection='3d')
surf = ax.plot_surface(X, Y, Z, cmap='viridis', edgecolor='none')
ax.set_xlabel('x'); ax.set_ylabel('y')
plt.tight_layout(); plt.show()
```

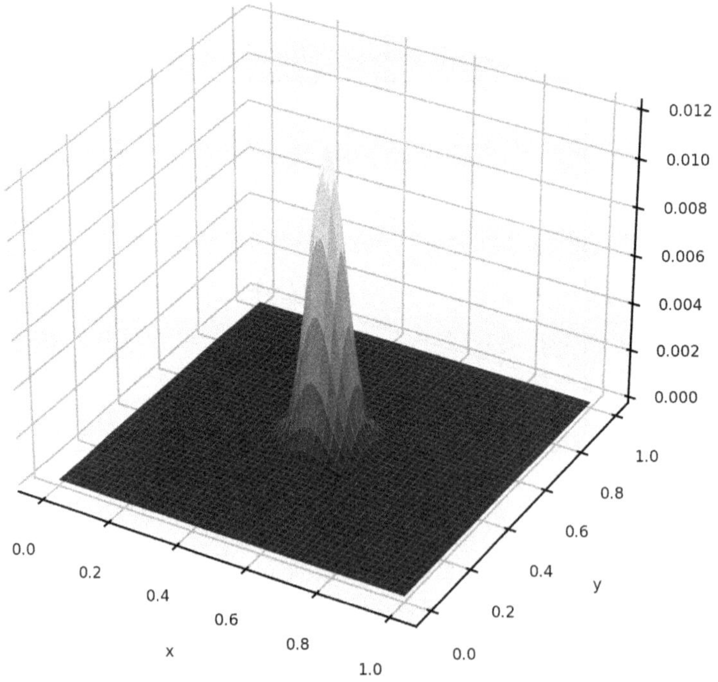

As shown, the force is sharply concentrated at the center of the plate at time zero.

5.8.2 Residual Normalization for Force Term

Since the pulse-like force function changes drastically over both the spatial domain and the time domain, the wave displacement u is expected to be a quite complicated function of (x, y, t). Thus, this case study presents several challenges, making it critical to carefully pose and model the problem in order to successfully train a high-quality PINN model. One important technique is to scale the force term using the scaling constant C shown in Eq.(5.19). This is achieved through **residual normalization**, described in detail below.

From the wave equation (5.1), it is clear that the wave displacement u is proportional to the amplitude of the force term $f(x, y, t)$. Therefore, we can scale the force in any way we choose, as long as the same scaling is also applied to u.

Consider the trivial case of $u = 0$. The residual at any point in the space-time domain in this trivial case becomes simply $f^2(x, y, t)$. Since the optimization in a PINN model is based on minimizing the squared residual, we want the baseline of the mean residual to be approximately unity for effective training. Therefore, the integral of $f^2(x, y, t)$ over the space-time domain, normalized by the volume of that domain, should equal 1:

$$\frac{1}{V_{xyt}} \int_{[0,1]^2 \times [0,t_{\text{end}}]} f^2(x, y, t)\, \mathrm{d}x\, \mathrm{d}y\, \mathrm{d}t = 1 \tag{5.20}$$

where V_{xyt} denotes the volume of the space-time domain.

Since the wave displacement u is generated by the force (i.e., if $f(x, y, t) = 0$, then $u(x, y, t) = 0$), the force acts as the cause. The purpose of residual normalization is to ensure that this causality is properly bounded, allowing the neural network to work effectively with the physical relationship and produce a high-quality solution.

Using Eqs.(5.19) and (5.20), the scaling constant C can be determined. The Python function to perform this computation is as follows:

```
def C_fxyt(t_end, f_t, f_xy, x_end = 1, y_end = 1):
    # Compute the L2 of time variation part of the force, f_t:
    integrand_t = lambda t: f_t(t)**2
    I_f_t2, I_err = si.quad(integrand_t, 0, t_end) # use scipy.integrate

    # Compute the L2 of spatial variation part of the force, f_xy:
    x_end = 1; y_end = 1
    integrand_xy = lambda x, y,: f_xy(x, y)**2
    I_f_xy2, I_err= si.quad(lambda x: si.quad(integrand_xy, 0, x,
                                        args=(x,))[0], 0, y_end)
    # The volume of the space-time domain:
    D_volume = (x_end-0)*(y_end-0)*(t_end-0)
    D_mean_fL2 = I_f_t2*I_f_xy2/D_volume    # domain mean L2 of the force
    C = np.sqrt(1/D_mean_fL2)              # scaling constant for the force
    return C
```

```
# Example of usage:
t_end = 1.0
C = C_fxyt(t_end, f_t, f_xy, x_end = 1, y_end = 1)
print(f"Coefficient of the force: {C}")
```

```
Coefficient of the force: 82.80972646233496
```

Using this function and the foregoing code, we found:

- When $t_{\text{end}} = 0.3$, the scaling constant is computed as $C = 45.35675516425333$.
- When $t_{\text{end}} = 3.0$, the scaling constant is computed as $C = 143.4306535936451$.

5.8.3 PINN Model for Waves Excited by Pulse-Like Force

We now proceed to build and train a PINN model for wave propagation in a unit square plate subjected to a prescribed pulse force. All four edges of the plate are fixed ($u = 0$), and the plate is initially at rest ($f_{IC} = 0$).

Since we use the `PINN_Wave2D_DBC_ICIV` class, all initial and boundary conditions are inherently satisfied through the trial solution formulation. As a result, only the PDE residual loss, `loss_PDE`, needs to be minimized during training.

To accurately capture the sharp variations of the applied force, the sampling points must be sufficiently dense. This necessitates a large number of collocation points to faithfully represent the underlying physics. While the samples-neurons theorem provides a minimum bound on the number of required sampling points to ensure a solution, the pulse-force scenario demands a significantly larger number. In other words, the number of sampling points is dictated not only by theoretical bounds but also by the physical nature of the problem.

For example, to spatially resolve the force on a 100×100 grid, a total of $10,000$ sampling points is needed. If a neural network with 208 neurons is used, the recommended number of sampling points, `N_sampl`, should be roughly 70 times the P^* value. With this carefully considered sampling strategy, we are ready to construct the PINN model for this problem.

A training snippet is provided below. It is self-explanatory and includes appropriate comments. In this case study, a time domain of $[0, 0.7]$ is used.

```
gr.set_seed_th(88)                                  # for reproducibility

t_end = 0.7                      # the end time; t_char= 0.707 ≈ 1/np.sqrt(2)
C = C_fxyt(t_end, f_t, f_xy, x_end = 1, y_end = 1)
def f_th(x, y, t):
    '''Pulse Force Term Function in the Wave PDE in torch'''
    return C*torch.exp(-(τ*(t-τ0))**2)*\
                                torch.exp(-(α*(x-ξ0))**2-(α*(y-η0))**2)

def f_IC(x, y): return 0*x + 0*y           # for the zero initial condition
def f_DBC(x, y): return torch.sin(np.pi*x)*torch.sin(np.pi*y) # true DBC

Nn, NL = 50, 4                                        # a new PINN model
# DBC, IC & IV are enforced in class:
pinn = PINN_Wave2D_DBC_ICIV(Nn, NL, f_DBC, f_IC, my_actvtn)
P_star, P_total = print_model_parameters(pinn, print_table=False)
N_sample = P_star
mPs = 70                        # use mPs*P_star collocaion points in the domain

# Generate the data:
xyt_clctn, xyt_DBC, u_DBC, xyt_IC, u_IC, xyt_IV, u_IV = sampleData(
```

```
                                N_clctn=mPs*N_sample, N_DBC=1, N_IC=mPs*N_sample,
                                                  x_end=1., y_end=1., t_end=t_end)

print(f"Sampled {xyt_clctn.shape[0]} collocation points in the domain.")
print(f"Is N_sample  P* ? {N_sample >= P_star}")

use_DBC = False                           # DBCs is enforced at the class level
use_NBC = False                                      # this problem has no NBCs
LossTerm = 'PDE_only'
Trainer  = trainerPDE_ICs_DBC
Residual = residual_wave
losses, totalTime = train_t_adam_lbfgs(pinn, Trainer, Residual, f_th,
          xyt_clctn, w_s_th, xyt_DBC, u_DBC, xyt_IC, u_IC, xyt_IV, u_IV,
                 epochs1=201, epochs2=11, use_lbfgs=True, adapt_every=0)
# executedexecuted in 59m 5s, finished 15:14:18 2025-06-28
```

```
P* (Total pseudo-dimension of the affine space) = 208
Total trainable parameters: 7901
Sampled 14560 collocation points in the domain.
Is N_sample  P* ? True
==================================================
201 epoches training with Adam:
Epoch   0, Loss:1.30e+00, PDE:1.30e+00, DBC:1.25e-15, IC:0.0, IV:0.00e+00
Epoch 100, Loss:1.22e+00, PDE:1.22e+00, DBC:1.24e-15, IC:0.0, IV:0.00e+00
Epoch 200, Loss:1.19e+00, PDE:1.19e+00, DBC:1.24e-15, IC:0.0, IV:0.00e+00
Time taken by adam: 51.637353 seconds

==================================================
11 epoches fine-tuning with LBFGS:
Epoch   0, Loss:1.27e-02, PDE:1.27e-02, DBC:1.25e-15, IC:0.0, IV:0.00e+00
Epoch  10, Loss:1.65e-04, PDE:1.65e-04, DBC:1.25e-15, IC:0.0, IV:0.00e+00

Training complete.
Time taken by lbfgs: 4407.823619 seconds
Total time: 4459.460972 seconds
Ratio (time_lbfgs/time_adam) of time taken per epoch:1559.780928
```

```
# # Convert to TorchScript and save to file for future use:
# example_input = torch.rand(1, 3)
# traced_pinn = torch.jit.trace(pinn, example_input)
# traced_pinn.save("PulseWave50_4net70m_d7.pt")         # trained and saved

# Load the tained model before using it:
loaded_pinn = torch.jit.load("PulseWave50_4net70m_d7.pt")
loaded_pinn.eval();
```

Because the wave function is expected to be highly complex due to the concentrated force that varies rapidly in time, the pinn model is constructed with 4 hidden layers, each containing 50 neurons. This results in a total parameter count of $P^* = 208$.

The number of randomly sampled collocation points is set to N_sample = 70 * P_star, yielding 14,560 points in total. The training process took approximately one hour and appears to have been successful.

5.8.4 Surface Plots of the Solution

The following code snippets generate surface plots of the solution obtained from the trained PINN model.

```
def data6slices4plot_XYT(pinn, hole, exact_u_xyt=None,
                         x0=0, y0=0, ra=0.2, rb=0.1,
                         N_plot=100, t_start=0.0, t_end=1.0):
    '''
    Generate 6 time slices of u_pred (and optionally u_exact) for
    a trained `pinn` over a 2D domain with a hole.
    Returns X, Y, list of u_pred_full (and optionally exact).
    Parameters: pinn-- trained PINN model
                hole-- function to apply mask on hole
                exact_u_xyt-- (optional) exact solution function
                t_start-- starting time
                t_end-- ending time
                N_plot--grid resolution per axis
    '''
    x = y = np.linspace(0, 1, N_plot)
    X, Y  = np.meshgrid(x, y)
    X_flat = X.flatten()
    Y_flat = Y.flatten()

    mask = ~hole(X_flat, Y_flat, x0=x0, y0=y0, ra=ra, rb=rb)
    idx  = np.where(mask)

    u_pred_list = []
    u_exact_list = []

    t_values = np.linspace(t_start, t_end, 6)
    pinn.eval()

    for t_obs in t_values:
        T = t_obs * np.ones_like(X_flat)
        XYT = np.stack([X_flat, Y_flat, T], axis=1)
        XYT_masked = XYT[mask]

        XYT_tensor = torch.tensor(XYT_masked, dtype=torch.float32)
        with torch.no_grad():
            u_pred_flat = pinn(XYT_tensor).detach().numpy().flatten()

        u_pred_full = np.full(X.shape, np.nan)
        u_pred_full.flat[idx] = u_pred_flat
        u_pred_list.append(u_pred_full)

        if exact_u_xyt is not None:
```

```
            u_exact_flat = exact_u_xyt(XYT_masked[:, 0], \
                                       XYT_masked[:, 1], t_obs)
            u_exact_full = np.full(X.shape, np.nan)
            u_exact_full.flat[idx] = u_exact_flat
            u_exact_list.append(u_exact_full)

    if exact_u_xyt is None:
        return X, Y, u_pred_list, t_values
    else:
        return X, Y, u_pred_list, t_values, u_exact_list
```

```
def plot2Dt_hole6slices(X, Y, u_pred, t_obs, dpi=300, zmin=0, zmax=1, \
                        vmin=0, vmax=1, n_surfaces=6, f_name='tmp'):
    '''
    Plot 2, 4, or 6 surface plots in 2 columns for a plate with a hole.
    Args: X, Y    -- 2D meshgrid arrays
          u_pred  -- array-like of shape (3×2=6) six 2D arrays
          t_obs   -- time value to display in titles
       vmin/vmax  -- color limits for consistency
    '''
    columns = 2; rows = int(n_surfaces/columns)

    width  = 8; height = 3*rows
    fig = plt.figure(figsize=(width, height), dpi=dpi)
    for i in range(n_surfaces):
        ax = fig.add_subplot(rows, columns, i+1, projection='3d')
        surf = ax.plot_surface(X, Y, u_pred[i], cmap='viridis',
                               edgecolor='none', vmin=vmin, vmax=vmax)
        # Set z-axis amplitude limits if provided
        if zmin is not None and zmax is not None:
            ax.set_zlim(zmin, zmax)
        #ax.set_title(f'Surface {i+1} at t={t_obs[i]:.4e}', fontsize=8)
        ax.set_title(f't={t_obs[i]:.4e}', fontsize=7)
        ax.set_xlabel('x')
        ax.set_ylabel('y')
        ax.set_zlabel('u')

    plt.tight_layout()
    fig.savefig(f_name+'.png', dpi=500)
    plt.show()
```

These plotting routines were originally developed for plates with a hole. However, they can still be used to visualize results over the square domain by setting the radii of the hole to a very small value.

```
x0=0.5; y0=0.5; ra=1e-9; rb=1e-9       # extremely small ellipse in center
```

```
def is_inside_ellipse(x, y, x0=x0, y0=y0, ra=ra, rb=rb):
    '''Elliptical hole exclusion function'''
    return ((x-x0)**2 / ra**2 + (y-y0)**2 / rb**2) < 1

t_end = 0.7
# Data for plots:
X, Y, u_pred_list, t_values = data6slices4plot_XYT(loaded_pinn,
    is_inside_ellipse, exact_u_xyt=None, x0=x0, y0=y0, ra=ra, rb=rb,
                                N_plot=200, t_start=0.1, t_end=t_end)
# plot 6 surfaces:
mima = .04
plot2Dt_hole6slices(X, Y, u_pred_list, t_values, dpi=500,
            zmin=-mima, zmax=mima, vmin=-mima,vmax=mima, n_surfaces=6,
                            f_name='imagesPN/PulseWave50_4net70m_d7')
```

Figure 5.1 shows six 3D surface plots of the wave generated by the pulse force at $t = 0.7$.

These six 3D surface plots clearly illustrate how the wave is generated by the pulse force and propagates toward all four edges. At $t = 0.7$, the wave reflects from the fixed boundaries, resulting in a change in sign. All these physical behaviors are consistent with the expected wave dynamics.

Since no exact solution is available for this type of complex problem, alternative methods are required to assess the quality of the obtained solution.

5.8.5 Prediction Confidence

We evaluate the residual error relative to the governing PDE. Inference using a trained PINN model is extremely fast, enabling the use of a large number of evaluation points to compute high-resolution residual estimates. As discussed in Chapter 2, although we generally cannot guarantee that the obtained solution exactly satisfies the wave equation, we can assess the **confidence level** of the trained PINN model.

This is achieved by computing the residual of the trained model over a large, independently sampled test dataset, and applying empirical statistical measures to estimate the likelihood that the model satisfies the PDE within a specified error tolerance.

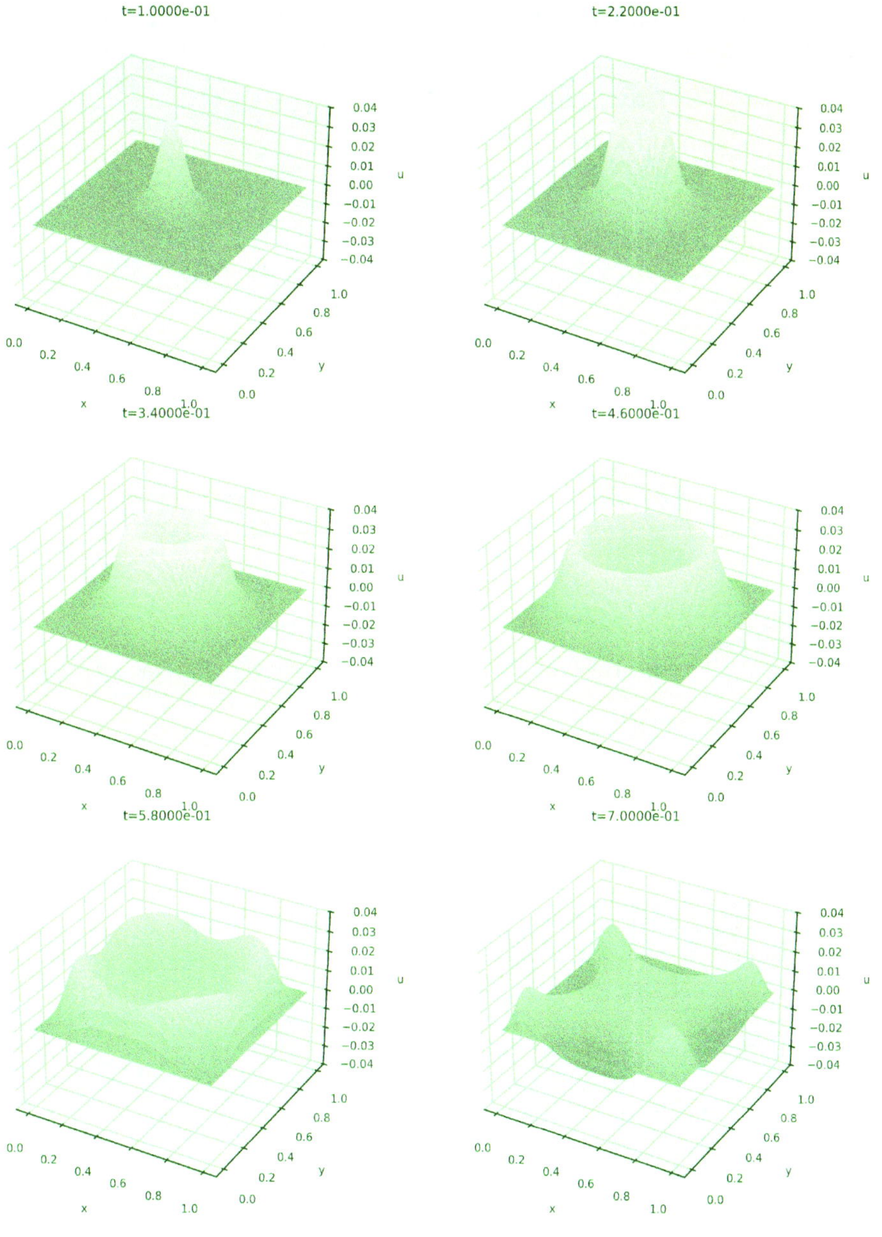

Figure 5.1: Six 3D surface plots of the wave generated by the pulse force at $t = 0.7$

The following Python script performs this evaluation. It includes a residual function that computes the PDE residuals using a test dataset independent of the training set. For the wave equation, we use the `residual_wave` function introduced earlier. The script first generates the independent dataset and then evaluates the residuals using the same function employed during training:

```
# Randomly sample data points for error (residual) evaluation of pinn:
gr.set_seed_th(8)              # use a different seed for random sampling
mPs = 20         # use a statistically sufficiently large number of points

t_end = 0.7
C = C_fxyt(t_end, f_t, f_xy, x_end = 1, y_end = 1)
def f_th(x, y, t):                                        # the pulse force
    return C*torch.exp(-(τ*(t-τ0))**2)*\
                                torch.exp(-(α*(x-ξ0))**2-(α*(y-η0))**2)

xyt_test, _, _, _, _, _, _ = sampleData(N_clctn=mPs*N_sample, N_DBC=1,
                     N_IC=mPs*N_sample, x_end=1., y_end=1., t_end=t_end)

print(f"Sampled {xyt_test.shape[0]} points for error evaluation.")

residual_pred = residual_wave(loaded_pinn, f_th, xyt_test)
print(f"L2 mean residual of trained pinn: {residual_pred[0]:.4e}")
```

```
Sampled 4120 points for error evaluation.
L2 mean residual of trained pinn: 8.5484e-01
```

The following Python function computes the confidence level of a trained PINN model.

```
# ---Prediction confidence evaluation using residual_fun() ----
def pinn_confidence(pinn, f_th, residual_fun, xyt_test, tol=1e-3):
    # Evaluate residual:
    mean_res, res_tensor = residual_fun(pinn, f_th, xyt_test)
    res_vals = res_tensor.detach().numpy().flatten()

    # Compute statistic properbility:
    L2_residual = np.mean(res_vals)     # = mean_res
    proportion_within_tol = np.mean(res_vals < tol)          # likelihood

    print(f"L2 mean residual: {L2_residual:.4e}")
    print(f"Proportion of points with residual < {tol}: "
          f"{proportion_within_tol:.4f}")

    return L2_residual, proportion_within_tol
```

```
# evaluate the confidence of `pinn` model for a given tolerance:
tol = 1e-3
_= pinn_confidence(loaded_pinn, f_th, residual_wave, xyt_test, tol=tol)
```

```
L2 mean residual: 8.5484e-01
Proportion of points with residual < 0.001: 0.8260
```

The results show that, within the space-time domain $[0,1]^2 \times [0,0.7]$, approximately 82.6% of the evaluation points exhibit a residual less than 10^{-3}. Naturally, if a larger tolerance is chosen, the computed confidence level will increase. Therefore, this metric is **subjective** and should be interpreted with appropriate context.

This case study demonstrates the successful training of PINN models for a fairly complex wave propagation problem, achieving at least good sub-optimal solutions. The models leveraged techniques developed in earlier chapters, as well as new methods introduced here. Among these, the **residual normalization** technique is particularly critical.

Readers can appreciate the importance of this technique by manually setting $C = 1$ in the provided code and observing how the training either fails to converge or results in significantly reduced accuracy for certain models. This emphasizes that **normalization** is an essential step in training any machine learning model—not just PINNs (see Ref. [12]).

Moreover, the method of normalization should be tailored to the specific nature of the problem being addressed. For the force-driven wave propagation problem considered here, the squared PDE residual is normalized under the assumption $u = 0$. This approach effectively quantifies the **initial causality** introduced by the applied body force, ensuring that the residual is appropriately scaled for the network to learn from physically meaningful signals during training.

5.9 Remarks

This chapter presents various PINN models for hyperbolic PDEs, which govern different types of wave phenomena, using the standard wave equation as a representative example. The training data is generated via collocation sampling guided by the samples-neurons theorem. The following key remarks can be made:

1. The partial derivative with respect to time is second-order in wave equations; thus, both initial conditions (ICs) and initial velocity conditions (IVs) must be specified.
2. Dirichlet boundary conditions (DBCs), ICs, and IVs can be enforced as additional terms in the total composite loss. However, it is often preferable to enforce them at the PINN class level for improved efficiency and accuracy.
3. The **residual normalization** technique is proposed and shown to be effective for PINN models addressing complex problems, such as wave propagation induced by a pulse-like force.
4. If the functional form of the solution is known or partially known, incorporating it into the activation function can improve the accuracy and generalization capability of the trained PINN model.
5. The training time for first-order optimizers such as `adam` scales approximately **linearly** with the number of sampling points. In contrast, quasi-second-order optimizers like `lbfgs` may require significantly more time, with performance strongly influenced by the nonlinearity and conditioning of the loss function.

The techniques for developing PINN models discussed in this chapter are, in principle, applicable to a wide variety of problems, including but not limited to acoustic waves, elastic waves, electromagnetic waves, and even quantum mechanics (via complex-valued wave equations). Nevertheless, constructing and training a high-quality PINN model remains a challenging task.

Chapter 6

PINN for Eigenvalue Equations

Let us import the necessary external modules or dependencies for later use in this chapter.

```
# Place cursor in this cell, and press Ctrl+Enter to import dependencies.
import os
import sys                                 # for accessing the computer system
sys.path.append('../grbin/')                 # Add in the path to your system

from commonImports import *          # Import dependencies from '../grbin/'
import grcodes as gr                       # Import the module of the author
importlib.reload(gr)                     # When grcodes is modified, reload it
import time
import math

init_printing(use_unicode=True)          # For latex-like quality printing
np.set_printoptions(precision=4, suppress=True)   # Digits in print-outs
```

Eigenvalue problems are ubiquitous in science and engineering. They are closely related to wave propagation problems in the wave number domain [22], and to vibration problems in the frequency domain [20]. Numerical methods have been extensively developed and are particularly suitable for handling problems with complex geometries. Among the most widely used are the finite element method (FEM) [21], the smoothed finite element method (S-FEM) [20], and meshfree methods [11].

Building on the techniques developed in the previous chapters, this chapter focuses on constructing **Physics-Informed Neural Network (PINN)** models for **eigenvalue equations**, with a particular emphasis on the **Helmholtz equation**, which represents the eigenvalue problem for the **Laplace operator**.

This problem is governed by second-order partial differential equations (PDEs) involving an unknown scalar function, called the **eigenfunction**, defined over a spatial domain, and an unknown scalar value, called the **eigenvalue**.

6.1 Governing PDE for the 2D Helmholtz Equation

We begin by considering the standard form of the 2D Helmholtz equation, given as follows:

$$\nabla^2 u(x,y) = -\underbrace{k^2}_{\lambda} u(x,y), \qquad (x,y) \in \Omega \tag{6.1}$$

where

- $u(x,y)$ often termed the **eigenfunction**, which is the unknown function.

- k is the wavenumber (for wave-propagation problems), and $\lambda = k^2$ is often called the **eigenvalue**, which is the unknown scalar.

In this chapter, the problem domain is the unit square $\Omega = (0,1)^2$.
We assume that Eq.(6.1) is subject to a Dirichlet boundary condition (DBC), $\partial\Omega_{\text{DBC}}$:

$$u(x,y)|_{\partial\Omega_{\text{DBC}}} = 0 \tag{6.2}$$

In general, there could also be Neumann boundary conditions (NBCs), as discussed in the previous chapters, because the spatial partial derivatives in the wave equation are of second-order. In this chapter, however, we will not repeat the discussion from the previous chapter on how to handle NBCs. The example problems studied here will be set without NBCs so that our focus can remain on the DBC together with the PDE.

6.2 Essential Properties of Eigenvalue Equations

1. **Equivalence of the Laplace operator and a scalar**
   Equation (6.1) states that the Laplacian of a function equals a negative scalar multiple of the function itself. Thus, the Laplacian ∇^2 acting on the function is equivalent to scaling it by the negative scalar $-\lambda$.

2. **Non-uniqueness**
   If any pair (u, λ) satisfies Eq.(6.1), then the pair (Cu, λ) also satisfies Eq.(6.1) for any scalar C. This non-uniqueness is typically resolved through **normalization**, which will be employed later in our PINN models.

3. **The characteristic pair**
   Not every function admits such a special pairing with a scalar. A function u and its corresponding scalar λ that satisfy Eq.(6.1) form what we call an **eigen-pair**. The function is the **eigenfunction**, and the scalar is the **eigenvalue**. The term *eigen* originates from the German word meaning *characteristic*.
   An eigen-pair is also referred to as an **eigenmode**. The eigenfunction is often called the **mode shape**, and the eigenvalue is related to the system's **natural frequency**.

4. **Infinite number of eigen-pairs**
   There exists an infinite number of characteristic pairs that satisfy Eq.(6.1). It is practically impossible to compute all of them numerically; fortunately, most applications require only a few of the lowest eigenvalues and their eigenfunctions, as these lowest modes usually dominate system behaviour [21].

5. **Orthogonality**
   In general, eigenfunctions corresponding to distinct eigenvalues are orthogonal to one another. Eigenfunctions sharing the same eigenvalue span an **eigenspace** , and any function within this space is orthogonal to eigenfunctions associated with different eigenvalues.

We will demonstrate and exploit this property later.

6.3 Challenges to PINN Models

The features and properties of eigenvalue problems pose significant challenges when building a PINN model to recover the eigenvalues and eigenfunctions satisfying Eq.(6.1).
Major challenges include:

1. **Single PDE, but multiple unknowns**
   Both the eigenvalue and the eigenfunction are unknown, rendering the problem inherently nonlinear.
2. **Multiple (infinite) solutions of eigen-pairs**
   The existence of infinitely many valid eigen-pairs makes it difficult to isolate the desired solution.
3. **Combination of challenges 1 and 2**
   The coexistence of multiple unknowns and multiple solutions complicates PINN training and convergence to a specific eigenmode.

In this chapter, we demonstrate how a PINN can be constructed and trained to find several of these solutions, and we discuss the difficulties encountered along the way.

6.4 Analytical Method

Many standard analytical methods can be used to solve the Helmholtz equation, including the method of separation of variables (discussed in the following section) and the Green's function method. PINNs offer an alternative approach to solving Eq.(6.1), producing a single—but possibly complicated—eigenfunction along with its corresponding eigenvalue, provided the model is properly trained.

To construct effective PINN models for the eigenvalue problem, it is essential to first understand the underlying mathematical structure. Therefore, we begin with an analytical solution approach in the next section.

6.4.1 Method of Separation of Variables

Equation (6.1) can be solved using the method of separation of variables. This approach assumes that the solution $u(x, y)$ can be expressed as the product of two unknown functions:

$$u(x, y) = X(x)Y(y) \tag{6.3}$$

where $X(x)$ is a function of x only, and $Y(y)$ is a function of y only. On the four edges of the domain, the Dirichlet boundary condition (DBC) specified in Eq.(6.2) becomes:

$$\begin{aligned}
u(0, y) = 0 &\Rightarrow X(0)Y(y) = 0 \Rightarrow X(0) = 0 \\
u(1, y) = 0 &\Rightarrow X(1)Y(y) = 0 \Rightarrow X(1) = 0 \\
u(x, 0) = 0 &\Rightarrow X(x)Y(0) = 0 \Rightarrow Y(0) = 0 \\
u(x, 1) = 0 &\Rightarrow X(x)Y(1) = 0 \Rightarrow Y(1) = 0
\end{aligned} \tag{6.4}$$

Substituting Eq.(6.3) into Eq.(6.1), we obtain:

$$X''(x)Y(y) + Y''(y)X(x) = -\lambda X(x)Y(y) \tag{6.5}$$

Now, let us assume that $\lambda = \lambda_x + \lambda_y$. Substituting into Eq.(6.5) gives:

$$X''(x)Y(y) + Y''(y)X(x) = -\lambda_x X(x)Y(y) - \lambda_y X(x)Y(y) \tag{6.6}$$

Regrouping terms, we have:

$$\left(X''(x) + \lambda_x X(x)\right) Y(y) = -\left(Y''(y) + \lambda_y Y(y)\right) X(x) \tag{6.7}$$

Since this equation must hold for all admissible $X(x)$ and $Y(y)$, the expressions inside the parentheses must each be equal to zero. This leads to the following two ordinary differential equations:

$$X''(x) + \lambda_x X(x) = 0 \qquad \text{with } X(0) = X(1) = 0 \tag{6.8}$$

and

$$Y''(y) + \lambda_y Y(y) = 0 \qquad \text{with } Y(0) = Y(1) = 0 \tag{6.9}$$

Here, we have used the Dirichlet boundary conditions (DBCs) specified in Eq.(6.4).

The original 2D Helmholtz partial differential equation has now been separated into two independent ordinary differential equations (ODEs), each representing a one-dimensional eigenvalue problem with fixed ends. The eigenvalues and eigenfunctions for these ODEs are determined as follows:

For $X(x)$:

$$\lambda_x = (m\pi)^2, \qquad X_m(x) = \sin(m\pi x), \qquad m = 1, 2, 3, ... \tag{6.10}$$

and for $Y(y)$:

$$\lambda_y = (n\pi)^2, \qquad Y_n(y) = \sin(n\pi y), \qquad n = 1, 2, 3, ... \tag{6.11}$$

Combining these two solutions, the eigenvalues for the original 2D Helmholtz equation become:

$$\lambda_{m,n} = (m\pi)^2 + (n\pi)^2 = \pi^2(m^2 + n^2) \tag{6.12}$$

where m and n are positive integers ($m, n \in \{1, 2, 3, ...\}$). The corresponding eigenfunctions are:

$$u_{m,n}(x, y) = \sin(m\pi x)\sin(n\pi y) \tag{6.13}$$

Clearly, there are infinitely many eigen-pairs.

6.4.2 Orthogonality of Eigenfunctions

Consider the inner product of two eigenfunctions:

$$\langle u_{m,n}, u_{p,q}\rangle = \int_0^1 \int_0^1 \sin(m\pi x)\sin(n\pi y)\sin(p\pi x)\sin(q\pi y)\,dx\,dy \tag{6.14}$$

This expression represents the standard L^2 inner product of two eigenfunctions.
By the orthogonality of sine functions over the interval $[0, 1]$, this double integral evaluates to **zero** unless $m = p$ and $n = q$, in which case it equals $\frac{1}{4}$:

$$\langle u_{m,n}, u_{p,q}\rangle = \begin{cases} \frac{1}{4}, & \text{if } m = p \text{ and } n = q \\ 0, & \text{otherwise} \end{cases} \tag{6.15}$$

This is the orthogonality property of any two eigenfunctions $u_{m,n}$ and $u_{p,q}$ for the Helmholtz equation defined on the unit square domain.

The following SymPy code snippet demonstrates that the eigenfunctions given in Eq.(6.13) are **mutually orthogonal** over the domain $[0, 1]^2$ with respect to the standard L^2 inner product:

```
# Define symbols
x, y = sp.symbols('x y')
m, n, p, q = sp.symbols('m, n, p, q', integer=True, positive=True)

# Define the eigenfunctions
u_mn = sp.sin(m * sp.pi * x) * sp.sin(n * sp.pi * y)  #*2,     normalized
u_pq = sp.sin(p * sp.pi * x) * sp.sin(q * sp.pi * y)  #*2,     normalized

# Compute the inner product over the unit square via double integral:
inner_product = sp.integrate(sp.integrate(u_mn * u_pq, (x, 0, 1)),
                                                        (y, 0, 1))
# Simplify the result
inner_product_simplified = sp.simplify(inner_product)
inner_product_simplified
```

$$\begin{cases} 0 & \text{for } m \neq p \vee n \neq q \\ \frac{1}{4} & \text{otherwise} \end{cases}$$

The preceding code confirms Eq.(6.15).

6.4.3 Normalization of Eigenfunctions

As demonstrated, the inner product of an eigenfunction with itself—i.e., its norm—is $\frac{1}{4}$. In practice, we typically normalize the eigenfunctions so that their norm is unity. This can be achieved by simply multiplying Eq.(6.13) by 2, leading to:

$$u_{m,n}(x, y) = 2\sin(m\pi x)\sin(n\pi y) \tag{6.16}$$

These eigenfunctions are now mutually **orthonormal**, which can also be confirmed using the preceding code. Normalization will be used later during PINN model training to ensure that the norm of the eigenfunction is controlled, which improves numerical stability.

6.5 PINN Model for Eigenvalue Solutions

Let us now construct a PINN model to solve the eigenvalue problem described in Eq.(6.1). In this study, our goal is to compute the first two eigenvalues and their corresponding eigenfunctions.

The following snippet defines the computational device for training the PINN model.
It ensures that a CUDA-enabled GPU is used for acceleration if available; otherwise, the CPU is used:

```
# Device management: ensure CUDA is available if you plan to use GPU
device = torch.device('cuda' if torch.cuda.is_available() else 'cpu')
gr.device = device                              # inject device to gr module
torch.cuda.empty_cache()                        # Clear previous model state
print(f"Using device: {device}")

# user defined activation
def my_actvtn(x):
    return torch.tanh(x) #sin(x) #
```

```
Using device: cpu
```

6.5.1 Multiple Solutions

As shown in the previous section, the exact solutions for the first two lowest eigenvalues and their corresponding eigenfunctions are given by:

$$u_{1,1}(x,y) = \sin(\pi x)\sin(\pi y), \quad \text{with } \lambda_{1,1} = 2\pi^2 \approx 19.7392 \tag{6.17}$$

$$\begin{aligned} u_{1,2}(x,y) &= \sin(\pi x)\sin(2\pi y), \quad \text{with } \lambda_{1,2} = 5\pi^2 \approx 49.3480 \\ u_{2,1}(x,y) &= \sin(2\pi x)\sin(\pi y), \quad \text{with } \lambda_{2,1} = 5\pi^2 \approx 49.3480 \end{aligned} \tag{6.18}$$

Note that the eigenfunctions $u_{1,2}(x,y)$ and $u_{2,1}(x,y)$ are distinct functions but share exactly the same eigenvalue, which has multiplicity 2. An eigenvalue associated with multiple eigenfunctions is said to be **degenerate**. These two eigenfunctions, $u_{1,2}(x,y)$ and $u_{2,1}(x,y)$, are orthogonal (or can be orthogonalized), and any linear combination of the two is also an eigenfunction corresponding to the same eigenvalue.

This subtle mathematical feature can influence the behavior of the PINN solution, and we will examine its implications in detail in the upcoming example section.

6.5.2 PINN Class for Eigenvalue Problems

A PINN class for eigenvalue problems is defined in the following code cell. It is largely similar to the one introduced in Chapter 2; the primary difference is that the eigenvalue λ is **unknown** and treated as a trainable parameter—just like the neural network weights $\hat{\mathbf{w}}$.

Therefore, λ is explicitly defined and initialized within the class template. The code snippets below include inline comments and are thus self-explanatory.

```
class EigenPINN(nn.Module):
    '''
    Define the PINN class with DBCs enforced automatically.
    '''
    def __init__(self, layers):
        super().__init__()
        self.net = nn.ModuleList()
        for i in range(len(layers) - 1):
            self.net.append(nn.Linear(layers[i], layers[i+1]))
            # Initialize weights with Xavier/He for better stability
            if i < len(layers)-2: # Don't apply to the last layer's bias
                nn.init.xavier_normal_(self.net[-1].weight)
                nn.init.zeros_(self.net[-1].bias)

        self.activation = torch.tanh
        # Initialize lambda as close as possible to expected values:
        self.lam = nn.Parameter(torch.tensor([15.],dtype=torch.float64))

    def forward(self, x):
        x1, x2 = x[:, 0:1], x[:, 1:2]
        u_hat = x
        for i in range(len(self.net) - 1):
            u_hat = self.activation(self.net[i](u_hat)) # output from NN

        u_hat = self.net[-1](u_hat)
        # Apply hard DBCs
        return x1 * (1 - x1) * x2 * (1 - x2) * u_hat
```

In the EigenPINN class, the Dirichlet boundary condition (DBC) is enforced exactly using a trial solution of the form:

$$u(x, y) = x(1-x)\,y(1-y)\cdot\mathcal{N}_{\hat{\mathbf{w}}} \tag{6.19}$$

This is the same approach used in Chapter 2. We only need to train the neural network $\mathcal{N}_{\hat{\mathbf{w}}}$, and the DBC is automatically satisfied by the construction of $u(x, y)$.

6.5.3 Sampling Point Generation

Since the DBCs are enforced at the class level, it is sufficient to sample points from the **interior** of the square domain. This can be easily achieved using the following Python function:

```
def generate_points(n_Ω):
    '''Generate sampling points in 2D domain Ω.
    n_Ω: number of smapling points in Ω.
    x_Ω: coordinates of the smapling points in Ω.
    '''
```

```
    x_Ω = torch.rand(n_Ω, 2, dtype=torch.float64)
    return x_Ω
```

6.5.4 Residual Evaluation for Eigenvalue Equations

The following snippet computes the residual for eigenvalue equations. It is quite similar to the residual function used for the Poisson equation discussed in Chapter 2. The main difference is the inclusion of the **unknown eigenvalue**, which is treated as a trainable parameter within the PINN model.

```
def eigen_residual(model, x):
    x.requires_grad_(True)                    # Ensure x requires gradients
    u = model(x)

    # Compute first-order gradients
    grad_u = torch.autograd.grad(u, x, torch.ones_like(u), \
                                    create_graph=True)[0]

    # Compute second-order gradients (Laplacian)
    du_xx = torch.autograd.grad(grad_u[:, 0], x, \
            torch.ones_like(grad_u[:, 0]), create_graph=True)[0][:, 0:1]
    du_yy = torch.autograd.grad(grad_u[:, 1], x, \
            torch.ones_like(grad_u[:, 1]), create_graph=True)[0][:, 1:2]

    residual = du_xx + du_yy + model.lam.clamp(min=1e-2) * u
    # model.lam is from the model class, a unknown, and trainable
    return residual, u
```

6.5.5 Enforcing Orthogonality

Since we aim to compute two eigenvalues and their corresponding eigenfunctions, we need to train **two separate PINN models**—one for each eigen-pair. The strategy is to first solve for the **fundamental (lowest) mode**, and then for the **second mode**.

Because eigenfunctions associated with different eigenvalues are orthogonal, we must leverage this property to guide the second PINN model toward the correct solution. In this case, when training the PINN for the second mode, we utilize the eigenfunction obtained from the first mode and explicitly enforce **orthogonality** with respect to it. Normalization is also applied to the eigenfunction during this process to ensure **numerical stability**.

The following Python function implements this procedure:

```
def eigen_residual_orthogonal(model, x, prev_models_list):
    '''
    Compute the eigenfunction using `model` that is orthogonal to the
    eigenfunctions trained previously and given in `prev_models_list`.
    It return the normalized eigenfunction u and its residual
    '''
    x.requires_grad_(True)
    u_raw = model(x)
```

```
    # Project u_raw onto the space orthogonal to all previous eigenfuns:
    u = u_raw
    for prev_model in prev_models_list:
        # Evaluate the previous eigenfunction using its model at the
        # CURRENT x points:
        u_prev_at_x = prev_model(x).detach()        # detaches the result

        # Normalize u_prev_at_x before projection for stability.
        # This ensures the projection coefficient is well-behaved.
        norm_u_prev_at_x = torch.norm(u_prev_at_x)
        if norm_u_prev_at_x > 1e-12:                # Avoid division by zero
            u_prev_at_x_normalized = u_prev_at_x / norm_u_prev_at_x
        else:
            u_prev_at_x_normalized = u_prev_at_x # If norm is too small,
                                             # leave as is (likely near zero)

        # Calculate projection coefficient: the dot product~ with the
                                                          # current u_raw
        proj_coeff = torch.sum(u_raw*u_prev_at_x_normalized)/(torch.sum(
               u_prev_at_x_normalized * u_prev_at_x_normalized) + 1e-12)
        # Subtract the projected component:
        u = u - proj_coeff * u_prev_at_x_normalized

    grad_u = torch.autograd.grad(u, x, torch.ones_like(u), \
                                                 create_graph=True)[0]
    du_xx = torch.autograd.grad(grad_u[:, 0], x, \
            torch.ones_like(grad_u[:, 0]), create_graph=True)[0][:, 0:1]
    du_yy = torch.autograd.grad(grad_u[:, 1], x, \
            torch.ones_like(grad_u[:, 1]), create_graph=True)[0][:, 1:2]

    residual = du_xx + du_yy + model.lam.clamp(min=1e-2) * u
    return residual, u
```

6.5.6 Code to Train the PINN Model

The following Python function trains the PINN model created. It is largely the same as that introduced in Chapter 2, and we will use a two-stage optimization strategy . The first-order optimizer `adam` is used for initial training, and the quasi-second-order optimizer `lbfgs` is used for refinement. The differences are as follows:

1. Orthogonalization is performed on the eigenfunctions during the training process.

2. Normalization is enforced using an L2 regularization term that penalizes deviation of the eigenfunction norm from 1. Therefore, we introduce a new `loss_norm` term, which will be monitored. A weight parameter `w_norm` is used to control the contribution of `loss_norm` in the composite loss. The `loss_norm` is computed as:

   ```
   loss_norm = (torch.mean(u_Ω**2) - 1.0)**2
   ```

3. Another L2 regularization term is applied to the model weights to prevent overfitting, which is standard in machine learning models. A very small regularization parameter is used for the PINN models.

4. The total loss is composed as:

```
loss = loss_pde + w_norm*loss_norm + 1e-6 * l2_reg
```

```
from tqdm import tqdm          # For displying the training progress bar

def train_mode(model, x_Ω, prev_models_list=None, use_ortho=False, \
                    adam_epochs=5000, lbfgs_max_iter=1000, w_norm=100.):
    '''
    Function to train the PINN `model`.
    x_Ω:     sampled points~ in the domain.
    w_norm: weight for the norm loss: loss_norm
    '''
    if prev_models_list is None:
        prev_models_list = []     # Initialize empty list if not provided

    optimizer = torch.optim.Adam(model.parameters(), lr=5e-4)

    # Use tqdm for initial training:
    for epoch in tqdm(range(adam_epochs), desc="Adam Training"):
        optimizer.zero_grad()
        # Compute the residual and the eigenfunction values u_Ω:
        if use_ortho:
            # Pass prev_models_list
            res, u_Ω = eigen_residual_orthogonal(model, x_Ω, \
                                                prev_models_list)
        else:
            res, u_Ω = eigen_residual(model, x_Ω)

        loss_pde = torch.mean(res**2)          # compute the loss for PDE

        # Apply regularization enforce normalization of u_Ω
        # Using L2 norm to penalize the norm that deviates from 1
        loss_norm = (torch.mean(u_Ω**2) - 1.0)**2

        #Use a small L2 regularization on weights to prevent overfitting
        l2_reg = 0.0
        for param in model.parameters():
            l2_reg += torch.norm(param, 2)

        loss = loss_pde + w_norm*loss_norm + 1e-6 * l2_reg  # add L2 reg

        loss.backward()
        optimizer.step()

        # print key variables to monitor the training progress:
        if epoch % (adam_epochs // 2) == 0 or epoch == adam_epochs - 1:
            tqdm.write(f"[Adam] Epoch {epoch:>4d} |"
                       f"λ≈{model.lam.item():.6f} |"
                       f"PDE = {loss_pde.item():.2e} | "
```

```
                             f"Norm = {loss_norm.item():.2e} |"
                             f"u_max = {u_Ω.max().item():.3e}")

    print("Refining with L-BFGS...")
    optimizer_lbfgs = torch.optim.LBFGS(model.parameters(), \
                                 max_iter=lbfgs_max_iter, tolerance_grad=1e-9,
                   tolerance_change=1e-9, line_search_fn="strong_wolfe")

    call_count = [0]
    def closure():
        optimizer_lbfgs.zero_grad()
        if use_ortho:
            res, u_Ω = eigen_residual_orthogonal(model, x_Ω, \
                                                      prev_models_list)
        else:
            res, u_Ω = eigen_residual(model, x_Ω)
        loss_pde = torch.mean(res**2)
        loss_norm = (torch.mean(u_Ω**2) - 1.0)**2

        l2_reg = 0.0
        for param in model.parameters():
            l2_reg += torch.norm(param, 2)

        loss = loss_pde + w_norm*loss_norm + 1e-6 * l2_reg

        loss.backward()
        call_count[0] += 1

        # print key variables to monitor the training progress:
        if call_count[0] % (lbfgs_max_iter//2)==0 or call_count[0] == 1\
                                   or call_count[0] == lbfgs_max_iter:
            tqdm.write(f"[L-BFGS-{call_count[0]}]"
                       f"λ≈{model.lam.item():.6f} |"
                       f"PDE = {loss_pde.item():.2e} | "
                       f"Norm = {loss_norm.item():.2e} |"
                       f"u_max = {u_Ω.max().item():.3e}")
        return loss

    optimizer_lbfgs.step(closure)
    return model
```

6.6 Case Study: PINN Models for Lowest eigenmodes

This study builds PINN models to find the lowest two eigenmodes of the Helmholtz equation defined on a unit square with all four edges fixed.

6.6.1 Training of PINN Models

The following snippet constructs and trains two PINN models for this problem. One is the `base_model`, which computes the eigenvalue and eigenfunction of the fundamental mode. The other is the `higher_model`, which computes the eigenvalue and eigenfunction of the second mode.

The initial guesses for these eigenvalues are chosen to be approximately 10% above the true values, based on the assumption that we typically have some rough prior knowledge about the eigenvalue problem to be solved.

```
gr.set_seed_th(seed=88)                       # Set seeds for reproducibility

# Analytical eigenvalues for a unit square are λ_mn= ^2 (m^2 + n^2)
# (1,1) mode: λ_11= ^2 (1^2 + 1^2)=2 ^2≈19.7392
# (2,1) mode: λ_21= ^2 (2^2 + 1^2)=5 ^2≈49.3480
# (1,2) mode: λ_12= ^2 (1^2 + 2^2)=5 ^2≈49.3480 (symmetric to 2,1)
# (2,2) mode: λ_22= ^2 (2^2 + 2^2)=8 ^2≈78.9568

n_interior_points = 10000
x_Ω = generate_points(n_interior_points)

print("Training base mode (1,1)...")
base_model = EigenPINN([2, 60, 60, 60, 1]).double()             # NL=3
base_model.lam = nn.Parameter(torch.tensor([22.0], dtype=torch.float64))
base_model = train_mode(base_model, x_Ω, adam_epochs=5000, \
                                        lbfgs_max_iter=1000, w_norm=1e3)

# plot_model(base_model, "Mode (1,1)")
print(f"Final λ for mode (1,1): {base_model.lam.item():.6f}")

print("\nTraining higher mode ...")  # increase one layer
higher_model = EigenPINN([2, 60, 60, 60, 60, 1]).double()       # NL=4
higher_model.lam = nn.Parameter(torch.tensor([55.], dtype=torch.float64))

# Pass the trained base_model object in the list
higher_model = train_mode(higher_model, x_Ω,\
                          prev_models_list=[base_model], use_ortho=True,
                      adam_epochs=5000, lbfgs_max_iter=1000, w_norm=2e3)
# plot_model(higher_model, usually a combied Mode (1,2) and Mode (2,1)):
print(f"Final λ for the second mode: {higher_model.lam.item():.6f}")

# executed in 58m 3s, finished 12:02:16 2025-07-01
```

```
Training base mode (1,1)…

Adam Training:   0%|  1/5000 [00:00<13:44,  6.06it/s]

[Adam] Epoch 0 |λ≈22.0005 |PDE=1.45e-03 | Norm=1.0e+00 |u_max=-1.646e-09

Adam Training:  50%| 2502/5000 [07:54<08:02,  5.18it/s]

[Adam] Epoch 2500 |λ≈20.497 |PDE=5.89e-01 | Norm=1.14e-06 |u_max=-1.267e-05
```

```
Adam Training:
100%| 5000/5000 [16:01<00:00,  5.20it/s]

[Adam] Epoch 4999 |λ≈19.847 |PDE=1.30e-02 | Norm=2.91e-09 |u_max=-1.261e-05
Refining with L-BFGS...
[L-BFGS-1]λ≈19.847 |PDE=1.29e-02 | Norm=3.11e-09 |u_max=-1.261e-05
[L-BFGS-500]λ≈19.739  |PDE=1.08e-05 | Norm=8.09e-15 |u_max=-1.255e-05
Final λ for mode (1,1): 19.739

Training higher mode ...

Adam Training:  0%| 1/5000 [00:00<1:02:16,  1.34it/s]

[Adam] Epoch  0 |λ≈54.9995 |PDE=1.79e-03 | Norm=1.00e+00 |u_max=1.507e-03

Adam Training:  50%| 2501/5000 [11:39<16:16,  2.56it/s]

[Adam] Epoch 2500 |λ≈53.612 |PDE=6.06e+01 | Norm=1.99e-03 |u_max=2.166e+00

Adam Training:
100%|  5000/5000 [22:54<00:00,  3.64it/s]

[Adam] Epoch 4999 |λ≈53.1732|PDE=2.01e+01 | Norm=4.20e-05 |u_max=2.142e+00
Refining with L-BFGS...
[L-BFGS-1]λ≈53.1732|PDE=2.01e+01 | Norm=4.20e-05 |u_max=2.142e+00
[L-BFGS-500]λ≈50.879|PDE=1.28e+01 | Norm=8.92e-06 |u_max=2.139e+00
[L-BFGS-1000]λ≈49.356 |PDE=9.02e+00 | Norm=4.64e-06 |u_max=2.145e+00
Final λ for the second mode: 49.1599
```

6.6.2 Plots of Eigenfunctions

The training has completed. Let us plot the results using the following function.

```python
def plot_model(model, title="Eigenfunction", f_name='tmp'):
    '''
    Plot the eigenvalue and eigenfunction using a trained `model`.
    '''
    x_plot = np.linspace(0, 1, 100)
    y_plot = np.linspace(0, 1, 100)
    X, Y = np.meshgrid(x_plot, y_plot)
    XY = torch.tensor(np.vstack([X.ravel(), Y.ravel()]).T, \
                                                    dtype=torch.float64)
    u_pred = model(XY).detach().numpy().reshape(100, 100)

    # Normalize the eigenfunction for plotting consistency
    u_pred = u_pred / np.sqrt(np.sum(u_pred**2)*(1/100)**2 * (1/100)**2)

    fig = plt.figure(figsize=(7, 3), dpi=500)

    # Contour Plot
    ax1 = fig.add_subplot(1, 2, 1)
    contour = ax1.contourf(X, Y, u_pred, 20, cmap='viridis')
    fig.colorbar(contour, ax=ax1, label="u(x,y)")
    ax1.set_title(f"Contour Plot: {title}, λ≈{model.lam.item():.4f}")
```

```
    ax1.set_xlabel("x")
    ax1.set_ylabel("y")

    # Surface Plot
    ax2 = fig.add_subplot(1, 2, 2, projection='3d')      # 3D projection
    surface = ax2.plot_surface(X, Y, u_pred, cmap='viridis', rstride=1,\
                                        cstride=1, antialiased=False)
    fig.colorbar(surface, ax=ax2, shrink=0.7, aspect=15, label="u(x,y)")
    ax2.set_title(f"Surface Plot: {title}, λ≈{model.lam.item():.4f}")
    ax2.set_xlabel("x")
    ax2.set_ylabel("y")

    plt.tight_layout()
    plt.savefig(f_name+'.png', dpi=500)
    plt.show()
```

The following snippet plots the first eigenvalue and its corresponding eigenfunction from the `base_mode` trained for the Helmholtz equation:

```
plot_model(base_model, title="Eigenfunction", f_name='imagesPN/
  ↪plateMode1st')
```

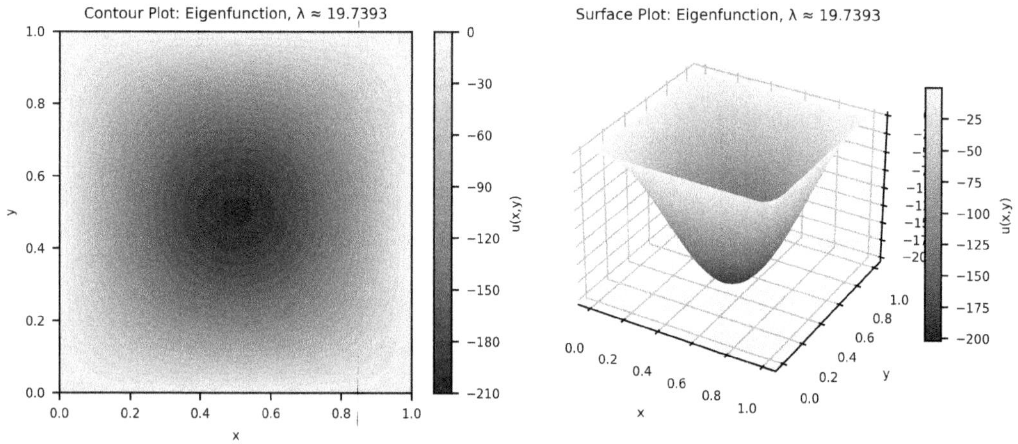

The eigenvalue obtained from the trained `base_model` is 19.739205. Compared with the exact solution of 19.739209, the PINN result is extremely accurate. The corresponding eigenfunction also exhibits the correct shape for the base mode.

The following snippet plots the second eigenvalue and its corresponding eigenfunction from the `higher_mode` model trained for the second mode.

```
plot_model(higher_model, title="Eigenfunction", f_name='imagesPN/
  ↪plateMode2nd')
```

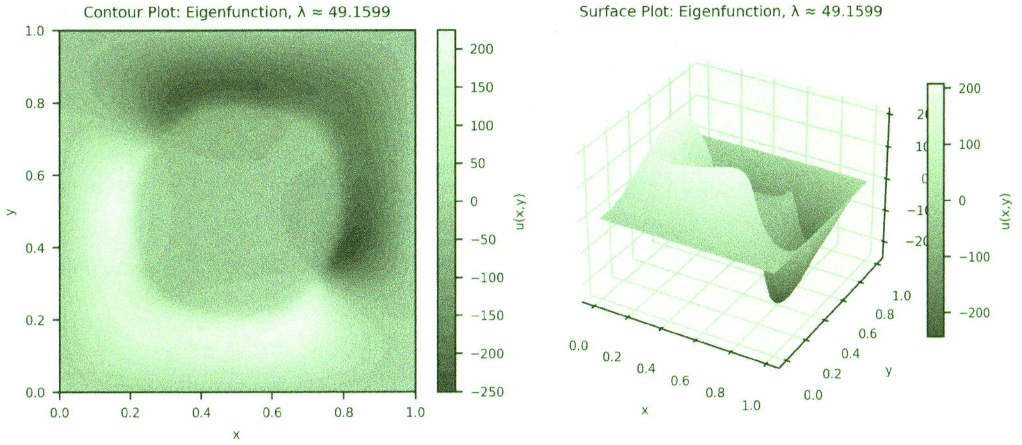

6.6.3 Result Examination

The eigenvalue from the trained `higher_model` is 49.16. Compared with the exact solution of 49.35, the PINN result has about 0.4% error, which is considered accurate.

The eigenfunction obtained is clearly different from the known exact solutions, $u_{1,2}(x, y)$ and $u_{2,1}(x, y)$, which are respectively given in the second and third equations in Eq.(6.17). Eigenfunction $u_{1,2}(x, y)$ is symmetric with respect to $y = 0.5$ and antisymmetric with respect to $x = 0.5$, while eigenfunction $u_{2,1}(x, y)$ is antisymmetric with respect to $y = 0.5$ and symmetric with respect to $x = 0.5$. However, the eigenfunction shown in the plots above is neither symmetric nor antisymmetric with respect to $x = 0.5$ or $y = 0.5$.

This discrepancy arises from the nature of the eigenvalue $5\pi^2 \approx 49.35$, which is a **degenerate eigenvalue**. A degenerate eigenvalue corresponds to multiple (in this case, two) linearly independent eigenfunctions. This is physically meaningful: the energy associated with either $u_{1,2}(x, y)$ or $u_{2,1}(x, y)$ is the same, although the eigenfunctions themselves are different.

Degeneracy implies that these two eigenfunctions span an **eigenspace**. Any function within this eigenspace is also a valid eigenfunction, and such functions can be formed as **linear combinations** of $u_{1,2}(x, y)$ and $u_{2,1}(x, y)$. The eigenfunction obtained in the plot is one such linear combination. Since it is not purely aligned with either $u_{1,2}$ or $u_{2,1}$, the antisymmetry properties are not preserved. Therefore, the PINN solution is still **mathematically correct**, even though it may not exactly match either exact mode individually. Some approximation error is also expected due to the training and numerical nature of the PINN model.

Readers are encouraged to experiment further by training models for **higher modes**. This can be done by modifying the code presented in this section.

6.7 Remarks

This chapter presents various PINN models along with accompanying code for solving the Helmholtz equation to compute eigenvalues and eigenfunctions within a unit square domain with all edges fixed. The training data is obtained via collocation sampling, using a sufficient number of points to satisfy the samples-neurons theorem. A case study is provided to compute the lowest two eigenvalues and their corresponding eigenfunctions. The following key remarks can be drawn:

1. An eigenvalue problem involves **two unknowns**: the eigenvalue and the eigenfunction. It is essentially a **nonlinear** problem. The eigenfunction is represented by the neural network in a PINN model, while the eigenvalue is treated as a **trainable parameter** within the network. Upon convergence, this parameter becomes the computed eigenvalue.

2. Eigenvalue problems admit **multiple (in fact, infinite)** solutions. The **orthogonality** property of eigenfunctions is exploited to compute higher modes.

3. **Normalization** of the eigenfunction is crucial for numerical stability and should be enforced during training.

In general, training high-quality PINN models to compute higher eigenmodes can be **challenging**. In this case study for the second mode, multiple training attempts were necessary, and the following adjustments were explored:

1. Modifying network capacity, including the number of layers (depth) and neurons per layer (width).

2. Changing the number of collocation points .

3. Adjusting training strategies, including the number of epochs and learning rates for `adam`, and iteration counts for `lbfgs`.

4. Tuning the weight `w_norm` associated with `loss_norm`.

5. Providing different initial guesses for the eigenvalue.

Chapter 7

Afterword

7.1 PINN Models for Inverse Problems

The Physics-Informed Neural Network (PINN) framework developed in this chapter can be readily extended to tackle **inverse problems**, which remain among the most challenging and impactful issues in contemporary science and engineering [19]. Such extensions require the following adaptations:

- Treat each **unknown physical parameter** in the governing PDE as a trainable model parameter, analogous to the treatment of the eigenvalue in the present chapter.
- When several scalar parameters are unknown, define one `torch.nn.Parameter` object for each parameter and include all of them in the residual formulation.
- When an unknown parameter **varies spatially**, replace the scalar with a dedicated neural-network sub-model that maps (x, y, t) to the parameter's value. This "parameter network" is then called inside the residual evaluation .

A detailed treatment of inverse problems lies beyond the scope of this volume.

7.2 Dimensionality Considerations

Although all illustrative examples in this book are two-dimensional, the accompanying code can be adapted straightforwardly:

- To formulate a PINN for 1D problem, replace (x, y, t) with (x, t).
- To address a PINN for 3D problem, extend the coordinate tuple to (x, y, z, t).

7.3 Extension to Nonlinear PDEs

Extending a PINN to nonlinear PDEs is conceptually straightforward. The overarching workflow for model construction remains identical for linear, time-dependent, and nonlinear equations; only the algebraic form of the residual changes to reflect the specific PDE. This flexibility illustrates one of the principal strengths of PINNs: once the infrastructure is in place, the same codebase can accommodate a broad range of equations.

That said, nonlinear problems typically present **greater training difficulty**. They exhibit:

- Increased sensitivity to weight initialization,
- More intricate, highly non-convex loss landscapes,
- A heightened need for regularization and tailored optimization strategies ,
- Greater challenges due to non-uniqueness issues.

7.4 Remarks on Training PINN Models

This introductory text emphasizes practical strategies for building and training PINNs on relatively simple benchmark problems. Numerous more sophisticated applications can be found in the literature. Readers are encouraged to consult recent works from the author's group [1–4, 10, 25, 31].

These works address highly nonlinear solid mechanics problems, including large deformation and contact mechanics involving vector field functions; the utilization of radial basis function networks for local approximation; the application of PINNs to inverse problems; the treatment of material interfaces in the mechanics of composite materials; and the development of specially designed activation functions (dual-nested activation) for complex heat transfer problems . TensorFlow codes for some of these problems are also provided.

Much earlier work [19] by the author's group may also be found useful in terms of the use of a genetic algorithm together with a neural network (not PINN per se) for finding the global optimum of loss functions.

In the author's view, the **central challenge** in high-quality PINN construction stems from the **representation capacity of neural networks**: a network approximates a single, highly nonlinear function, rendering the composite loss landscape correspondingly complex. Locating a **global optimum** in this high-dimensional parameter space is therefore inherently demanding and computationally intensive.

Accordingly, the choice of an effective **optimizer**, or a more sophisticated **optimization strategy**, is critical for advancing the range of problems to which PINNs can be applied. Equally important are careful network design, systematic data handling, and judicious weighting of the multiple terms in the composite loss function.

Bibliography

[1] J. Bai, H. Jeong, C.P. Batuwatta-Gamage, S. Xiao, Q. Wang, C.M. Rathnayaka, L. Alzubaidi, G.R. Liu, and Y. Gu. An introduction to programming physics-informed neural network-based computational solid mechanics. *International Journal of Computational Methods*, 20(10):2350013, 2023.

[2] J. Bai, G.R. Liu, L. Alzubaidi, X.Q. Feng, and Y. Gu. Physics-informed radial basis network (pirbn) for solving nonlinear partial differential equations. In *The 14th International Conference of Computational Methods (ICCM2023)*, 2023.

[3] J. Bai, G.R. Liu, A. Gupta, L. Alzubaidi, X.Q. Feng, and Y. Gu. Physics-informed radial basis network (pirbn): A local approximating neural network for solving nonlinear partial differential equations. *Computer Methods in Applied Mechanics and Engineering*, 415:116290, 2023.

[4] J. Bai, G.R. Liu, T. Rabczuk, Y. Wang, X.Q. Feng, and Y. Gu. A robust radial point interpolation method empowered with neural network solvers (rpim-nns) for nonlinear solid mechanics. *Computer Methods in Applied Mechanics and Engineering*, 429:117159, 2024.

[5] A. P. Boresi and R. J. Schmidt. *Advanced Mechanics of Materials.* John Wiley & Sons, New York, 2002.

[6] Jie Gao, Wujun Sun, and Changyou Chen. Error-aware adaptive sampling for physics-informed neural networks. *arXiv preprint arXiv:2206.02980*, 2022.

[7] Sina Heydari and Thomas Goldstein. Softadapt: Techniques for adaptive loss weighting of neural networks with multi-task learning. *arXiv preprint arXiv:1912.12355*, 2019.

[8] Alex Kendall, Yarin Gal, and Roberto Cipolla. Multi-task learning using uncertainty to weigh losses for scene geometry and semantics. In *Proceedings of the IEEE Conference on Computer Vision and Pattern Recognition*, pages 7482–7491, 2018.

[9] Arjun Krishnapriyan, Amir Gholami, Shideh Zhang, Robert Kirby, and Michael W. Mahoney. Characterizing possible failure modes in physics-informed neural networks. *Advances in Neural Information Processing Systems*, 34:26548–26560, 2021.

[10] Mitra Ramakrishna Lanka and G. R. Liu. A novel neural network method for solid mechanics problems of heterogeneous bars. *Engineering Applications of Artificial Intelligence*, 156:111084, 2025.

[11] G.R. Liu. *Mesh free methods: moving beyond the finite element method.* CRC press, Taylor and Francis Group, New York, 2010.

[12] G.R. Liu. *Machine Learning with Python: Theory and Applications.* World Scientific, 2023.

[13] G.R. Liu. *Mechanics of Materials: formulations and solutions with Python.* World Scientific, 2024.

[14] G.R. Liu. *Solid Mechanics with Python: Cartesian Coordinates.* World Scientific, New Jersey, 2024.

[15] G.R. Liu. *Solid Mechanics with Python: PolarCartesian Coordinates.* World Scientific, New Jersey, 2024.

[16] G.R. Liu. Calculus: a practical course with python, 2025.

[17] G.R. Liu. *Vector Calculus: Formulations, Applications, and Python Codes.* World Scientific, New Jersey, 2025.

[18] G.R. Liu and Y. Gu. An introduction to meshfree methods and their programming, 2005.

[19] G.R. Liu and X Han. *Computational inverse techniques in nondestructive evaluation.* CRC Press, Taylor and Francis Group, New York, 2003.

[20] G.R. Liu and TT Nguyen. *Smoothed finite element methods.* CRC Press, New York, 2010.

[21] GR Liu and Siu Sin Quek. *The finite element method: a practical course.* Butterworth-Heinemann, London, 2013.

[22] GR Liu and ZC Xi. *Elastic waves in anisotropic laminates.* CRC press, New York, 2001.

[23] Lu Lu, Xuhui Meng, Zhiping Mao, and George Em Karniadakis. Deepxde: A deep learning library for solving differential equations. *SIAM Review*, 63(1):208–228, 2021. Includes residual-based adaptive sampling (RAR).

[24] Lu Lu, Xuhui Meng, Zongyi Mao, and George Em Karniadakis. Deepxde: A deep learning library for solving differential equations. *SIAM Review*, 63(1):208–228, 2021.

[25] Y.J. Lu, S.Y. Duan, and G.R. Liu. Pre-conditioned physics-informed neural network for inverse problems. *International Journal of Computational Methods*, page 2441010, 2025.

[26] Maziar Raissi, Paris Perdikaris, and George Em Karniadakis. Physics-informed neural networks: A deep learning framework for solving forward and inverse problems involving nonlinear partial differential equations. *Journal of Computational Physics*, 378:686–707, 2019.

[27] Jiaxuan Wang, Hanwen Teng, Luning Meng, and George Em Karniadakis. Self-adaptive physics-informed neural networks using a soft attention mechanism. *arXiv preprint arXiv:2009.04544*, 2020.

[28] Shuhang Wang, Yifan Teng, and Paris Perdikaris. Understanding and mitigating gradient flow pathologies in physics-informed neural networks. *SIAM Journal on Scientific Computing*, 43(5):A3055–A3081, 2021.

[29] Sifan Wang, Yujun Teng, and Paris Perdikaris. Understanding and mitigating gradient pathologies in physics-informed neural networks. *SIAM Journal on Scientific Computing*, 43(5):A3055–A3081, 2021.

[30] Sifan Wang, Xinling Yu, and Paris Perdikaris. When and why pinns fail to train: A neural tangent kernel perspective. *Journal of Computational Physics*, 449:110768, 2022.

[31] Tianhao Wang, Guirong Liu, Eric Li, and Xu Xu. Adaptive deep physics-informed neural network with dual-nested activation for solving complex partial differential equations. *Computer Methods in Applied Mechanics and Engineering*, 444:118125, 2025.

[32] Yongle Zheng, Zhe Zhong, Ji Wang, and Shengyong Cai. Clw-pinn: Curriculum loss weighting based physics-informed neural network for solving time-dependent pdes. *Computer Methods in Applied Mechanics and Engineering*, 400:115539, 2022.

Index

www.ingramcontent.com/pod-product-compliance
Lightning Source LLC
LaVergne TN
LVHW062034101126
852251LV00025B/333